Methods in Molecular Biology

Series Editor
John M. Walker
School of Life and Medical Sciences
University of Hertfordshire
Hatfield, UK

For further volumes:
http://www.springer.com/series/7651

For over 35 years, biological scientists have come to rely on the research protocols and methodologies in the critically acclaimed *Methods in Molecular Biology* series. The series was the first to introduce the step-by-step protocols approach that has become the standard in all biomedical protocol publishing. Each protocol is provided in readily-reproducible step-by-step fashion, opening with an introductory overview, a list of the materials and reagents needed to complete the experiment, and followed by a detailed procedure that is supported with a helpful notes section offering tips and tricks of the trade as well as troubleshooting advice. These hallmark features were introduced by series editor Dr. John Walker and constitute the key ingredient in each and every volume of the *Methods in Molecular Biology* series. Tested and trusted, comprehensive and reliable, all protocols from the series are indexed in PubMed.

Cerebrospinal Fluid (CSF) Proteomics

Methods and Protocols

Second Edition

Edited by

Enrique Santamaría and Joaquín Fernández-Irigoyen

Clinical Neuroproteomics Unit, Navarrabiomed, Pamplona, Navarra, Spain

Humana Press

Editors
Enrique Santamaría
Clinical Neuroproteomics Unit
Navarrabiomed
Pamplona, Navarra, Spain

Joaquín Fernández-Irigoyen
Clinical Neuroproteomics Unit
Navarrabiomed
Pamplona, Navarra, Spain

ISSN 1064-3745 ISSN 1940-6029 (electronic)
Methods in Molecular Biology
ISBN 978-1-0716-4461-4 ISBN 978-1-0716-4462-1 (eBook)
https://doi.org/10.1007/978-1-0716-4462-1

Cover Illustration Caption: Cover design by María Arraiza from Communication and Design Unit (Navarrabiomed-Spain).

This Humana imprint is published by the registered company Springer Science+Business Media, LLC, part of Springer Nature.
The registered company address is: 1 New York Plaza, New York, NY 10004, U.S.A.

Preface

Due to the complexity of the molecular organization of the brain, characterization of protein profiles within specific brain structures and cerebrospinal fluid (CSF) forms an essential part of unearthing the molecular perturbation associated with neuropsychiatric disorders and neurodegenerative diseases. However, even though CSF is considered the best compartment for the analysis of protein and peptide biomarkers to study brain disorders and to support CNS drug development, it presents intrinsic characteristics that hamper conducting in-depth routine analysis in a straightforward manner. Although we are in the early stages of a continuous learning curve, the application of low and high-throughput quantitative -omics is emerging as a powerful toolbox to profile CSF molecular profiles in normal and pathological states, increasing our understanding of human brain biology and pushing the detection and validation of consistent CSF biomarkers in the neurology field. As a continuation of the first edition of the book published in 2019, the themes discussed in this second edition will be mainly focused on mass spectrometry and different protein methods used in neurobiological disciplines during recent years.

Specifically, this book will bring a collection of detailed protocols written by experienced professionals around the world to cover a variety of experimental workflows used in the characterization of the CSF molecular landscape. Chapters will include CSF sample preparation methods, shotgun and targeted proteomic workflows, as well as methods to characterize specific posttranslational modifications to achieve finer molecular resolution in proteomic studies of CSF. Moreover, additional chapters will focus on experimental strategies based on other -omics applied to the CSF lipidome and metabolome. Finally, several chapters will describe different bioinformatic pipelines useful to analyze and integrate the molecular information derived from high-throughput -omics experiments performed in CSF. All of these approaches are useful to quantify CSF molecular profiles, detect potential biomarker candidates, and identify post-translational modifications and are beginning to shed new light on the translation, integration, and interpretation of the metabolic regulation that occurs in brain disorders at the CSF level.

We believe this book will be a useful resource for graduate students or junior postdoctoral fellows interested in starting a journey into CSF molecular profiles, as well as established researchers seeking valuable insight into the growing utility of proteomics and metabolomics in neuroscience.

Pamplona, Navarra, Spain

Enrique Santamaría
Joaquín Fernández-Irigoyen

Contents

Part VI Bioinformatics Applied to Cerebrospinal Fluid

Contributors

FRANCISCO AMADO • *LAQV-REQUIMTE, Department of Chemistry, University of Aveiro, Aveiro, Portugal*

VERONICA G. ANANIA • *Department of Translational Medicine, Genentech, Inc., South San Francisco, CA, USA*

ELENA ANAYA-CUBERO • *Clinical Neuroproteomics Unit, Proteomics Platform, Navarrabiomed, Hospitalario Universitario de Navarra (HUN), Universidad Pública de Navarra (UPNA), IdiSNA, Navarra Institute for Health Research, Pamplona, Spain*

POL ANDRÉS-BENITO • *Cognition and Behavior Study Group, Institut de Recerca Biomèdica de Lleida (IRBLleida), Lleida, Spain*

NURIA BARBARROJA • *Cobiomic Bioscience. EBT University of Cordoba/IMIBIC, Cordoba, Spain; Rheumatology Unit/Reina Sofia Hospital/University of Cordoba/IMIBIC, Cordoba, Spain*

BRENO JOSÉ ALENCAR PIRES BARBOSA • *Academic Area of Neuropsychiatry, Medical Sciences Center, Federal University of Pernambuco, Recife, Brazil*

PAULO BASTOS • *RISE-Health, Department of Surgery and Physiology, Faculty of Medicine, University of Porto, Porto, Portugal*

GEORGY BEREZHNOY • *Core Facility Metabolomics & Werner Siemens Imaging Center, Medical Faculty University of Tübingen, Tübingen, Germany*

ALIONA BOGDANOVA • *Max Planck Institute of Molecular Cell Biology and Genetics, Dresden, Germany*

MIREIA BOLUDA-NAVARRO • *Health in Code (HiC), Valencia, Spain*

PAZ CARTAS-CEJUDO • *Clinical Neuroproteomics Unit, Proteomics Platform, Navarrabiomed, Hospitalario Universitario de Navarra (HUN), Universidad Pública de Navarra (UPNA), IdiSNA, Navarra Institute for Health Research, Pamplona, Spain*

SHILAN CHEN • *Clinical Biobank Center, Microbiome Medicine Center, Department of Laboratory Medicine, Guangdong Provincial Clinical Research Center for Laboratory Medicine, Zhujiang Hospital, Southern Medical University, Guangzhou, China; Neurosurgery Center, The National Key Clinical Specialty, Engineering Research Center of Diagnostic and Therapeutic Technology and Devices for Cerebrovascular Diseases, Ministry of Education, Guangdong Provincial Key Laboratory on Brain Function Repair and Regeneration, Zhujiang Hospital Institute for Brain Science and Intelligence, Zhujiang Hospital, Southern Medical University, Guangzhou, China*

IDOIA COMPANY-MARÍN • *Department of Experimental Medicine, Lleida Biomedical Research Institute, University of Lleida, Lleida, Spain*

ADRIANA CORTÉS • *Clinical Neuroproteomics Unit, Proteomics Platform, Navarrabiomed, Hospitalario Universitario de Navarra (HUN), Universidad Pública de Navarra (UPNA), IdiSNA, Navarra Institute for Health Research, Pamplona, Spain*

YUANYUAN DENG • *State Key Laboratory of Digital Medical Engineering, School of Biological Science and Medical Engineering, Southeast University, Nanjing, Jiangsu, China*

RAMÓN DÍAZ-PEÑA • *Proteomics Platform, Clinical Neuroproteomics Unit, Navarrabiomed, Hospital Universitario de Navarra (HUN), Universidad Pública de Navarra (UPNA), IdiSNA, Pamplona, Spain*

IGNACY RZAGALINSKI • *Max Planck Institute of Molecular Cell Biology and Genetics, Dresden, Germany*

ENRIQUE SANTAMARÍA • *Clinical Neuroproteomics Laboratory, Proteomics Unit, Navarrabiomed, Hospital Universitario de Navarra (HUN), Universidad Pública de Navarra (UPNA), IdiSNA, Pamplona, Spain; Clinical Neuroproteomics Unit, Proteomics Platform, Navarrabiomed, Hospital Universitario de Navarra (HUN), Universidad Pública de Navarra (UPNA), IdiSNA, Pamplona, Spain; Proteomics Platform, Clinical Neuroproteomics Unit, Navarrabiomed, Hospital Universitario de Navarra (HUN), Universidad Pública de Navarra (UPNA), IdiSNA, Pamplona, Spain; Proteomics Platform, Clinical Neuroproteomics Unit, Navarrabiomed, Complejo Hospitalario de Navarra (CHN), Universidad Pública de Navarra (UPNA), IdiSNA, Pamplona, Spain; Clinical Neuroproteomics Unit, Proteomics Platform, Navarrabiomed, Hospitalario Universitario de Navarra (HUN), Universidad Pública de Navarra (UPNA), IdiSNA, Navarra Institute for Health Research, Pamplona, Spain*

ANDREJ SHEVCHENKO • *Max Planck Institute of Molecular Cell Biology and Genetics, Dresden, Germany*

JOY SOLOMON • *Department of Chemistry and Biochemistry, Texas Tech University, Lubbock, TX, USA*

JANNIK SPRENGEL • *Core Facility Metabolomics & Werner Siemens Imaging Center, Medical Faculty University of Tübingen, Tübingen, Germany*

SANJEEVA SRIVASTAVA • *Department of Biosciences and Bioengineering, Indian Institute of Technology Bombay, Mumbai, Maharashtra, India*

HAITAO SUN • *Clinical Biobank Center, Microbiome Medicine Center, Department of Laboratory Medicine, Guangdong Provincial Clinical Research Center for Laboratory Medicine, Zhujiang Hospital, Southern Medical University, Guangzhou, China; Neurosurgery Center, The National Key Clinical Specialty, Engineering Research Center of Diagnostic and Therapeutic Technology and Devices for Cerebrovascular Diseases, Ministry of Education, Guangdong Provincial Key Laboratory on Brain Function Repair and Regeneration, Zhujiang Hospital Institute for Brain Science and Intelligence, Zhujiang Hospital, Southern Medical University, Guangzhou, China; Key Laboratory of Mental Health of the Ministry of Education, Guangdong–Hong Kong–Macao Greater Bay Area Center for Brain Science and Brain-Inspired Intelligence, Southern Medical University, Guangzhou, China*

W. ANDY TAO • *Department of Biochemistry, Purdue University, West Lafayette, IN, USA; Department of Chemistry, Purdue University, West Lafayette, IN, USA; Institute for Cancer Research, Purdue University, West Lafayette, IN, USA*

CHRISTOPH TRAUTWEIN • *Core Facility Metabolomics & Werner Siemens Imaging Center, Medical Faculty University of Tübingen, Tübingen, Germany*

FÁBIO TRINDADE • *RISE-Health, Department of Surgery and Physiology, Faculty of Medicine, University of Porto, Porto, Portugal*

RATHNA VEERAMACHANENI • *KCAS Bioanalytical Sciences, Olathe, KS, USA*

TUANE CRISTINE R. G. VIEIRA • *Medical Biochemistry Institute Leopoldo de Meis, National Institute of Science, Technology, Structural Biology and Bioimaging, Federal University of Rio de Janeiro, Rio de Janeiro, Brazil*

RUI VITORINO • *RISE-Health, Department of Surgery and Physiology, Faculty of Medicine, University of Porto, Porto, Portugal; iBiMED, Department of Medical Sciences, University of Aveiro, Aveiro, Portugal*

LAURA WINCHESTER • *Department of Psychiatry, University of Oxford, Oxford, UK*

ZHUOYING XIE • *State Key Laboratory of Digital Medical Engineering, School of Biological Science and Medical Engineering, Southeast University, Nanjing, Jiangsu, China*

FABIOLA ZAMBOM-FERRARESI • *Department of Geriatric Medicine, Hospital Universitario de Navarra (HUN), Navarrabiomed, Universidad Pública de Navarra (UPNA), IdiSNA, Pamplona, Spain*

FABRICIO ZAMBOM-FERRARESI • *Department of Geriatric Medicine, Hospital Universitario de Navarra (HUN), Navarrabiomed, Universidad Pública de Navarra (UPNA), IdiSNA, Pamplona, Spain; CIBER of Frailty and Healthy Aging (CIBERFES), Instituto de Salud Carlos III, Madrid, Spain*

HANJUN ZHAO • *Department of Psychiatry, University of Oxford, Oxford, UK*

Part I

Introduction

Chapter 1

Recent Advances in Human Cerebrospinal Fluid Proteomics

Joaquín Fernández-Irigoyen and Enrique Santamaría

Abstract

Cerebrospinal fluid (CSF) proteomics has become an alternative that allows zooming-in where pathophysiological alterations are taking place, detecting protein mediators that might eventually be considered as potential biomarkers in neurological or psychiatric diseases. From a technological point of view, mass-spectrometry-based-proteomics as well as antibody/aptamer-based platforms allow the simultaneous monitoring of secreted and circulating proteins in a broad concentration range in a robust manner, generating innovative facets in biomarker validation. This chapter highlights recent discoveries in the field of CSF protein analysis, covering sample preparation methods for proteomic workflows, shotgun proteomics, subproteomics, and other omics applied to CSF.

Key words Proteomics, Human CSF, Mass spectrometry, Aptamers

1 Introduction

Given the segmented functions of the brain, the characterization of protein profiles within specific areas is essential to decipher the molecular basis for structure specialization as well as the proteostasis impairment associated with neurodegenerative and psychiatric diseases. The selection of a fluid in close proximity to a diseased organ may increase the probability of finding a biomarker panel originating from a pathological tissue. In that respect, CSF, which is in closest contact with the brain, represents a useful reservoir of potential clinically relevant biomarkers for neurological diseases [1, 2]. Several reports have demonstrated the presence of at least 4500 proteins in human CSF [3, 4]. In addition, studies focused on PTM characterizations have pointed out the presence of phosphorylated, glycosylated, oxidized, and carbonylated proteins in human CSF. However, despite cerebrospinal fluid (CSF) being considered the best compartment for the analysis of protein and peptide biomarkers to study neurological disorders and to support CNS drug development [5], the number of publications has certainly been constant during the last decade (Fig. 1). Human CSF

Enrique Santamaría and Joaquín Fernández-Irigoyen (eds.), *Cerebrospinal Fluid (CSF) Proteomics: Methods and Protocols*, Methods in Molecular Biology, vol. 2914, https://doi.org/10.1007/978-1-0716-4462-1_1,

in CSF proteomics is the deployment of Tandem Mass Tagging (TMT) in order to increase the accuracy of the quantitative protein data. This approach has allowed to employ tandem mass tag (TMT) quantification methods from CSF samples comprising patients with ALS [17], Japanese encephalitis [18], Alzheimer's disease [19, 20], cerebral venous sinus thrombosis [21], severe traumatic brain injury [22] and multiple sclerosis [23]. In some cases, technological combination has allowed the interlocking of aptamer-based and mass-spectrometry-based workflows. For example, proximal extension assay (PEA) and untargeted data-dependent mass spectrometry were combined to generate orthogonal lists of protein marker candidates for Syndromic CLN3-Batten, a fatal, neuropediatric disease caused by variants in the endolysosomal transmembrane CLN3 protein [24]. The combination of liquid chromatography mass spectrometry with multiple reaction monitoring/Rules-Based Medicine, tandem mass tag mass spectrometry, and SOMAscan pointed out the existence of amyloid-independent and -dependent CSF proteome changes with older age, representing physiological aging and initial Alzheimer's disease [25]. Chaudhary et al. integrated flow cytometry and mass spectrometry proteomics to evaluate how protein components from human CSF associate with Aβ fibrils to constitute plaque particles, establishing a new point of view to model AD-related pathological mechanisms [26]. Johansson et al. combined mass-cytometry and PEA assays to monitor CSF signaling molecules and surface markers with the aim of identifying a multiple sclerosis-specific immune cell population by deep immune phenotyping [27]. Mun et al. generated a catalog of peptides and proteins in human CSF in data-dependent mode on a quadrupole time-of-flight mass spectrometer coupled with a trapped ion mobility spectrometry (timsTOF) operated in parallel accumulation-serial fragmentation (PASEF), supporting the utility of this spectral library in future PRM experiments for future quantification of potential biomarkers in neurological disorders [28]. Another useful molecular resource in the Parkinson's disease field is the characterization of blood, brain and CSF through the SOMASCAN platform [29].

4 Cerebrospinal Fluid Subproteomics

One of the most relevant aspects of CSF protein analysis is the characterization of extracellular vesicles (EVs) [30]. These vesicular compartments carry lipids, nucleic acids and proteins having biological roles related to communication in the CNS. Recently, their potential as biomarkers for CNS diseases has stimulated the scientific community, increasing its interest in CSF EV research. However, this topic is challenging because of the invasive nature of CSF collection, the limited CSF sampling, and the low number of

EVs in CSF compared to other biofluids like plasma. From the proteomic point of view, multiple groups have decided to characterize the protein composition of CSF EVs in different biological contexts. Kangas et al. validated ultrafiltration combined with size-exclusion chromatography (UF-SEC) method for the isolation of EVs, performing proteomics from 0.5 mL of CSF [31]. Hou et al. applied label-free and PRM-based proteomics to detect key proteins related to leptomeningeal carcinomatosis at the level of total CSF as well as at the exosomal level [32]. Rather HA et al. demonstrated that EVs from CSF may potentially serve as biomarkers to evaluate the transgenerational adverse effects due to prenatal cocaine exposure. For that, they performed a CSF EV characterization based on the surface expression of specific tetraspanins, concentration (particles/mL), size distribution, and cargo proteins by mass spectrometry [33]. Chatterjee et al. pointed out that CSF EVs may be used as potential AD biomarkers, playing a role in immune system regulation during the disease progression. In this study, CSF EVs were isolated by ultracentrifugation and analyzed by untargeted quantitative mass spectrometry-based proteomics and ELISA [34]. Targeted proximity extension assays (PEA) have also been applied to CSF EVs. For example, Sjoqvist et al. analyzed CSF EVs using this approach, demonstrating its feasibility using different EV isolation kits [35].

Multiple studies have been focused on the characterization of specific protein subsets in CSF such as endo-lysosomal proteins or apolipoproteins [36, 37] or on the enrichment of specific posttranslational modified CSF proteins by phosphorylation [8], glycosylation [38–41], sulphydration [42], glutathionylation [43] or citrullination [44]. Finally, different mass spectrometry-based workflows have been optimized to detect and quantify specific protein groups [45–47], neuropeptide proteoforms [48], interactomes [49], cytokines [50] and conformational biomarkers [51–53].

5 Other Omics Applied to Cerebrospinal Fluid

Other omics technologies such as transcriptomics, metabolomics and metagenomics are currently being used in combination with proteomics to deeply characterize the molecular landscape of CSF in a variety of biological scenarios. Borkowski et al. developed a pipeline to integrate CSF metabolomics with CSF proteomic revealing novel associations between lipid mediators and central nervous system vascular and energy metabolism [54]. Lilley and colleagues have established a comparative multi-omic database of signatures derived from a human population of 30 individuals (<75 years) of serum and cerebrospinal fluid (obtained by the combination of proteomics and metabolomics) with the aim to be

22. Shultz SR, Shah AD, Huang C, Dill LK, Schittenhelm RB, Morganti-Kossmann MC, Semple BD (2022) Temporal proteomics of human cerebrospinal fluid after severe traumatic brain injury. J Neuroinflammation 19(1):291. https://doi.org/10.1186/s12974-022-02654-0
23. Shi Y, Ding Y, Li G, Wang L, Osman RA, Sun J, Qian L, Zheng G, Zhang G (2021) Discovery of novel biomarkers for diagnosing and predicting the progression of multiple sclerosis using TMT-based quantitative proteomics. Front Immunol 12:700031. https://doi.org/10.3389/fimmu.2021.700031
24. Dang Do AN, Sleat DE, Campbell K, Johnson NL, Zheng H, Wassif CA, Dale RK, Porter FD (2023) Cerebrospinal fluid protein biomarker discovery in CLN3. J Proteome Res 22(7): 2493–2508. https://doi.org/10.1021/acs.jproteome.3c00199
25. Wesenhagen KEJ, Gobom J, Bos I, Vos SJB, Martinez-Lage P, Popp J, Tsolaki M, Vandenberghe R, Freund-Levi Y, Verhey F, Lovestone S, Streffer J, Dobricic V, Bertram L, Alzheimer's Disease Neuroimaging I, Blennow K, Pikkarainen M, Hallikainen M, Kuusisto J, Laakso M, Soininen H, Scheltens P, Zetterberg H, Teunissen CE, Visser PJ, Tijms BM (2022) Effects of age, amyloid, sex, and APOE epsilon4 on the CSF proteome in normal cognition. Alzheimers Dement (Amst) 14(1):e12286. https://doi.org/10.1002/dad2.12286
26. Chaudhary H, Meister SW, Zetterberg H, Lofblom J, Lendel C (2020) Dissecting the structural organization of multiprotein amyloid aggregates using a bottom-up approach. ACS Chem Neurosci 11(10):1447–1457. https://doi.org/10.1021/acschemneuro.0c00110
27. Johansson D, Rauld C, Roux J, Regairaz C, Galli E, Callegari I, Raad L, Waldt A, Cuttat R, Roma G, Diebold M, Becher B, Kuhle J, Derfuss T, Carballido JM, Sanderson NSR (2021) Mass cytometry of CSF identifies an MS-associated B-cell population. Neurol Neuroimmunol Neuroinflamm 8(2). https://doi.org/10.1212/NXI.0000000000000943
28. Mun DG, Budhraja R, Bhat FA, Zenka RM, Johnson KL, Moghekar A, Pandey A (2023) Four-dimensional proteomics analysis of human cerebrospinal fluid with trapped ion mobility spectrometry using PASEF. Proteomics 23(10):e2200507. https://doi.org/10.1002/pmic.202200507
29. Winchester L, Barber I, Lawton M, Ash J, Liu B, Evetts S, Hopkins-Jones L, Lewis S, Bresner C, Malpartida AB, Williams N, Gentlemen S, Wade-Martins R, Ryan B, Holgado-Nevado A, Hu M, Ben-Shlomo Y, Grosset D, Lovestone S (2023) Identification of a possible proteomic biomarker in Parkinson's disease: discovery and replication in blood, brain and cerebrospinal fluid. Brain Commun 5(1):fcac343. https://doi.org/10.1093/braincomms/fcac343
30. Sandau US, Magana SM, Costa J, Nolan JP, Ikezu T, Vella LJ, Jackson HK, Moreira LR, Palacio PL, Hill AF, Quinn JF, Van Keuren-Jensen KR, McFarland TJ, Palade J, Sribnick EA, Su H, Vekrellis K, Coyle B, Yang Y, Falcon-Perez JM, Nieuwland R, Saugstad JA, International Society for Extracellular Vesicles Cerebrospinal Fluid Task F (2024) Recommendations for reproducibility of cerebrospinal fluid extracellular vesicle studies. J Extracell Vesicles 13(1):e12397. https://doi.org/10.1002/jev2.12397
31. Kangas P, Nyman TA, Metsahonkala L, Burns C, Tempest R, Williams T, Karttunen J, Jokinen TS (2023) Towards optimised extracellular vesicle proteomics from cerebrospinal fluid. Sci Rep 13(1):9564. https://doi.org/10.1038/s41598-023-36706-z
32. Hou L, Chen X, Qiu G, Qi X, Zou Y, He J, Bu H (2023) Cerebrospinal fluid exosomal protein alterations via proteomic analysis of NSCLC with leptomeningeal carcinomatosis. J Neuro-Oncol 164(2):367–376. https://doi.org/10.1007/s11060-023-04428-x
33. Rather HA, Mishra S, Su Y, Kumar A, Singh S, Misra BB, Lee J, Furdui CM, Hamilton LR, Gould RW, Nader SH, Nader MA, Deep G (2022) Mass spectrometry-based proteome profiling of extracellular vesicles derived from the cerebrospinal fluid of adult rhesus monkeys exposed to cocaine throughout gestation. Biomol Ther 12(4). https://doi.org/10.3390/biom12040510
34. Chatterjee M, Ozdemir S, Kunadt M, Koel-Simmelink M, Boiten W, Piepkorn L, Pham TV, Chiasserini D, Piersma SR, Knol JC, Mobius W, Mollenhauer B, van der Flier WM, Jimenez CR, Teunissen CE, Jahn O, Schneider A (2023) C1q is increased in cerebrospinal fluid-derived extracellular vesicles in Alzheimer's disease: a multi-cohort proteomics and immuno-assay validation study. Alzheimers Dement 19(11):4828–4840. https://doi.org/10.1002/alz.13066
35. Sjoqvist S, Otake K, Hirozane Y (2020) Analysis of cerebrospinal fluid extracellular vesicles by proximity extension assay: a comparative study of four isolation kits. Int J Mol Sci 21(24). https://doi.org/10.3390/ijms21249425

36. Lowe AJ, Sjodin S, Rodrigues FB, Byrne LM, Blennow K, Tortelli R, Zetterberg H, Wild EJ (2020) Cerebrospinal fluid endo-lysosomal proteins as potential biomarkers for Huntington's disease. PLoS One 15(8):e0233820. https://doi.org/10.1371/journal.pone.0233820
37. Hu Y, Meuret C, Martinez A, Yassine HN, Nedelkov D (2021) Distinct patterns of apolipoprotein C-I, C-II, and C-III isoforms are associated with markers of Alzheimer's disease. J Lipid Res 62:100014. https://doi.org/10.1194/jlr.RA120000919
38. Cho BG, Gutierrez Reyes CD, Mechref Y (2021) N-Glycomics of cerebrospinal fluid: method comparison. Molecules 26(6). https://doi.org/10.3390/molecules26061712
39. Chen Z, Wang D, Yu Q, Johnson J, Shipman R, Zhong X, Huang J, Yu Q, Zetterberg H, Asthana S, Carlsson C, Okonkwo O, Li L (2022) In-depth site-specific O-glycosylation analysis of glycoproteins and endogenous peptides in Cerebrospinal Fluid (CSF) from healthy individuals, Mild Cognitive Impairment (MCI), and Alzheimer's Disease (AD) patients. ACS Chem Biol 17(11): 3059–3068. https://doi.org/10.1021/acschembio.1c00932
40. Chen Z, Yu Q, Yu Q, Johnson J, Shipman R, Zhong X, Huang J, Asthana S, Carlsson C, Okonkwo O, Li L (2021) In-depth site-specific analysis of N-glycoproteome in human cerebrospinal fluid and glycosylation landscape changes in Alzheimer's disease. Mol Cell Proteomics 20:100081. https://doi.org/10.1016/j.mcpro.2021.100081
41. Furukawa JI, Hanamatsu H, Yokota I, Hirayama M, Ando T, Kobayashi H, Ohnishi S, Miura N, Okada K, Sakai S, Yuyama K, Igarashi Y, Ito M, Shinohara Y, Sakamoto N (2021) Comprehensive Glycomic approach reveals novel low-molecular-weight blood group-specific Glycans in serum and cerebrospinal fluid. J Proteome Res 20(5): 2812–2822. https://doi.org/10.1021/acs.jproteome.1c00056
42. Greco V, Neri C, Pieragostino D, Spalloni A, Persichilli S, Gastaldi M, Mercuri NB, Longone P, Urbani A (2021) Investigating different forms of hydrogen sulfide in cerebrospinal fluid of various neurological disorders. Metabolites 11:3. https://doi.org/10.3390/metabo11030152
43. Srivastava D, Kukkuta Sarma GR, Dsouza DS, Muralidharan M, Srinivasan K, Mandal AK (2019) Characterization of residue-specific glutathionylation of CSF proteins in multiple sclerosis – a MS-based approach. Anal Biochem 564-565:108–115. https://doi.org/10.1016/j.ab.2018.10.015
44. Mitra S, Harvey-Jones K, Kraev I, Verma V, Meehan C, Mintoft A, Norris G, Campbell E, Tucker K, Robertson NJ, Hristova M, Lange S (2023) The extracellular vesicle Citrullinome and signature in a piglet model of neonatal seizures. Int J Mol Sci 24(14). https://doi.org/10.3390/ijms241411529
45. Johnson ECB, Bian S, Haque RU, Carter EK, Watson CM, Gordon BA, Ping L, Duong DM, Epstein MP, McDade E, Barthelemy NR, Karch CM, Xiong C, Cruchaga C, Perrin RJ, Wingo AP, Wingo TS, Chhatwal JP, Day GS, Noble JM, Berman SB, Martins R, Graff-Radford NR, Schofield PR, Ikeuchi T, Mori H, Levin J, Farlow M, Lah JJ, Haass C, Jucker M, Morris JC, Benzinger TLS, Roberts BR, Bateman RJ, Fagan AM, Seyfried NT, Levey AI, Dominantly Inherited Alzheimer N (2023) Cerebrospinal fluid proteomics define the natural history of autosomal dominant Alzheimer's disease. Nat Med 29(8):1979–1988. https://doi.org/10.1038/s41591-023-02476-4
46. Bergstrom S, Remnestal J, Yousef J, Olofsson J, Markaki I, Carvalho S, Corvol JC, Kultima K, Kilander L, Lowenmark M, Ingelsson M, Blennow K, Zetterberg H, Nellgard B, Brosseron F, Heneka MT, Bosch B, Sanchez-Valle R, Manberg A, Svenningsson P, Nilsson P (2021) Multi-cohort profiling reveals elevated CSF levels of brain-enriched proteins in Alzheimer's disease. Ann Clin Transl Neurol 8(7): 1456–1470. https://doi.org/10.1002/acn3.51402
47. Xu M, Lu M, Zhang W, Jin Q, Chen Y (2021) Simultaneous detection of six isoforms of tau protein in human cerebrospinal fluid by multidimensional mass spectrometry-based targeted proteomics. J Proteome Res 20(5): 2299–2307. https://doi.org/10.1021/acs.jproteome.0c00826
48. Quinn JP, Ethier EC, Novielli A, Malone A, Ramirez CE, Salloum L, Trombetta BA, Kivisakk P, Bremang M, Selzer S, Fournier M, Das S, Xing Y, Arnold SE, Carlyle BC (2023) Cerebrospinal fluid and brain proteoforms of the granin neuropeptide family in Alzheimer's disease. J Am Soc Mass Spectrom 34(4): 649–667. https://doi.org/10.1021/jasms.2c00341
49. Ferreira A, Timmerman E, Staes A, Vuylsteke M, De Muynck L, Gevaert K (2023) Protein interactors of 3-O sulfated heparan sulfates in human MCI and age-matched control cerebrospinal fluid. Sci

Data 10(1):121. https://doi.org/10.1038/s41597-023-02009-1

50. Li Y, Yang K, Zhang F, Wang J, Shen H, Liu M, Guo J, Wang J (2022) Identification of cerebrospinal fluid biomarker candidates for anti-N-methyl-D-aspartate receptor encephalitis: high-throughput proteomic investigation. Front Immunol 13:971659. https://doi.org/10.3389/fimmu.2022.971659
51. Wang B, Zhong X, Fields L, Lu H, Zhu Z, Li L (2023) Structural proteomic profiling of cerebrospinal fluids to reveal novel conformational biomarkers for Alzheimer's disease. J Am Soc Mass Spectrom 34(3):459–471. https://doi.org/10.1021/jasms.2c00332
52. Shuken SR, Rutledge J, Iram T, Losada PM, Wilson EN, Andreasson KI, Leib RD, Wyss-Coray T (2022) Limited proteolysis-mass spectrometry reveals aging-associated changes in cerebrospinal fluid protein abundances and structures. Nat Aging 2(5):379–388. https://doi.org/10.1038/s43587-022-00196-x
53. Mackmull MT, Nagel L, Sesterhenn F, Muntel J, Grossbach J, Stalder P, Bruderer R, Reiter L, van de Berg WDJ, de Souza N, Beyer A, Picotti P (2022) Global, in situ analysis of the structural proteome in individuals with Parkinson's disease to identify a new class of biomarker. Nat Struct Mol Biol 29(10): 978–989. https://doi.org/10.1038/s41594-022-00837-0
54. Borkowski K, Seyfried NT, Arnold M, Lah JJ, Levey AI, Hales CM, Dammer EB, Blach C, Louie G, Kaddurah-Daouk R, Newman JW (2023) Integration of plasma and CSF metabolomics with CSF proteomic reveals novel associations between lipid mediators and central nervous system vascular and energy metabolism. Sci Rep 13(1):13752. https://doi.org/10.1038/s41598-023-39737-8
55. Lilley LM, Sanche S, Moore SC, Salemi MR, Vu D, Iyer S, Hengartner NW, Mukundan H (2022) Methods to capture proteomic and metabolomic signatures from cerebrospinal fluid and serum of healthy individuals. Sci Rep 12(1):13339. https://doi.org/10.1038/s41598-022-16598-1
56. Hooshmand K, Xu J, Simonsen AH, Wretlind A, de Zawadzki A, Sulek K, Hasselbalch SG, Legido-Quigley C (2024) Human cerebrospinal fluid sample preparation and annotation for integrated lipidomics and metabolomics profiling studies. Mol Neurobiol 61(4):2021–2032. https://doi.org/10.1007/s12035-023-03666-4
57. Stojanovic F, Taktek M, Khieu NH, Huang J, Jiang S, Rennie K, Chakravarthy B, Costain WJ, Cuperlovic-Culf M (2021) NMR analysis of the correlation of metabolic changes in blood and cerebrospinal fluid in Alzheimer model male and female mice. PLoS One 16(5):e0250568. https://doi.org/10.1371/journal.pone.0250568
58. Prieto-Fernandez E, Lopez-Lopez E, Martin-Guerrero I, Barcena L, Gonzalez-Lopez M, Aransay AM, Lozano JJ, Benito J, Falcon-Perez JM, Garcia-Orad A (2020) Variability in cerebrospinal fluid MicroRNAs through life. Mol Neurobiol 57(10):4134–4142. https://doi.org/10.1007/s12035-020-02011-3
59. Kang Y, Ji X, Guo L, Xia H, Yang X, Xie Z, Shi X, Wu R, Feng D, Wang C, Chen M, Zhang W, Wei H, Guan Y, Ye K, Zhao G (2021) Cerebrospinal fluid from healthy pregnant women does not harbor a detectable microbial community. Microbiol Spectr 9(3): e0076921. https://doi.org/10.1128/Spectrum.00769-21

Part II

Cerebrospinal Fluid Subproteomics

Chapter 2

Using Real-Time Quaking-Induced Conversion (RT-QuIC) in Clinical Practice for Creutzfeldt-Jakob Disease Diagnostics

Tuane Cristine R. G. Vieira and Breno José Alencar Pires Barbosa

Abstract

Prion diseases, a group of fatal neurodegenerative disorders, present significant diagnostic challenges due to their subtle clinical manifestations, which often mimic those of other treatable neurological conditions. This diagnostic ambiguity underscores the necessity for reliable and specific diagnostic tools. Real-Time Quaking-Induced Conversion (RT-QuIC) has emerged as a groundbreaking technique for diagnosing these diseases using cerebrospinal fluid (CSF) samples. This chapter outlines a comprehensive and standardized protocol for employing RT-QuIC to detect the pathological form of the prion protein in CSF, offering a powerful tool for clinical diagnosis. It provides step-by-step instructions for patient examinations, laboratory RT-QuIC analysis, and data interpretation. By adhering to these protocols, researchers and clinicians can achieve high sensitivity and specificity in diagnosing prion diseases, facilitating timely and accurate patient management.

Key words RT-QuIC, Prion diseases, Cerebrospinal fluid, Protein purification, Neurodegenerative disorders, Diagnostic assay

1 Introduction

Neurodegenerative diseases, characterized by the progressive degeneration of the nervous system, represent a significant burden on global health. These disorders, which include Alzheimer's disease, Parkinson's disease, and prion diseases, often have overlapping symptoms, such as cognitive decline, behavioral changes, and psychiatric symptoms, that complicate differential diagnosis [1]. Additionally, the clinical manifestations of neurodegenerative diseases can also resemble those of treatable neurological conditions, further complicating accurate diagnosis [2]. Reliable biomarkers are critically needed to facilitate early and precise diagnosis, enabling timely intervention and improved patient outcomes.

Enrique Santamaría and Joaquín Fernández-Irigoyen (eds.), *Cerebrospinal Fluid (CSF) Proteomics: Methods and Protocols*, Methods in Molecular Biology, vol. 2914, https://doi.org/10.1007/978-1-0716-4462-1_2,

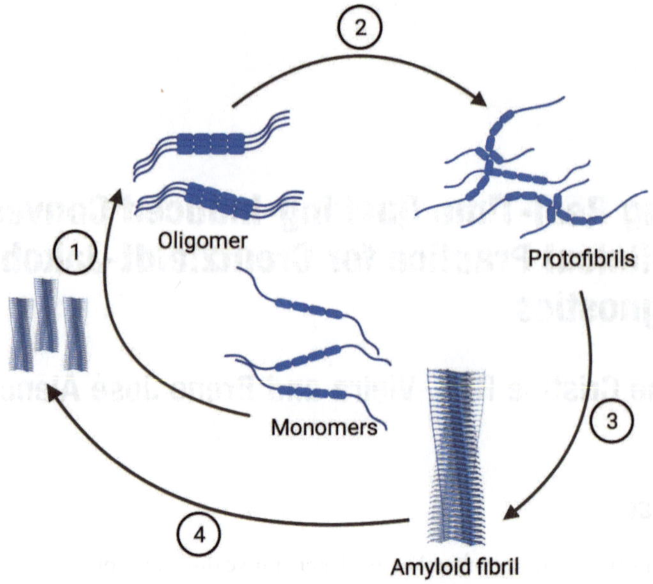

Fig. 1 The stages of amyloid fibril formation, which is central to the RT-QuIC assay mechanism. Protein monomers can misfold and aggregate to form oligomers (1). These oligomers will continue to aggregate, forming protofibrils, intermediate structures exhibiting early fibrillar properties (2). Protofibrils will mature into amyloid fibrils that are highly structured and stable (3), which can be fragmented (4) and further seed the formation of new oligomers, continuing the cycle

Cerebrospinal fluid (CSF) is a valuable medium for biomarker discovery and clinical diagnostics in neurodegenerative diseases. As a clear, colorless body fluid found in the brain and spinal cord, CSF directly interacts with the brain's extracellular space, reflecting its biochemical environment [3]. This proximity makes CSF an excellent source for detecting pathological changes associated with neurodegenerative conditions. Moreover, CSF sampling is minimally invasive compared to brain biopsies, enhancing its practicality and acceptability in clinical settings.

Real-Time Quaking-Induced Conversion (RT-QuIC) is a groundbreaking technique that has gained prominence for detecting misfolded proteins, which are hallmark features of several neurodegenerative diseases [2, 4]. RT-QuIC operates on the principle of amplifying small amounts of disease-associated protein by inducing its conversion into a detectable form (Fig. 1). This method offers exceptional sensitivity and specificity, making it highly suitable for diagnosing prion diseases using CSF samples [5]. As various disease control entities recommend, RT-QuIC is used as a paraclinical assay to diagnose Creutzfeldt-Jakob Disease (CJD) in conjunction with clinical symptoms [6, 7].

In summary, the need for reliable biomarkers in neurodegenerative diseases, the advantages of CSF as a sample medium, and the principles and applicability of RT-QuIC collectively underscore the potential of this technique in clinical diagnostics. This chapter will bring a standardized protocol for employing RT-QuIC to detect prion diseases, providing a crucial tool for improving patient management and outcomes.

2 Materials

All solutions should be prepared using ultrapure water and high-grade analytical reagents. Ensure solutions are freshly prepared and filtered using a 0.22 μm membrane.

2.1 Heterologous Protein Expression

1. PET plasmids containing the PrP sequence can be used without attaching to any Tag. Sequences from different species can be used in PrP RT-QuIC, but the hamster sequence from amino acids 90–231 is the most applied substrate for IQ-CSF.
2. BL21 *E. coli*.
3. Luria Bertani (LB) medium.
4. Antibiotic according to the plasmid or *E. coli* chosen.
5. 1 M solution of IPTG (Isopropyl β-D-1-thiogalactopyranoside).

2.2 Protein Purification

1. Bug Buster Master Mix 1x.
2. Ni-NTA beads.
3. SnakeSkin Dialysis Tubing 7 K.
4. Filter, Millex-GP, 0.22 μM.
5. Denaturing Buffer 1: 0.1 M sodium phosphate, 8 M guanidine HCl, and 10 mM Tris Base, pH 8.0.
6. Denaturing Buffer 2: 0.1 M sodium phosphate, 6 M guanidine HCl, and 10 mM Tris Base, pH 8.0.
7. Refolding Buffer: 0.1 M of sodium phosphate and 10 mM Tris Base, pH 8.0.
8. Elution Buffer: 0.1 M of sodium phosphate, 10 mM of Tris Base, and 0.5 M of imidazole, pH 5.8.
9. Dialysis buffer: 10 mM sodium phosphate, pH 5.8.

2.3 RT-QuIC Materials

1. Black, transparent, flat bottom 96 well plates and sealers.
2. 100 KDa filters.
3. SDS/PBS Buffer: 0.1% (wt/vol) SDS in PBS. Keep the buffer stock at 4 °C.

4. 5x PBS, pH 6.9, buffer. Keep the buffer stock at 4 °C.
5. 100 mM EDTA. Keep the buffer stock at 4 °C.
6. 2 M NaCl.
7. 1 mM ThT. Keep the 1 mM stock in the dark at room temperature.

3 Methods

3.1 Patient Examination and Sample Collection

Given the CSF's crucial role in maintaining central nervous system (CNS) homeostasis such as brain maintenance, clearance of metabolites, and hydrodynamic stabilization its examination is a critical method for diagnostic and therapeutic purposes in Neurological clinical practice and research. This fluid is typically collected through a standard medical procedure called lumbar puncture (LP). LP is vital for diagnosing bacterial, fungal, mycobacterial, and viral CNS infections and, in specific circumstances, aids in diagnosing subarachnoid hemorrhage, demyelinating diseases, inflammatory disorders, and CNS malignancies. For neurodegenerative CNS disorders, an LP for CSF assessment should be performed in particular scenarios, as detailed below.

3.2 Indications for CSF Sampling in Suspected Neurodegenerative Cases [8, 9]

1. rapidly progressive dementias (i.e., Cognitive decline leading to severe functional impairment in less than 1–2 years) to rule out inflammatory, infectious, or neoplastic conditions mimicking degenerative disorders. In this scenario, additional testing includes (1) testing for 14-3-3 protein, which could be elevated in prion diseases but has poor specificity; (2) screening for autoantibodies for autoimmune encephalopathies in suspected cases; and (3) assessing Alzheimer's disease biomarkers such as amyloid-beta, phospho-tau, and total tau profiles.
2. early-onset dementias (i.e., Cognitive decline in patients younger than 65 years old) to rule out inflammatory, infectious, or neoplastic conditions mimicking degenerative disorders.
3. atypical dementia syndromes (i.e., Language-predominant cognitive decline) since, in these scenarios, the dementia's etiology could be either due to Alzheimer's disease or frontotemporal dementia. Whenever available, Alzheimer's disease biomarkers should be performed for improved diagnostic accuracy.

3.3 Step-by-Step Procedure for CSF Collection [10]

1. LP should be performed by trained physicians in a proper medical facility. For most patients, we recommend the use of the lateral decubitus position as it facilitates access to the cerebrospinal fluid (CSF) space and permits an accurate measurement of intraspinal opening and closing pressures.

2. The site of needle insertion is determined visually and confirmed by palpation; a direct line joining the highest points of the iliac crests is a guide to the fourth lumbar vertebral body. The spinal needle can be safely inserted into the subarachnoid space at the L3–4, L4–5, or L5-S1 interspaces since they are below the termination of the spinal cord in most patients.
3. The overlying skin should be cleaned with alcohol and a disinfectant such as povidone-iodine or chlorhexidine (0.5% in 70% alcohol). After the skin is cleaned and allowed to dry, a sterile drape with an opening over the lumbar spine is placed on the patient. The Healthcare Infection Control Practices Advisory Committee and the Centers for Disease Control and Prevention (CDC) recommend face masks for individuals who place a catheter or inject material into the spinal canal.
4. Insert a styleted 20- or 22-gauge spinal needle at a slightly cephalad angle. If a cutting tip needle is used, insert the needle with the bevel parallel to the long axis of the spine to reduce the risk of post-dural puncture headache (PDPH).
5. Once CSF flows well from the needle hub, advance the needle no further.
6. Allow CSF to flow by gravity into sterile plastic tubes. Do not aspirate CSF, as this may increase the risk of bleeding. Typically, 8–16 mL of CSF is collected (2–4 mL in each of four tubes) and sent for analysis.
7. The stylet may be replaced before removing the spinal needle.

3.4 Handling and Storage of CSF Samples Before Analysis [11]

1. For proper RT-QuIC analysis, a volume of at least 0.5 mL of clear and colorless CSF is required.
2. The CSF sample should be collected into a routine universal container.
3. If the lumbar puncture is traumatic, we advise that the CSF sample is collected in a series of containers until it is clear. We advise that the RCC is performed on the clearest sample.
4. Ask the laboratory to freeze the CSF sample. It should be frozen within 2–3 h. A normal −20 °C freezer is acceptable, but if the lab has a −70 °C/80 °C, that is better.

3.5 PrP Purification

The protocol detailed here is suitable for purifying both full-length and truncated PrP.

3.5.1 Competent BL21 *E. coli* Cells Transformation and Culturing

1. Chill cell samples on ice for 15 min (~50 μL of cell suspension in conical tubes).
2. Add 1 μL of plasmid solution (5–10 ng) to the bacterial samples (controls are treated similarly without plasmid).

8. Measure ThT fluorescence every ~15 min with settings: excitation at 450 nm, emission at 480 nm, bottom read, 20 flashes, 20 μs integration time. Adjust gain with a negative control initially.

3.7 Data Interpretation and Analysis

The RT-QuIC assay measures Thioflavin T (ThT) fluorescence, which increases as prion aggregates form. For analyzing fluorescence data:

3.7.1 Fluorescence Data Analysis

1. Baseline Correction: Adjust the fluorescence readings to correct for any baseline drift by subtracting the initial fluorescence values from all subsequent readings.
2. Threshold Determination: Establish a fluorescence threshold that distinguishes between positive and negative reactions. This threshold is typically set based on the fluorescence intensity of negative control wells.
3. Kinetic Analysis: Plot the fluorescence intensity (media of at least four replicates) against time. Analyze the reaction's kinetics, focusing on the ThT response. Higher ThT responses indicate higher prion seeding activity.
4. Qualitative Comparison: Compare the fluorescence signal of test samples to those of positive and negative controls. The area under the curve (AUC) can also be used for comparisons.

3.7.2 Interpretation of Results

1. Positive Samples: Samples exhibiting fluorescence intensity exceeding the established threshold will be considered positive if 2/4, 2/4, and 4/4 of wells test positive (Fig. 2).
2. Negative Samples: Samples exhibiting fluorescence intensity that do not show a significant increase in fluorescence or remain close to the baseline will be considered negative if 4/4 of the wells test negative (Fig. 2).
3. Indeterminate Results: Wells with fluorescence intensities near the threshold or if 1/4 of wells test positive may require repeat testing or additional confirmatory assays to determine their status (Fig. 2).
4. Controls: To validate the assay's performance, ensure the positive controls consistently exhibit high fluorescence and the negative controls remain low.

4 Notes

1. You can store a glycerol culture of the transformed bacteria at −80 °C by aliquoting it with 25% glycerol and keeping it in a −80 °C freezer.

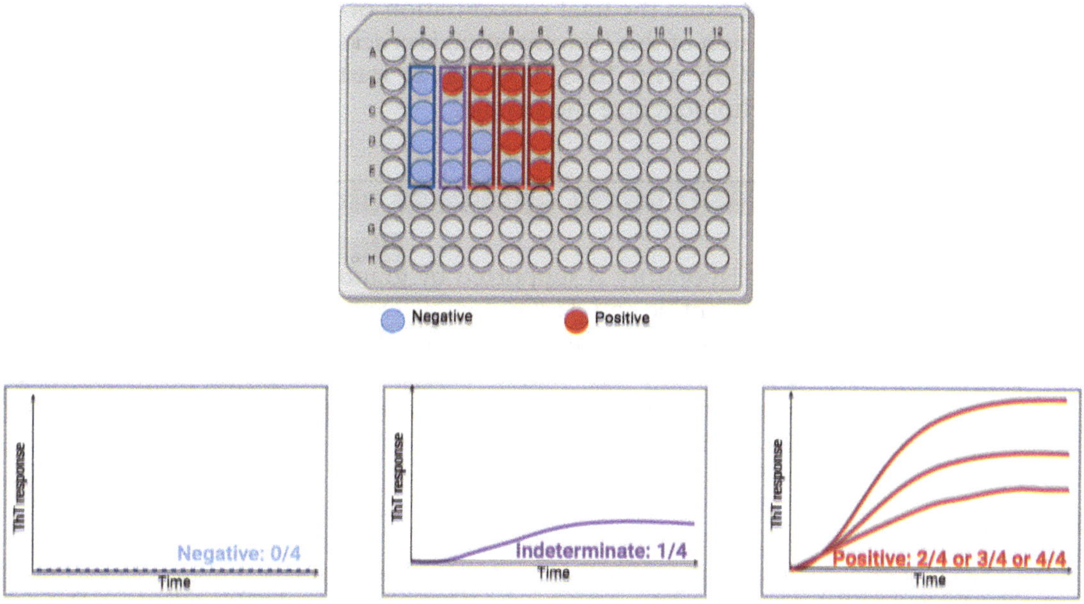

Fig. 2 The possible results of an RT-QuIC assay used to detect the presence of PrP aggregates. The top section shows a microplate with 96 wells, where each well is either negative (blue) or positive (red) for the aggregated protein. The bottom section presents three graphs showing the ThT response over time, corresponding to different assay outcomes: Negative (0/4): no significant increase in ThT response, indicating no protein aggregate detected. Indeterminate (1/4): a slight rise in ThT response in one of four replicates suggests an ambiguous result. Positive (2/4, 3/4, or 4/4): a significant increase in ThT response in two, three, or all four replicates, indicating the presence of misfolded proteins

2. This method typically produces approximately 8 g of cell paste. You can proceed to protein purification or store cell pellets at −20 °C for two to 3 months.
3. Ensure that the pellet is fully resuspended. You can also resuspend the pellet using Denaturing Buffer 1 and sonicate it in an ice bath. You should sonicate it for 30 s on and 30 s off cycles until the pellet changes consistency. After this step, proceed to **step 12**. This protocol is straightforward, but it necessitates the availability of a sonicator. Furthermore, it should be noted that the Ni-NTA superflow resin beads may exhibit increased impurities in the subsequent steps and are not recommended for reuse.
4. This step can be stopped, and the inclusion body pellet can be kept at −20 °C for two to 3 months.
5. The typical yield is approximately 20–30 mg when using the Ni-NTA superflow resin beads for the first time. The resin can be cleaned according to the manufacturer's protocol and reused, but it's important to note that with each reuse, its performance will decrease, and it will yield less protein.

mixture is loaded onto a specialized cartridge containing a large number of microwells, typically more than 235,000. Each microwell is sized to admit just one bead, allowing an ultrasensitive detection of the analyte. To create a liquid-tight seal and ensure that each bead occupies a single microwell, oil is added to the cartridge. This step helps isolate individual beads and prevents cross-contamination.

The fluorescence emitted by the beads is measured using a fully or semi-automated system equipped with a fluorescent microscope. Because each microwell can hold only one bead, the number of microwells with fluorescence is proportional to the amount of target protein in the sample. This enables single molecule detection and the ability to detect very low concentrations of biomarkers, making it particularly useful in the early detection of diseases or monitoring disease progression [3].

A graphical explanation of the method can be seen in Fig. 1.

There are two different systems for bead-bound protein detection: (1) a fully automated SIMOA bead-based immunoassay platform designed to multiplex up to four biomarkers in a single assay with unrivaled sensitivity; and (2) a semi-automated compact analyzer that offers simple workflows with minimal user intervention for running bead-based assays [4].

The SIMOA software calculates the fluorescent measurements as an average number of enzymes per bead (AEB), which can be further translated to concentration units using a calibration curve. At low concentrations of target protein, the reaction environment includes many more beads than protein molecules; hence, following the Poisson distribution, most beads carry no or only one enzyme. Therefore, at low concentrations, AEB is determined by the digital count of positive beads, defined as those emitting fluorescence. At higher concentrations of a target protein, more enzymes can be bound to a single bead; the AEB is computed based on the averaged analog signal output from all the beads present on a microarray [5].

A key advantage of SIMOA is that it can detect proteins in the femtomolar range (typically ranging from 10 fg/mL to 1 pg/mL) [3].

SIMOA technology relies on its high sensitivity in detecting individual beads, enabling the precise quantification of target biomolecules even at very low concentrations. This high sensitivity renders it a valuable tool for assessing the diagnostic applicability of classical biomarkers within Cerebrospinal Fluid (CSF), blood plasma or saliva [1]. Moreover, its procedural time is shorter compared to traditional ELISA methods [6].

Among other strengths of SIMOA is that, in addition to single-plex assays, multiplex assays can be run, allowing for simultaneous measurement of multiple analytes of interest in a single test, all while utilizing minimal sample volume [7]. This multiplexing

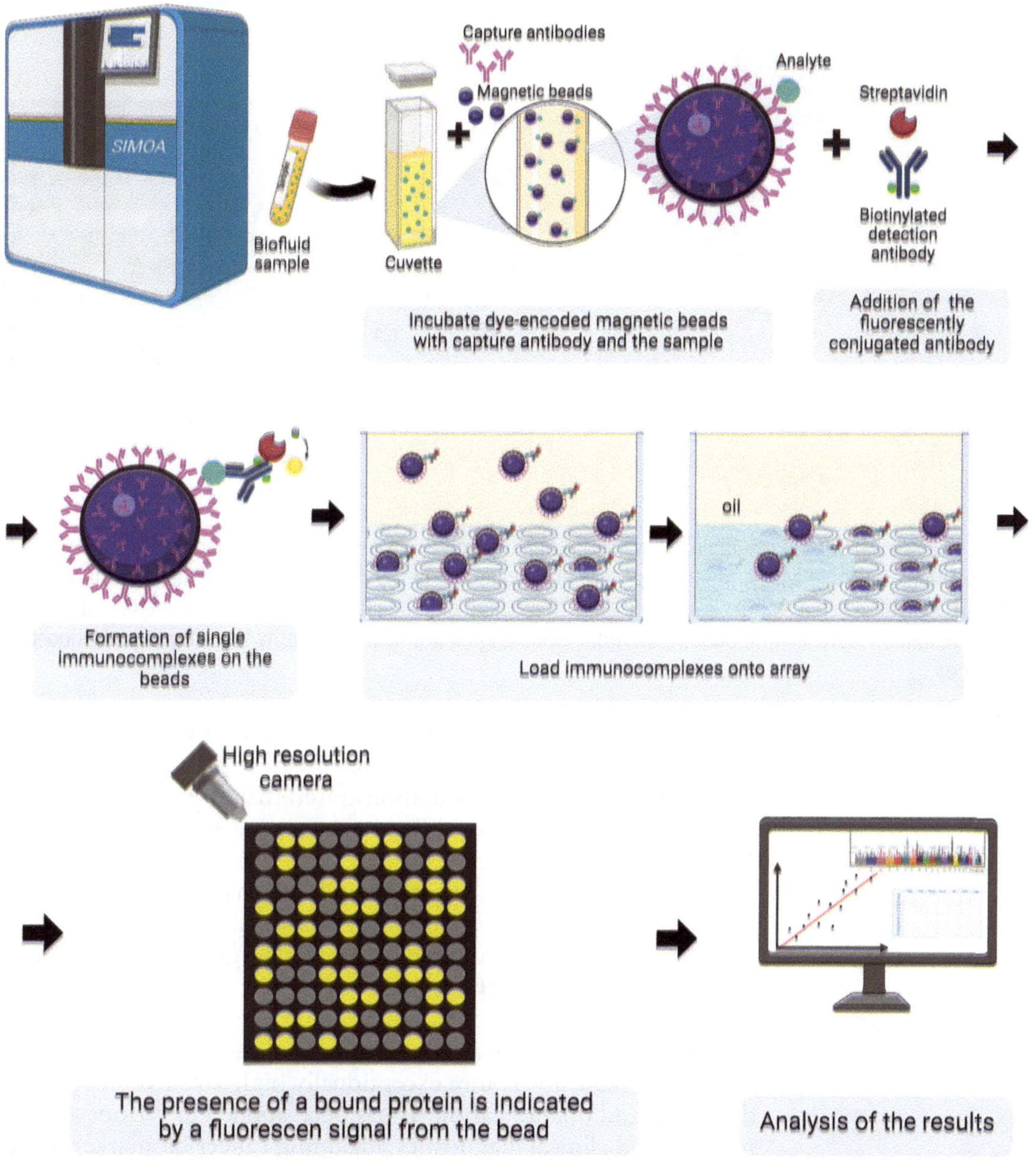

Fig. 1 *Schematic illustration of Single Molecule Array (SIMOA) workflow.* Capture antibodies are coupled to microbeads and then incubated with the biological sample and biotinylated detection antibodies to form an immunocomplex. Following hybridization, microbeads suspended in the substrate are loaded onto an array of femtoliter-size wells. After loading, the wells are sealed with oil. This results in an array of isolated reaction chambers, each of which contains either none or one microbead. If the immunocomplex is present on the microbead, it generates a fluorescent product, resulting in a detectable fluorescent signal. The array is then imaged and analyzed to determine the total number of microbeads to calculate the average enzyme per bead

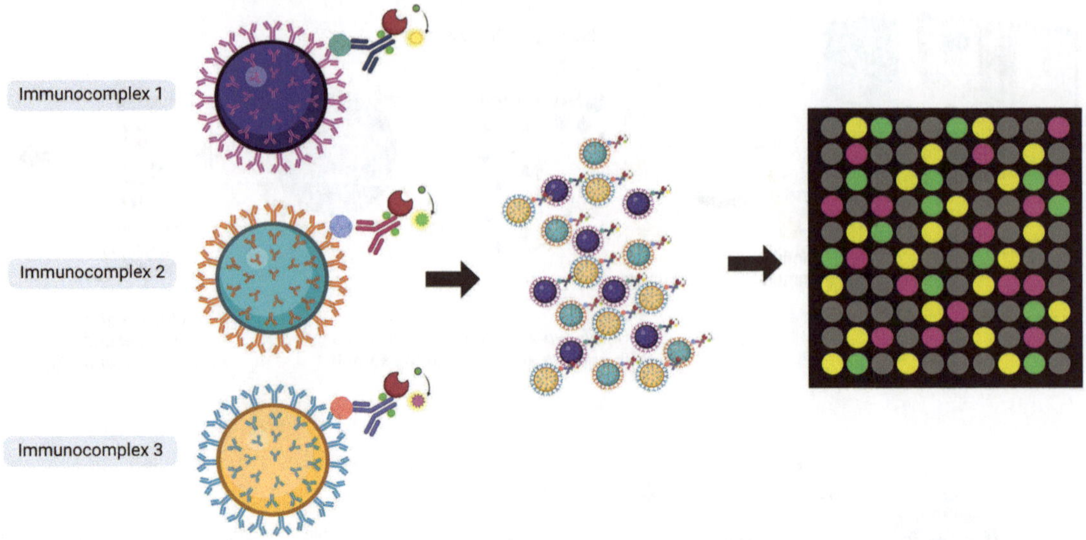

Fig. 2 *Multiplexed SIMOA*. Multiplexed measurements involve the simultaneous quantification of multiple analytes within a single assay. Multiple capture antibodies are immobilized on microbeads, each targeting a different analyte. When a sample is applied, each microbead captures its respective target analyte, generating subpopulations. Fluorescent images at different wavelengths are used to identify the different immunocomplex types

capability is especially advantageous considering the multifactorial nature of many neurodegenerative diseases, such as Alzheimer's disease, where research has demonstrated that assessing a combination of markers enhances diagnostic accuracy [8] (Fig. 2) (*see* **Note 1**).

While SIMOA achieves the highest sensitivity compared to standard ELISA methods, this technology also has drawbacks as the currently limited number of protein targets, the higher cost associated with SIMOA kits, and the specific detection instrument [9].

Overall, SIMOA is a powerful technology for researchers and diagnosticians who require exceptionally high sensitivity in protein detection and quantification. It has been applied in various areas of research and clinical diagnostics, including cancer biomarker detection, neurodegenerative disease research (such as Alzheimer's and Parkinson's disease), and cardiovascular disease diagnostics. Its high sensitivity and precision have made it a valuable tool for advancing our understanding of various diseases and improving patient care.

2 Materials

SIMOA Assay Kits contain ready-to-use reagents for running 96 tests of target analyte(s) on any of the different SIMOA

Analyzers using single-molecule array technology. Depending on the biomarkers of study, the SIMOA Assay Kit is different, with the reagents and the antibodies varying, and so on; the protocol is specific for each biomarker.

2.1 SIMOA Assay Kits

All SIMOA Assay Kits include:

1. *Bead Reagent kit*: It includes capture antibody, paramagnetic beads, Bead Conjugation Buffer with protein stabilizers, Bead Wash Buffer, Bead Blocking Buffer, and Bead Diluent Buffer.
2. *Detector Reagent kit*: It includes detector antibody, Biotinylation Reaction Buffer with protein stabilizers and Biotin vial.
3. *SβG Reagent*: Streptavidin-β-galactosidase (SβG) in buffer with protein stabilizers.
4. *RGP Reagent*: Resorufin β-D-galactopyranoside (RGP) in buffer with a surfactant.
5. *Sample Diluent*: Buffer with protein stabilizers; it is a heterophilic blocker.
6. *Quanterix 96-well plate.*
7. *Calibrators*: Antigen in buffer with protein stabilizers.
8. *Controls*: Antigen in buffer with protein stabilizers.

2.2 Other Materials

1. 1x PBS.
2. Amicon Centrifugal Filters.
3. EDC (1-ethyl-3-[3-dimethylaminopropyl] carbodiimide hydrochloride).
4. HulaMixer orbital shaker.
5. Coulter Counter Z2: includes a Z-Pac clean solution.
6. Polypropylene tubes.
7. SIMOA Analyzer (Quanterix).
8. SIMOA System Wash Buffer 1: specific of the analyzer (Quanterix) *.
9. SIMOA System Wash Buffer 2: specific of the analyzer (Quanterix) *.
10. SIMOA Sealing Oil (Quanterix) *.
11. SIMOA Cuvettes (Quanterix) **.
12. SIMOA disposable Pipettor tips (Quanterix) **.
13. SIMOA Discs (Quanterix) **.

    * SIMOA System Wash Buffers and the Sealing Oil are sold with the Analyzer (Quanterix).

    ** Cuvettes, Pipettor Tips and Discs are sold as consumable unit "SIMOA Disc Kit 2.0" (Quanterix).

3 Methods

3.1 Bead Coupling

The bead coupling protocol is based on the coupling protocols developed by Quanterix. The specific manual for the biomarker studied will be included in the Assay Kit, and as it will be optimized for the given antibody and assay, it can be slightly different.

3.1.1 Preparation of the Capture Antibody

1. Thaw 100 μg of capture antibody to 500 μL using 1x PBS to get a final concentration of 0.2 mg/mL. Mix it for 10 min with gentle rotation.
2. Place the 500 μL solution in an Amicon Ultra Centrifugal Filter of 0.5 mL and centrifuge at 14,000× *g* for 5 min.
3. Discard the flow through and add 450 μL of Bead Conjugation Buffer.
4. Centrifuge at 14,000× *g* for 5 min.
5. Discard the flow through and add 450 μL of Bead Conjugation Buffer.
6. Centrifuge at 14,000× *g* for 5 min.
7. Discard the flow-through and invert the filter into a new 1.5 mL tube.
8. Centrifuge at 1000× *g* for 2 min.
9. Add 25 μL of Bead Conjugation Buffer to each side of the filter (total volume, 50 μL).
10. Centrifuge at 1000× *g* for 2 min.
11. Measure the concentration of the antibody in solution using a NanoDrop spectrophotometer. Blank the NanoDrop with Bead Conjugation Buffer.
12. Transfer the volume of the antibody solution to a new 1.5 mL tube and label it as *Capture Antibody Solution (CA).*
13. Place it on ice or at 4 °C until use.

3.1.2 Wash and Activation of the Beads

1. Let sit EDC at room temperature for 10 min.
2. Discard the supernatant and remove the tube with the paramagnetic beads from the Dynamag magnet.
3. Add 200 μL of Bead Wash Buffer, vortex for 3 s at high speed and place back on the magnet.

   Repeat twice.
4. Add 200 μL of Bead Conjugation Buffer to the tube, vortex for 3 s at high speed and place back on the magnet.

   Repeat twice.
5. Add 300 μL of Bead Conjugation Buffer to the tube.

6. Reconstitute the EDC in 100 μL of Bead Conjugation Buffer and vortex for 10 s.
7. Add the appropriate amount of EDC to the tube of beads (indicated in SIMOA Assay Kit instructions).
8. Mix the tube with the beads on a HulaMixer with rotation and intermittent shaking for 30 min at the appropriate temperature for each biomarker (indicated in SIMOA Assay Kit instructions).
9. Spin the tube with the beads before placing it on the Dynamag magnet.
10. Discard the supernatant. Remove the tube from the Dynamag magnet, add 200 μL of Bead Conjugation Buffer, and vortex for 3 s.
11. Add 300 μL of the *Capture Antibody Solution* to the tube and vortex for 3 s.
12. Put the tube in the HulaMixer with rotation and intermittent shaking for 2 h at the appropriate temperature for conjugation.
13. Remove the beads from the HulaMixer and spin the tube before placing it on the Dynamag magnet.
14. Remove the supernatant, place it in a tube labeled *Supernatant*, and add 200 μL of Bead Wash Buffer. Vortex for 3 s. Place back the beads tube on the Dynamag magnet.
15. Remove the supernatant, place it in a tube labeled *Wash*, and add 200 μL of Bead Wash Buffer. Vortex for 3 s. Place back the beads tube on the Dynamag magnet.
16. Discard the supernatant and add 200 μL of Bead Wash Buffer. Vortex for 3 s. Place back the beads tube on the Dynamag magnet.

    Repeat twice.
17. Discard the supernatant and add 200 μL of Bead Blocking Buffer. Vortex for 3 s.
18. Put the beads in the HulaMixer with rotation and intermittent shaking for 30 min at room temperature for blocking.
19. Spin the tube with the beads before placing it on the Dynamag magnet.
20. Discard the supernatant. Remove the tube from the Dynamag magnet and add 200 μL of Bead Wash Buffer. Vortex for 3 s. Place back the beads tube on the Dynamag magnet.
21. Discard the supernatant. Remove the tube from the Dynamag magnet and add 200 μL of Bead Diluent Buffer. Vortex for 3 s.

3.3.2 Calculation of the Volume of Detector Antibody to Be Used per Assay

1. The total volume of a *Biotinylated Detection Antibody* to be inserted in the instrument is:

$$\mathrm{Vol_{BDA}}(\mu\mathrm{L}) = M \times 30\ \mu\mathrm{L} + 700\ \mu\mathrm{L}$$

2. The concentration of the Detection antibody necessary to add to each sample is indicated in the SIMOA kit: *[DA]*.

   The amount of *Biotinylated Detection Antibody* to spike into each sample is *(X)*:

$$X\ (\mu\mathrm{L}) = \frac{[DA] \times \mathrm{Vol_{BDA}}}{[\mathrm{Concentration\ of\ BDA}]}$$

3.3.3 Calculation of the Volume of SβG Solution to Be Used per Assay

1. The total volume of *SβG Solution* to be inserted in the instrument is:

$$\mathrm{Vol_{S\beta G}}(\mu\mathrm{L}) = M \times 110\ \mu\mathrm{L} + 700\ \mu\mathrm{L}$$

2. The SβG concentration necessary depends on the biomarker of study, as it is indicated in the SIMOA kit: $[\mathrm{S\beta G_{theoretical}}]$.

   The SβG stock concentration (SβG Reagent) is generally 50 nM (but if not, adjust the next equation). The amount of SβG Reagent to add to the total volume of solution in the Analyzer is:

$$\mathrm{S\beta G\ Reagent}\ (\mu\mathrm{L}) = \frac{[\mathrm{S\beta G_{theoretical}}] \times \mathrm{Vol_{S\beta G}}}{[50,000\ \mathrm{pM}]}$$

3.3.4 Calculation of the Amount of RGP

The instrument uses 25 μL of RGB per measure:

$$\mathrm{RGP\ Reagent}\ (\mu\mathrm{L}) = 25\ \mu\mathrm{L} \times M$$

3.4 Preparation of the Instrument

1. Turn on the Analyzer and initialize the SIMOA software.
2. Perform pre-run maintenance by going to the "*Maintenance*" tab and clicking "Run Task." This takes up to 20 min.
3. Download the Assay Definition from the customer portal and install it on the SIMOA Analyzer prior to performing the assay.
4. Import the assay definition: "Custom assay" – "Import."

3.5 Setting Up Assay

1. Allow Calibrator and Control Reagents to come to room temperature.

   Do not heat vials to accelerate thawing.
2. When the Calibrator and Control Reagents are fully thawed, mix them by multiple gentle inversions or vortexing.

3.6 Sample Preparation

Sample preparation for SIMOA assays using cerebrospinal fluid involves specific steps to ensure accurate and reliable results.

3.6.1 Collection and Handling

The CSF volume taken can influence the concentration of biomarkers. Most molecules and cell numbers have a rostro-caudal concentration gradient [10, 11]. This means that small volumes of CSF may reflect the composition of the lumbar sac, whereas large volumes of the rostral spinal composition. Therefore, comparing biomarker concentrations from different CSF volumes can lead to erroneous results. Thus, a standard volume of 12 mL of CSF should be collected.

Because of the concentration gradient, the best location of puncture is vertebral body L3-L5.

Then, a removal of bloody CSF samples should be done as blood proteins lead to the suppression of proteomic patterns in CSF (*see* **Note 2**).

3.6.2 Aliquoting and Storage

Collect CSF samples into sterile polypropylene tubes and centrifuge for 10 min at 400× *g* at room temperature.

Once processed, aliquot and store the supernatant in 0.5 mL low-binding tubes and immediately freeze at −80 °C pending analysis.

Avoid repeated freeze–thaw cycles, as this can affect the integrity of the proteins in the CSF.

3.6.3 Thawing Samples

When you are ready to perform the SIMOA assay, remove the required aliquots from the freezer. Thaw the samples at room temperature. Avoid using a microwave or heating samples rapidly, as this can denature proteins.

3.6.4 Assay Specifics

Follow the specific SIMOA assay protocol provided by the manufacturer, which will detail how to mix the CSF samples with the antibody-coated beads and detection reagents.

3.7 Calibration Curves

1. Each diluent is optimized to work for spike and recovery and dilution linearity in both cerebrospinal fluid and plasma.

   The calibration curves are written for the stock concentrations of the biomarkers of the SIMOA Assay Kit. Moreover, the Sample Diluent used depends on the Assay Kit too, and the mixing reaction is indicated in the kit.

   Serially dilute the biomarker of study at the indicated concentrations with Sample Diluent.

2. Plate in each well of a Quanterix supplied 96-well plate the required volumes of the diluted biomarker as well as of Calibrator and Control Reagents as indicated in the SIMOA Assay Kit Protocol to generate standard curves.

3.8 Loading Instrument and Run Assay

3.8.1 Load Volumes

1. Load the calculated volumes of beads (*ABC Solution*), detector antibody (*BDA Solution*), and *SβG solution* into the reagent cuvette and incubate for 35 min.
2. Select "Load Reagent" tab and select the reagent lane. Use the barcode reader to assign positions of reagents in the reagent rack. Insert the reagent rack.
3. The beads are then pelleted with a magnet and the supernatant is removed.
4. Add 100 μL of *SβG solution* and incubate for 5 min.
5. Add the calculated volume of RGP to the cuvette and load onto the array disc.
6. Use the barcode reader to assign RGP and insert the array disc.
7. Touch the "Done Loading Reagents" tab.

   The array disc is then sealed with oil automatically.

3.8.2 Assign Calibrators and Samples

1. Create the plate layout by selecting the "Setup Run" tab and assigning the "Batch Name" and the "Plate Barcode."
2. Assign calibrators by selecting the "Assign Calibrators" tab. Select the number of replicates per well.
3. Select the "Assign Sample" tab and highlight all wells that contain any controls or samples. Select "Assay" and assign the number of replicates per well.
4. Click of "List View" to confirm the selections. When confirmed, insert the plate rack and select "Done with Setup" (*see* **Note 3**).

3.8.3 Start Run

1. Go to the "System Resources" tab and select "All Resources."
2. Click "Start Run."
3. If you want to monitor the progress of the run, go to the "Current Run" tab (*see* **Note 4**).

3.9 Post-Run

1. Remove the sample plate, reagents and RGP.
2. Review the results and export them as a CSV file.
3. Perform post-run maintenance: select the "Maintenance" tab and check the "End of Day" task. Click "Run Task." It takes 15 min.
4. Shut down the software; turn off the instrument and the PC.

3.10 Data Analysis

Analysis of array images and AEB values calculations are done automatically by the software in the SIMOA Analyzer.

Digital counting of molecules is used to determine the concentration of target analytes using the Poisson distribution:

$$P_{\mathrm{AEB}} = e^{-\mathrm{AEB}} \times \left(\frac{\mathrm{AEB}^z}{z!}\right)$$

where "*z*" is the number of enzyme molecules associated with the beads (*see* **Note 5**).

The concentration of unknown samples is determined using the calibration curves. This analysis is automated using Quanterix-provided software on the SIMOA Analyzer. The limit of detection (LOD) is calculated by adding three times the standard deviation to the mean of the background signal.

Calibrator concentrations are adjusted to optimize the quantification of the biomarkers within the sample. The lower limit of quantitation (LLOQ) and upper limit of quantitation (ULOQ) are defined as the lowest and highest calibrators, respectively, and are measured in triplicate with coefficients of variation (CVs) less than 20%. The LLOQ also satisfies the criterion of having a signal at least 10 standard deviations above the zero calibrator (background). Finally, samples outside the range LLOQ – ULOQ are set to the respective limit in the figures and are excluded from the statistical analysis.

4 Notes

SIMOA is a powerful technology, but like any analytical method, it can face certain challenges or problems. Some of the major issues that can occur with SIMOA and how they can be addressed include:

1. Measurements using SIMOA technology can be conducted through two primary approaches: individual measurements and multiplexed measurements. Multiplexed measurements involve the simultaneous quantification of multiple analytes within a single assay. Indeed, multiplexed detection is achieved using optically distinct beads, which are coated with different dyes or variations in color intensities to generate unique bead subpopulations. Each of these optically distinct bead subpopulations is conjugated to a specific antibody targeting a particular target analyte. Using a specific multiplex SIMOA kit is crucial when conducting multiplexed assays. These kits are designed to ensure that the detection antibodies are uniquely tagged with different dyes, allowing for the differentiation and recognition of various biomarkers of interest within a single sample (Fig. 2).
2. Contamination of samples, particularly with blood, can introduce extraneous proteins and interfere with the quantification of target biomarkers. If contamination happens, preprocessing steps such as centrifugation or filtration can remove blood cells and large debris. Additionally, control samples can assess the extent of contamination.

3. Instrument drift or malfunction can affect data quality. To overcome it, regular instrument calibration, maintenance, and quality control checks are essential.
4. Interactions of non-target molecules, which result in signal production due to non-specific binding (NSB), produce a high background signal in SIMOA assays that can reduce the accuracy of measurements. NSB can lead to reduced specificity, making it difficult to distinguish between true positive signals and signals generated by non-specific interactions. Moreover, NSB can result in false-positive results, leading to inaccurate conclusions and misleading findings. If there is a high background signal, optimization of assay conditions, like washing steps, which can help to remove unbound molecules and reduce background noise. Another strategy could be using blocking agents such as bovine serum albumin or non-fat dry milk, which can reduce NSB by saturating available binding sites on surfaces. Finally, properly designed negative and quality controls can help assess and account for background signal sources.
5. Poisson noise, also known as shot noise, is inherent in counting discrete events, such as individual molecules, and is directly related to the number of measurements made (M). The Poisson noise of M measurements is proportional to the square root of M, leading to increased uncertainty when dealing with a lower number of molecules. In addition, multiplexed measurements can introduce additional complexity and sources of noise. To ensure minimal Poisson noise it is important to interrogate a sufficient number of beads, so in multiplex assays, there should be a huge number of beads from each distinct subpopulation. However, using too many bead subpopulations may introduce noise. Balancing the trade-offs between increasing the number of measurements to reduce Poisson noise and maintaining manageable complexity and noise levels in multiplexed assays is essential to achieve accurate and reliable single-molecule measurements. Because the Poisson noise is an intrinsic error that cannot be reduced, it should be considered in the context of the analytical requirements.

Acknowledgments

The Clinical Neuroproteomics Unit is a member of the Spanish Olfactory Network—ROE (grant RED2022-134081-T from the Spanish Ministry of Science & Innovation) and is supported by grants PID2023-152593OB-I00 funded by MCIU/AEI/10.13039/501100011033 / FEDER, UE to ES and JFI and 0011-1411-2023-000028 (from Government of Navarra—

Department of Economic and Business Development-S4). EA-C is supported by "Programa MRR Investigo 2023" in the framework of the European Union recovery and resilience facility.

References

1. Thangavelu B, Boutté AM (2021) Single molecule assay for ultrasensitive detection of Cathepsin B in human blood. ACS Omega 6: 9609–9616. https://doi.org/10.1021/acsomega.1c00180
2. Norman M, Ter-ovanesyan D, Trieu W, Norman M (2020) Simoa extracellular vesicle assays roey bead coupling protocol. pp 1–12
3. Costa OR, Verhaeghen K, Roels S et al (2018) An analytical comparison of three immunoassay platforms for subpicomolar detection of protein biomarker GAD65. PLoS One 13:1–10. https://doi.org/10.1371/journal.pone.0193670
4. Wilson DH, Rissin DM, Kan CW et al (2016) The Simoa HD-1 analyzer: a novel fully automated digital immunoassay analyzer with single-molecule sensitivity and multiplexing. J Lab Autom 21:533–547. https://doi.org/10.1177/2211068215589580
5. Teunissen CE, Kimble L, Bayoumy S et al (2023) Methods to discover and validate biofluid-based biomarkers in neurodegenerative dementias. Mol Cell Proteomics:100629. https://doi.org/10.1016/j.mcpro.2023.100629
6. Kuhle J, Barro C, Andreasson U et al (2016) Comparison of three analytical platforms for quantification of the neurofilament light chain in blood samples: ELISA, electrochemiluminescence immunoassay and Simoa. Clin Chem Lab Med 54:1655–1661. https://doi.org/10.1515/cclm-2015-1195
7. Rissin DM, Kan CW, Song L et al (2013) Multiplexed single molecule immunoassays. Lab Chip 13:2902–2911. https://doi.org/10.1039/c3lc50416f
8. Thijssen EH, La Joie R, Wolf A et al (2020) Diagnostic value of plasma phosphorylated tau181 in Alzheimer's disease and frontotemporal lobar degeneration. Nat Med 26:387–397. https://doi.org/10.1038/s41591-020-0762-2
9. De Meyer S, Schaeverbeke JM, Verberk IMW et al (2020) Comparison of ELISA- and SIMOA-based quantification of plasma Aβ ratios for early detection of cerebral amyloidosis. Alzheimers Res Ther 12:1–16. https://doi.org/10.1186/s13195-020-00728-w
10. Reiber H (2001) Dynamics of brain-derived proteins in cerebrospinal fluid. Clin Chim Acta 310:173–186. https://doi.org/10.1016/S0009-8981(01)00573-3
11. Martino G, Grimaldi LME, Moiola L et al (1990) Discontinuos distribution of IgG oligoclonal bands in cerebrospinal fluid from multiple sclerosis patients. J Neuroimmunol 30: 129–134. https://doi.org/10.1016/0165-5728(90)90096-6

Chapter 4

Luminex Multiplex Immunoassay for Proteomic Profiling of Cerebrospinal Fluid

Joana M. Huertas-Pons and Ana Quiroga-Varela

Abstract

The enzyme-linked immunosorbent assay (ELISA) is commonly used in research and diagnostics but has limitations, including its single-substance detection and large sample volume requirements. In contrast, the Luminex technique, a bead-based sandwich immunoassay combined with flow cytometry, overcomes these limitations by simultaneous measurement of multiple analytes in small volumes. Using dyed micro-beads coated with monoclonal antibodies, this method can detect 100–500 immune markers in cerebrospinal fluid, capturing multiple targets from a single sample. This chapter discusses key pre-analytical factors like sample collection and preservation that influence detection outcomes, highlighting the Luminex technique's adaptability for measuring diverse biomarkers and its advantages in diagnostic applications, reducing variability, and optimizing resources.

Key words Bead-based immunoassay, Multiplex immunoassay, Biomarker detection, Diagnostic applications, Cytokine profiling

1 Introduction

The enzyme-linked immunosorbent assay (ELISA) has traditionally served as the main method for identifying specific substances in biological samples, extensively used in both research and medical diagnostics. Nonetheless, ELISA has its limitations [1]. It is restricted to detecting one substance at a time and requires substantial quantities of samples and reagents to effectively capture the targeted antigen. Smaller sample volumes are crucial when working with limited portions of samples, such as specific components of blood or cerebrospinal fluid (CSF). Additionally, many ELISA protocols rely on enzyme-mediated signal amplification to achieve adequate sensitivity. However, this amplification process is not consistently linear, potentially leading to skewed results [1, 2].

The capacity to simultaneously analyze multiple substances, known as multiplexing, offers numerous benefits over traditional

Enrique Santamaría and Joaquín Fernández-Irigoyen (eds.), *Cerebrospinal Fluid (CSF) Proteomics: Methods and Protocols*, Methods in Molecular Biology, vol. 2914, https://doi.org/10.1007/978-1-0716-4462-1_4,

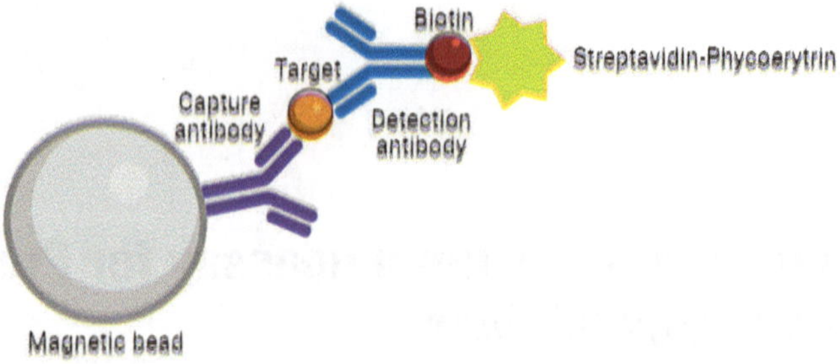

Fig. 1 Magnetic bead sandwich immunoassay. (Adapted from Luminex User Guide)

single-substance testing. This advanced approach is increasingly favored in many research and diagnostic settings due to its efficiency and comprehensive analytical capability. A prime example of multiplexing technology is Luminex® xMAP® Technology. This innovative system utilizes uniquely colored microspheres to identify and measure multiple substances within a single sample with high precision [3, 4]. The xMAP® technology represents a synthesis of several established scientific methods, including flow cytometry, microspheres, laser technology, digital signal processing, and conventional chemistry [2]. By integrating these techniques, Luminex® has created a powerful tool that enhances the accuracy and efficiency of substance detection and quantification in various applications.

The Luminex® technique enables the simultaneous measurement of various nucleic acids and proteins, including cytokines, enzymes, receptors, and ligands. Specific research applications include single nucleotide polymorphism (SNP) genotyping, profiling gene and microRNA expression, cytokine assays, and screening for genetic diseases [3, 5, 6].

The xMAP® assay setup consists of a suspension array where specific binding agents are covalently linked to the surfaces of internally colored microspheres (beads; Fig. 1). These binding reactions occur on the microsphere surfaces, forming an array of 100–500 members. Detection of analyte binding is facilitated by a reporter fluorophore, typically Phycoerythrin (PE) [4, 5]. Each set of beads is coated with a capture antibody tailored to a particular assay, enabling the capture and identification of a specific analyte. On the flow-based detection instrument equipped with dual lasers, the fluorescent intensity directly correlates with the amount of bound analyte, enabling the quantification of each specific analyte [6].

Multiple bead types have been developed, initially featuring 100 fluorescent color tones based on two internal dyes, primarily red and infrared. The addition of a third dye has expanded the array to encompass 500 different color tones. The original MicroPlex®

beads are 5.6 μm polystyrene microspheres functionalized with carboxyl groups for assay capture molecule attachment. Subsequently, MagPlex® beads, superparamagnetic 6.45 μm microspheres have been developed [5, 6].

The capability of adding multiple conjugated beads to each sample results in the ability to obtain multiple results from a single sample. Analyzing multiple factors simultaneously offers several advantages: it maximizes the use of limited samples by allowing data collection from just 25 μL or less of undiluted sample; it minimizes experimental variability since samples are processed only once, resulting in multiple data points derived from a single manipulation; it optimizes productivity by requiring minimal sample preparation and processing while generating high volumes of data; and it is economical, as examining multiple analytes in a single sample saves time and resources [7, 8].

2 Materials

2.1 CSF Collection and Sampling

Proper collection and handling of CSF samples are essential to preserve biomarker integrity and ensure reliable analysis in research studies [9].

Materials required:

- Sterile polypropylene tubes: For collecting CSF samples due to their low protein binding potential.
- Centrifuge: For processing CSF samples immediately after collection.
- Small polypropylene tubes: For aliquoting supernatant.
- −80 °C freezer: For long-term storage of CSF samples.
- Timer: To ensure timely processing of samples within 30 min.
- Personal protective equipment (PPE): Gloves, lab coat, and face shield for handling CSF samples to maintain sterility and safety.

2.2 Luminex® Technology

To conduct Luminex® assays effectively, a specific set of reagents and equipment is essential. This section details the components included in the Luminex® kit and outlines the necessary materials required for conducting assays, ensuring precise and reliable results [2, 6, 8].

Reagents included in the Luminex® kit:

- Standard for Luminex Assay.
- Standard diluent.
- Quality Controls.
- Assay Buffer.
- Wash Buffer (10X).

- Detection Antibodies.
- Streptavidin-Phycoerythrin.
- Set of one 96-well plate with 2 sealers.
- Mixing bottle.
- Magnetic Bead Antibody-Immobilized Magnetic Beads.

Materials required:

- Adjustable pipettes with tips suitable for dispensing 2–1000 μL and 2–200 μL (*see* **Note 1**).
- Luminex reader, available from different manufacturers, each offering their specialized software.
- MAGPIX® Drive Fluid, xMAP® Sheath Fluid PLUS or xMAP® Sheath Concentrate PLUS.
- Vortex mixer and sonicator.
- Titer plate shaker.
- Automatic plate washer for magnetic beads or handheld magnetic separation block.
- Dual-laser flow-based plate reader complete with software for data acquisition and analysis.

2.3 Beads

Several types of beads have been developed for Luminex® assays. Initially, the first generation featured 100 fluorescent color tones based on internal dyes, primarily red and infrared [5, 6]. Subsequent advancements introduced a third dye, expanding the array to encompass 500 different color tones [5].

The original MicroPlex® beads are 5.6 μm diameter polystyrene microspheres, functionalized with carboxyl groups for covalent attachment of assay capture molecules. They are available in 500 different color sets. In contrast, MagPlex® beads are superparamagnetic microspheres functionalized with surface carboxyl groups and are available in 500 different color sets [5].

3 Methods

3.1 CSF Collection

Fresh CSF is collected via lumbar puncture (LP) by inserting a sterile needle into the L3/L4 or L4/L5 intervertebral space. The CSF is then collected in sterile polypropylene tubes, yielding approximately 5–10 mL per procedure (Fig. 2). The timing between withdrawing CSF, centrifugation, and freezing is critical to minimize delays exceeding 30 min, thereby preserving cellular components and biomarkers effectively (*see* **Note 2**). Immediate storage at −80 °C is essential for long-term preservation and subsequent analysis [9] (*see* **Note 3**). Careful consideration of ethical guidelines is paramount when collecting CSF from healthy

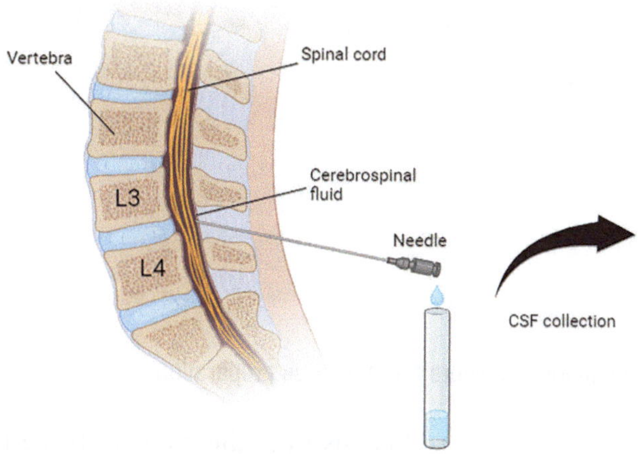

Fig. 2 Lumbar puncture procedure for CSF collection, essential for biomarker analysis in neurological research

donors, ensuring alignment with specific research objectives and ethical considerations.

3.2 Sample Preparation and Handling

Centrifuge CSF samples at 400× *g* for 10–15 min to preserve fragile cells, ensuring sample integrity. Alternatively, use 2000× *g* for specific purposes [9]. Immediately aliquot the cell-free supernatant into small polypropylene tubes to prevent degradation before freezing at −80 °C, preserving biomarker stability for subsequent analysis (*see* **Note 4**).

3.3 Luminex® Protocol

Ensure that all reagents are brought to room temperature (20–25 ° C) before using them in the assay. Thaw CSF samples and keep them on ice throughout preparation. Maintain all samples on ice until they are ready for use.

3.3.1 Sample Preparation

When working with frozen samples, thoroughly thaw them and ensure proper mixing by vortexing. To prevent low bead counts, centrifuge the samples before use in the assay for 5–10 min at a minimum of 10,000× *g* to remove any debris, lipids, or cells. Depending on the assay requirements, samples may need dilution in an assay buffer or sample diluent. Typically, the maximum required sample volume is 25 μL [8, 10].

3.3.2 Preparation of Reagents

Antibody-Immobilized Beads: When using individual antibody-bead vials, add 150 uL of each to a mixing bottle and adjust the final volume to 3.0 mL with the appropriate assay buffer.

Quality Controls: Before use, mix the quality control with 250 μL of standard diluent (deionized water) and invert the vial several times, and then vortex.

Standards:

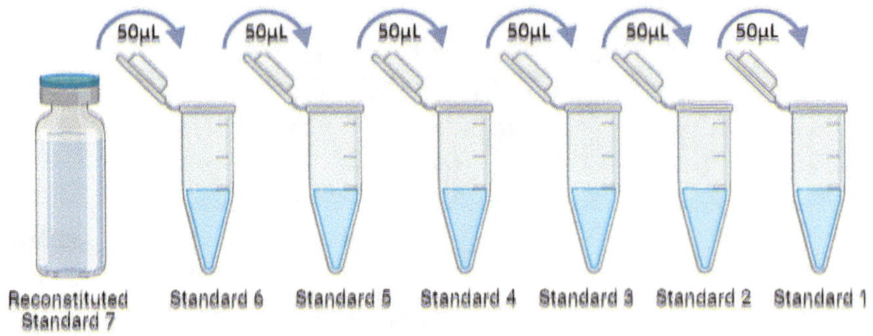

Fig. 3 Serial dilution procedure for standards preparation

- Reconstitute the standard by adding 250 μL of deionized water. Gently mix by inverting the vial multiple times and vortexing for 10 s. Let the vial stand for 5–10 min before use.
- Prepare serial dilutions of the standards in 6 polypropylene tubes (*see* Fig. 3): Begin by adding 150 μL of diluent to each tube. Transfer 50 μL of the reconstituted standard to Tube 6 and mix thoroughly. Transfer 50 μL from Tube 6 to Tube 5, ensuring proper mixing between transfers. Repeat this process sequentially from Tube 5 to Tube 1, maintaining thorough mixing (*see* **Note 5**).

3.3.3 Plate Layout

The arrangement of standards, controls, and samples on the assay plate is crucial in Luminex® assays to ensure accurate measurement and reliable data interpretation. An example plate layout is depicted in Fig. 4, illustrating the strategic placement of standards (S1–S7) for calibration, a blank (B) for background measurement, and quality controls (C1, C2) for validation purposes. The remaining 76 wells are allocated for distributing serum, plasma, or CSF samples, optimizing assay efficiency and experimental consistency (*see* **Note 6**).

3.3.4 Performing the Assay

1. Prepare the plate: Add 200 μL of wash buffer into all wells. Seal the plate with adhesive plastic sealer and incubate for 10–30 min at room temperature with continuous shaking (according to the manufacturer's protocol). Decant the wash buffer.
2. Add 25 μL of standards and controls to their respective wells.
3. Prepare sample wells: Add 25 μL of assay buffer to background and sample wells. For background, standards, and control wells, add 25 μL of the appropriate matrix solution (use assay buffer when working with CSF; *see* **Note 7**).
4. Add 25 μL of samples (diluted if necessary) to the appropriate sample wells.

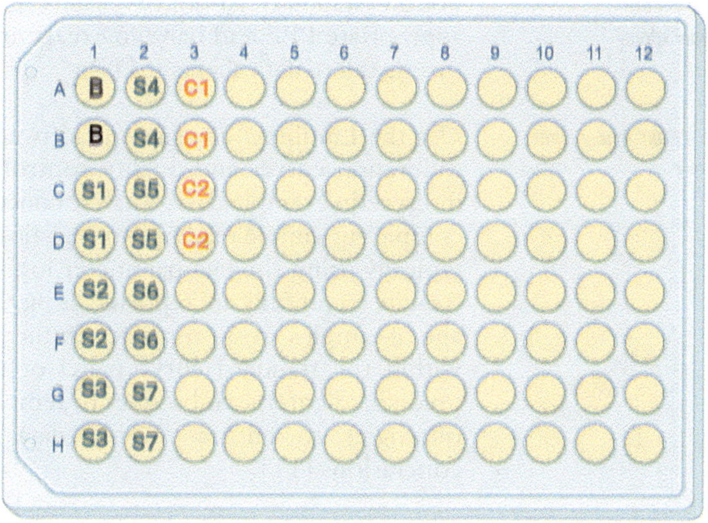

Fig. 4 Example 96-well plate layout demonstrating the arrangement of blanks (B), standards (S1→ S7), quality controls (C), and sample wells in a Luminex assay. In this format, 38 samples can be analyzed in duplicate

5. Dispense 25 μL of beads into each well, shaking the bottle to prevent settling (*see* **Note 8**).
6. Seal the plate with a plate sealer and cover with foil. Incubate overnight (16–18 h) at 2–4 °C with constant agitation on a plate shaker (500–600 rpm) or for 2 h at room temperature.
7. Wash plate: After incubation, aspirate the well contents and wash the plate three times with 200 μL of wash buffer either manually or using a plate washer.
8. Add 25 μL of detection antibodies per well. Incubate for 1 h on a plate shaker at room temperature (500–600 rpm). Do not aspirate after incubation.
9. While the plate is incubating, prepare the software protocol. Standard and control concentrations must be provided by the manufacturer.
10. Add 25 μL of Streptavidin-Phycoerythrin to each well. Incubate for 30 min on a plate shaker at room temperature. Then aspirate and wash three times with 200 μL of wash buffer (*see* **Note 9**).
11. Prepare for Reading: Dispense 150 μL of Sheath Fluid PLUS or Drive Fluid PLUS into all wells. Resuspend the beads on a plate shaker for 5 min (ensure thorough resuspension).
12. Run the Plate: Run the plate on a Luminex instrument using the corresponding software (*see* **Note 10**).

3.3.5 Safety Considerations

Follow standard laboratory safety protocols, including wearing appropriate PPE and handling reagents according to the manufacturer's safety data sheets (SDS).

3.3.6 Post-Assay Data Analysis

Use the Luminex® instrument software for data analysis, ensuring calibration curves and quality control results are verified prior to interpreting sample data. Analyte concentrations are determined by comparing sample readings to logistic curve-fitted standard curves, typically using the five-parameter logistic (5PL) regression model. Results commonly are reported as pg/mL ± SD, range, or percentage of samples, as appropriate. Outliers are identified automatically within the standard curve panel or manually selected based on predefined criteria. Subsequently, employ rigorous statistical methodologies to ensure reliable data interpretation and robust conclusions [11].

4 Notes

1. For convenience and optimal performance, it is highly recommended to use a multichannel pipette. Ensure correct sealing and pipette with the multichannel device, avoiding contact with the reagent in the plate to prevent cross-well contamination.
2. For gene expression analysis studies, it is recommended to use immediate stabilization methods such as RNAlater solution to preserve RNA integrity. CSF collected directly into RNAlater (at a 1:4 ratio of RNAlater to CSF) can be stored at ambient temperature or + 4 °C for up to 1 week. Alternatively, promptly freezing CSF samples at −80 °C is preferred to prevent dilution and ensure optimal preservation for subsequent RNA extraction and analysis [12].
3. Long-term storage at −80 °C is recommended to preserve biomarker integrity and ensure accurate analysis. Avoid prolonged storage at −20 °C and more than two freeze/thaw cycles to prevent degradation. Freeze samples immediately if not tested right away [9, 10].
4. To maintain biomarker concentrations and integrity, aliquot CSF samples into small volumes (e.g., 0.25, 0.5, and 1 mL) and store them at −80 °C in screw-capped vials with freeze-proof labels. Thaw samples completely and mix thoroughly before analysis to ensure accurate results. Be aware that hemolyzed, icteric, or lipemic samples may invalidate certain tests. [9, 10, 13].
5. The serially diluted standards must be utilized within 1 h of preparation. Any remaining standards, excluding the stock, can

be stored at −20 °C for up to 1 month, and at −80 °C for longer durations exceeding 1 month.

6. For optimal assay performance, remember to run standards, samples, and controls in duplicate when creating your plate layout.
7. The diluent serves as the background solution.
8. Ensure thorough mixing of the more concentrated standards before proceeding with the next dilution.
9. Any remaining mixed antibody-immobilized beads can be stored in the designated mixing bottle at 2–8 °C for up to 1 month.
10. Ensure to sonicate the Luminex® probe both before and after reading the plate to maintain cleanliness. Not doing this could lead to calibration or reading problems.

Acknowledgments

JMHP is supported by funds of the RICORS Program (RD21/0002/0063) from ISCIII, co-funded by the European Union. AQV is supported by funds from Fundación Francisco Soria Melguizo. Figures in this chapter were created with BioRender.com.

References

1. Baker HN, Murphy R, Lopez E, Garcia C (2012) Conversion of a Capture ELISA to a Luminex xMAP Assay using a Multiplex Antibody Screening Method. J Vis Exp. https://doi.org/10.3791/4084
2. Zhang Y, Birru R, Di YP (2014) Analysis of clinical and biological samples using microsphere-based multiplexing luminex system. In: Methods in molecular biology, pp 43–57
3. Dunbar SA (2006) Applications of Luminex® xMAP™ technology for rapid, high-throughput multiplexed nucleic acid detection. Clin Chim Acta 363:71–82. https://doi.org/10.1016/j.cccn.2005.06.023
4. Wild D, John R, Sheehan C (2013) Front matter. In: The immunoassay handbook. Elsevier, p iii
5. Dunbar SA, Hoffmeyer MR (2013) Microsphere-based multiplex immunoassays. In: The immunoassay handbook. Elsevier, pp 157–174
6. Faresjö M (2014) A useful guide for analysis of immune markers by fluorochrome (Luminex) technique. Methods Mol Biol 1172:87–96. https://doi.org/10.1007/978-1-4939-0928-5_7
7. Bio-Techne; R&D Systems (2024) Luminex assay users guide. In: Luminex assay users guid a clear easy guid to luminex mult immunoassays, pp 1–15
8. Merck Millipore (2024) Guide to MILLIPLEX® multiplexing
9. Teunissen CE, Petzold A, Bennett JL et al (2009) A consensus protocol for the standardization of cerebrospinal fluid collection and biobanking. Neurology 73:1914–1922
10. Srivastava A, Dixit AB, Tripathi M et al (2024) Quantification of neuroinflammatory markers in blood, cerebrospinal fluid, and resected brain samples obtained from patients. In: Methods in molecular biology
11. Breen EJ, Tan W, Khan A (2016) The statistical value of raw fluorescence signal in luminex

xMAP based multiplex immunoassays. Sci Rep 6. https://doi.org/10.1038/srep26996

12. Atkinson L, Lee JC, Lennon A et al (2023) Untargeted metagenomics protocol for the diagnosis of infection from CSF and tissue from sterile sites. Heliyon 9. https://doi.org/10.1016/j.heliyon.2023.e19854
13. Henry JB (2001) Book review: clinical diagnosis and management by laboratory methods. Lab Med 32:700–700. https://doi.org/10.1309/XL9U-4DYG-MJ76-XCC8

Chapter 5

Monitoring Cerebrospinal Fluid Inflammatory Mediators by Olink Target 48 Cytokine Panel

Fabricio Zambom-Ferraresi, Fabiola Zambom-Ferraresi, Maite Izco, Nicolás Martínez-Velilla, Leire Extramiana, Joaquín Fernández-Irigoyen, Enrique Santamaría, Adrían Llamas-Urbano, Yasin Hanaee, Julio M. Martinez-Moreno, Nuria Barbarroja, and Carlos Pérez-Sánchez

Abstract

The diagnosis and monitoring of neurological disorders are challenging due to their complexity and the difficulty of obtaining samples of damaged tissue while the patient is alive. Cerebrospinal fluid (CSF) is an excellent biofluid for conducting diagnostic and monitoring studies because it is in direct contact with the brain, allowing for the analysis of secreted proteins and peptides associated with various neurological disorders.

The Proximity Extension Assay, which combines antibodies with oligonucleotides to detect proteins via qPCR, is revolutionizing the field of biomarker identification. The Olink Target 48 Cytokine Panel is an optimal tool for evaluating the inflammatory profile of CSF samples, requiring only one microliter. The workflow involves incubation with antibody-linked oligonucleotides, extension, amplification through a PCR assay, and detection using microfluidic qPCR, enabling precise and accurate quantification of cytokine levels in pg/mL. This approach offers valuable insights into inflammatory processes within the CSF, serving as a robust tool for studying and understanding inflammation in neurological contexts.

Key words CSF, Cytokines, Neurobiology, Inflammation, Proximity extension assay

1 Introduction

The brain, one of the most complex organs in the human body and the main organ of the nervous system, remains largely unknown today. Consequently, our understanding of the various diseases and injuries that affect it is still limited. These conditions include physical injuries, neurodegenerative diseases, and cancer. However, within each of these categories, there are numerous diseases with different molecular abnormalities. Generally, these conditions

Enrique Santamaría and Joaquín Fernández-Irigoyen (eds.), *Cerebrospinal Fluid (CSF) Proteomics: Methods and Protocols*, Methods in Molecular Biology, vol. 2914, https://doi.org/10.1007/978-1-0716-4462-1_5,

result from small changes in various molecular factors that accumulate over time, leading to alterations in complex molecular processes and eventually giving rise to clinical symptoms.

Diagnosing and monitoring these pathologies are challenging, not only due to their complexity but also because it is nearly impossible to obtain samples of the damaged tissue while the patient is alive. This limitation has fueled growing interest in the use of other types of biological samples that can be obtained through less invasive procedures and still adequately reflect brain pathology. Unlike many other diseases, where blood is the ideal fluid for diagnostic and monitoring studies, the Central Nervous System (CNS) is isolated from the blood by the Blood-Brain Barrier (BBB) [1]. Cerebrospinal Fluid (CSF) is, therefore, the biofluid that best meets the necessary criteria for studying CNS diseases. This fluid surrounds and permeates the brain and spinal cord, placing it in direct contact with the brain and making it useful for studying the secreted protein and peptide composition associated with different neurological disorders [2]. Additionally, the rotation of the cerebrospinal fluid (CSF) occurs approximately 4–5 times per day [3], and because the process of obtaining samples is relatively fast and simple, patient follow-ups can be performed more easily.

Neurodegenerative diseases, such as multiple sclerosis (MS), amyotrophic lateral sclerosis (ALS), Alzheimer's disease (AD), and Parkinson's disease (PD), involve significant neuroinflammatory components [4–6], which in PD have been suggested to contribute to the disease pathology [7]. Furthermore, the persistent inflammatory response in AD patients, triggered by IFN-γ and TNF-α [8], is related to the deposition of amyloid-β (Aβ) [4]. Patients with spinal cord injuries (SCI) also exhibit acute inflammation in their brains [9]. Glioblastoma multiforme (GBM) accounts for 50% of malignant gliomas diagnosed in adulthood [10–12]. These tumor cells secrete various cytokines and growth factors (GF) into the environment, which inhibit the antitumor immune response and contribute to tumor expansion [13–21]. However, neuroinflammation and cytokine production do not always worsen pathology. In AD, Aβ plaque clearance is modulated by activated microglia and peripheral monocytes, whose entry into the CNS is stimulated by IFN-γ [22–24]. Both clinical and preclinical studies on the use of oncolytic viruses for CNS tumor treatment have shown an increase in inflammatory cells and TH1 cell-related cytokines following treatment, resulting in total or partial tumor remission and increased survival rates [25–28].

All cytokines, to varying degrees, are pleiotropic and redundant [29]. As seen in the conditions described above, the pro-inflammatory process and immune system activation can either improve or worsen the disease. This difference may partly be due to the combined expression of certain cytokines under specific conditions, stages of disease, or types of treatment. To fully understand the biological context of each disease or injury, and to identify effective and specific treatments, it is essential to study the expression patterns of the various cytokines present in the CSF together.

Current proteomic workflows, based on mass spectrometry, do not allow for easy detection of cytokines and growth factors in CSF [30]. In this chapter, we introduce a groundbreaking technology called Proximity Extension Assay (PEA), developed by the Swedish company Olink Proteomics, which has revolutionized proteomics research. This state-of-the-art method enables the simultaneous analysis of thousands of proteins from just a minimal volume (1 microliter) of any biofluid (blood, saliva, CSF, etc.), greatly advancing the potential of proteomic studies.

PEA (Proximity Extension Assay) combines the precision of antibody-based immunoassays with the efficiency of qPCR, offering exceptional levels of specificity, sensitivity, and dynamic range. This technique uses a dual-recognition immunoassay, where two antibodies labeled with unique DNA oligonucleotides bind to a target protein, bringing them into close proximity. This proximity allows the DNA oligonucleotides to hybridize, forming a unique double-stranded DNA "barcode" that correlates with the initial concentration of the target protein. This barcode is then amplified through PCR.

The combination of dual-antibody recognition and DNA-based measurement in PEA ensures an unmatched level of specificity, effectively addressing challenges commonly encountered in multiplex immunoassays, such as antibody cross-reactivity. This remarkable specificity sets PEA apart from other proteomic technologies, making it a revolutionary tool in the field. Results obtained through PEA can be further validated using techniques such as cytokine arrays, Western blotting, ELISA, Luminex, or Mass Spectrometry with multiple reaction monitoring (MRM), which has demonstrated a good correlation with PEA findings.

In this chapter, we provide a detailed procedure for thoroughly evaluating the Target 48 Cytokine panel, which simultaneously assesses 45 inflammatory mediators. It is important to note that other panel formats are available, tailored to the specific needs of researchers in various fields, including oncology, cardiovascular research, neurology, and metabolism studies (https://cobiomicbioscience.com/services/). This technology has significantly improved our understanding of the biology and progression of various diseases [31–33], enabled the identification of new biomarkers, and facilitated the characterization of the microenvironment and behavior of different cell types following specific treatments [34–36].

In conclusion, combining CSF analysis with Olink® Panels, specifically the Olink® Target 48 Cytokine panel, is one of the most effective approaches for thoroughly characterizing and monitoring the neuroinflammatory aspects of various CNS diseases. Here, we describe the methodology associated with this technology.

2 Materials

2.1 CSF Sample Obtention and Processing

1. Spinal needle of 22 gauge (22G).
2. Polypropylene or glass sterile tubes.
3. Centrifuge adapted for selected tubes that spin at 3000 rpm for 15 min at room temperature.

2.2 Incubation Step

Ensure the laboratory (pre-PCR room) is properly prepared for the incubation step by arranging pipettes and all required consumables in advance. Verify that the reagents and controls have fully acclimated to room temperature before starting the procedure.

The required materials are as follows:

1. Pipettes for necessary volumes.
2. Pipette tips: Sterile and suitable for your pipettes.
   - Three pipette tips of 10 μL.
   - Four pipette tips of 100/200 μL.
   - One pipette tip of 1000 μL.

   And 104 multichannel pipette tips of 10 μL.
3. Prepared and thawed 96-well plate with samples.
4. Reagents (vortex and spin all reagents before use):
   - Incubation Solution (Target 48 panel).
   - Frw and Rev. probes (Target 48 panel).
5. Controls (vortex and spin before use):
   - Sample and Negative Controls (Target 48 panel).
   - Calibrator (Target 48 panel).
6. Two 8-well strips.
7. One new 96-well plate.
8. One microcentrifuge tube (1–1.5 mL).
9. Two adhesive films.
10. One vortex mixer.
11. One plate mixer.
12. One microcentrifuge.
13. One plate spinner.
14. Refrigerator for overnight incubation at 4 °C.

2.3 Extension and Amplification Step

Ensure the laboratory (pre-PCR room) is properly prepared for the extension and amplification steps by arranging pipettes and all required consumables in advance. Allow the PEA solution to fully acclimate to room temperature before proceeding with the procedure.

The required materials are as follows:

1. Pipettes for necessary volumes.
2. Pipette tips: Sterile and suitable for your pipettes.
   - Two pipette tips of 1000 μL.
   - Two pipette tips of 200 μL.
   - And 8 multichannel pipette tips of 200 μL.
3. Reagents:
   - PEA Solution. Keep at room temperature for 15 min before use (Target 48 panel).
   - PEA Enzyme (Target 48 panel).
   - PEA Enhancer. At room temperature (Target 48 panel).
4. DNase- and RNase-free water of high purity.
5. One centrifugal tube (15 mL).
6. One multichannel pipette reservoir.
7. One adhesive film.
8. One vortex mixer.
9. One plate mixer.
10. One microcentrifuge.
11. One plate spinner.
12. PCR machine (Thermocycler).

2.4 Detection Step

Ensure the laboratory (post-PCR room, separate from the pre-PCR room) is fully prepared for the detection step by organizing pipettes and all necessary consumables in an orderly manner.

1. 48.48 Dynamic Array IFC chip and Two syringes with control line fluid for priming the chip.
2. One 8-well strip.
3. One 96-well plate.
4. One microcentrifuge tube (1.5 mL).
5. Two pipette tips (10 μL).
6. One pipette tip (100/200 μL).
7. Two pipette tips (1000 μL).
8. Three boxes + eight multichannel pipette tips (10 μL).

9. Reagents:
   - Detection Solution (Target 48 panel).
   - Detection Enzyme (Target 48 panel).
   - PCR Polymerase (Target 48 panel).
   - Primer Plate (Target 48 panel).
10. DNase- and RNase-free water of high purity.
11. Two adhesive films.
12. One plate mixer.
13. Extension products.
14. Olink Signature Q100 instrument.

3 Methods

3.1 CSF Sample Obtention and Processing (See Note1)

1. Lumbar puncture (LP) is the procedure used to access the subarachnoid space in the lumbar region. It is used to obtain CSF, measure CSF pressure, or administer intrathecal treatment. As an invasive procedure, several items must be achieved and/or taken into consideration such as patient preparation (helping to open up the intervertebral spaces for easier needle insertion (*see* **Note 2**), local anesthesia (*see* **Note 3**) and needle insertion (*see* **Note 4**).
2. CSF should be collected in polypropylene or glass sterile tubes In some cases, CSF is collected in three to four tubes for various analyses (e.g., microbiology, chemistry, cell count, and cytology).
3. Samples must be proceeded immediately or temporally stored at appropriate temperatures at 4 °C and processed as soon as possible. Do not leave the sample at 4 °C for more than 12 h.
4. For proteomics analysis, CSF must be centrifuged at 3000 rpm for 15 min at room temperature. The supernatant must be collected and aliquoted in volumes of 220 μL and stored at −80 °C until use.

3.2 Olink® Target 48 Procedure/Process

3.2.1 Incubation Step (See Note 5)

1. Thaw the samples.
2. After defrosting the Sample Control, Negative Control and Calibrator, vortex and briefly spin them. Then add 4 μL of each control to an 8-well strip in the following order: Sample Control (A, B, C), Negative Control (D, E) and Calibrator (F, G, H).
3. Prepare the Incubation Mix in a 1.5 mL tube by adding the following reagents, taking care to avoid foaming when pipetting. Vortex and spin each reagent before use:

- 160 μL of Incubation Solution
- 21 μL Frw-probes
- 21 μL Rev-probes

4. Vortex and spin down the Incubation Mix to ensure thorough mixing.
5. Transfer 23 μL of the Incubation Mix to each well of a new 8-well strip.
6. Using a multichannel pipette, transfer 3 μL of the Incubation Mix to the bottom of the wells in a new 96-well plate), which will be referred to as the Incubation Plate, by reversing pipetting (*see* **Notes 6** and 7) without changing the tips.
7. Mix the plate with samples using a plate mixer at 2000 rpm for 30 s, or vortex with a flat-bottom platform at full speed for 20–30 s.
8. Spin down the liquid at 400–1000× *g* for 1 min at room temperature.
9. Using a multichannel pipette, transfer 1 μL of each sample to the bottom of the wells of columns 1–5 in the Incubation plate by forward pipetting (*see* **Note 8**).
10. Transfer 1 μL from the prepared 8-well strip containing Calibrator, Sample Control, and Negative Control to column 6 of the plate by forward pipetting too (*see* **Note 8**). Change pipette tips between every column.
11. Incubate the plate overnight for 16–22 h at +4 °C, sealing the plate with adhesive plastic film and spinning at 400–1000× *g* for 1 min at room temperature (*see* **Notes 9** and **10**). During incubation, proteins will bind to their targets.
12. Prepare a tube of DNase- and RNase-free water of high purity for the extension step the next day and store it at +4 °C overnight.
13. Thaw the PEA Solution overnight at +4 °C and leave the PEA Enhancer at room temperature overnight.

3.2.2 Extension and Amplification Step (See Note 5)

After 16–22 h of incubation, specific DNA reporter sequences for each target protein are obtained through regular PCR during the extension and amplification steps.

1. Prepare the PCR machine by setting the PEA program to 50 °C and pausing it.
2. Prepare the Extension Mix in a 25 mL tube, adding the following reagents and avoiding foaming when pipetting. Vortex and spin each reagent before use (*see* **Notes 11** and **12**):
   - 4350 μL High Purity Water
   - 580 μL PEA Enhancer

- 580 μL PEA Solution (bring to room temperature before use)
- 58 μL PEA Enzyme

3. Vortex the Extension Mix. Then, spin down the Incubation Plate at 400–1000× *g* for 1 min at room temperature (*see* **Note 9**).
4. Start a 5-min timer and transfer 96 μL of Extension Mix to the upper parts of each well wall of the Incubation Plate by reversing pipetting (*see* **Notes 6** and 7) using a multichannel pipette without changing tips, ensuring it does not come into contact with the contents of the wells.
5. Seal the plate with an adhesive plastic film and vortex thoroughly using a plate mixer at 2000 rpm for 30 s before spinning it (*see* **Notes 10** and **13**).
6. Transfer the plate to the PCR thermocycler and resume the PEA program (*see* **Notes 14**).
7. Once the PCR PEA program is complete (approximately 1 h and 55 min), proceed to the detection step.

3.2.3 Detection Step (See Note 5)

After the extension and amplification steps, each biomarker is quantified using high-throughput real-time qPCR on the Olink Signature Q100 system during the final detection step.

1. Priming the IFC Chip:
   - Remove the protective film from the bottom of the chip and ensure that the valve is fully open with the seal pressed down and to the side (*see* **Note 15**).
   - Insert the syringe tip into the valve with the chip tilted, then slowly inject a syringe of control line fluid into each IFC accumulator.
   - On the Olink Signature Q100 instrument, press "Target 48" on the home screen, place the chip on the interface with the barcode facing forward, close the lid, and press start (*see* **Note 16**).
2. Preparing the Detection Mix:
   - While the IFC is being primed, prepare the Detection Mix in a 1.5 mL tube by adding the following reagents (vortex and spin down each reagent before use):
     - 275 μL Detection Solution
     - 116 μL High Purity Water
     - 3,9 μL Detection Enzyme
     - 1,5 μL PCR Polymerase

3. Vortex the Detection Mix and briefly spin down. Transfer 46 μL of the mix to each well of an 8-well strip.
4. Using a multichannel pipette without changing the tips, add 7.2 μL of the Detection Mix to each well of columns 1–6 in a new 96-well plate, labeled as the Sample Plate by reverse pipetting. (*see* **Notes 6** and 7).
5. Remove the Extension products from the thermocycler (*see* **Note 10**), vortex for 30 s and spin down the liquid for 3 min.
6. Using a multichannel pipette, transfer 2.8 μL from the extension products to the Sample Plate by forward pipetting (*see* **Note 8**), changing the tips between different columns.
7. Seal, vortex, and spin the Sample Plate at 400 g for 1 min at room temperature.
8. Remove the primed chip from the Olink Signature Q100 instrument. Vortex and spin down both the Sample Plate and Primer Plate.
9. Remove the adhesive film from the Primer Plate carefully, avoiding contamination between wells.
10. Position the chip with the barcode oriented towards the left side. Load the primers on the left side and the samples on the right side of the 48.48 Dynamic Array IFC according to the following instructions:
    - Using a multichannel pipette, load both primer and sample plates by reverse pipetting (*see* **Note 6**), transferring 5 μL of each well of columns 1–3 of the plate to columns 1–3 of the chip, starting with the first well of the first row.
    - Then transfer from columns 4–6 of the plate to columns 1–3 of the chip, starting with the first well of the second row.
11. After loading, search and remove any bubbles with a syringe needle, being careful to avoid contamination.
12. Place the chip into the Olink Signature Q100 instrument and press "Start Run" to initiate the detection process.
13. Finally, the data are uploaded into the NPX Signature Software which generates the absolute quantification in pg/mL.

    The technical process overview is illustrated in Fig. 1.

In conclusion, Olink Target 48 Cytokine analysis offers a highly effective method for assessing inflammation in CSF samples. The workflow-comprising incubation with antibody-linked oligonucleotides, extension and amplification through PCR, and detection using microfluidic qPCR—enables precise and accurate quantification of cytokine levels in pg/mL. This approach provides valuable insights into inflammatory processes within the CSF, making it a robust tool for studying and understanding inflammation in neurological contexts.

UE) and the Department of Economic and Business Development from Government of Navarra (Ref. 0011-1411-2023-000028) to ES. CPS and NB were supported by Ramón y Cajal Program of MCIN (RyC-2021-033828-I and RyC-2017-23437) and ALU by the Industrial PhD Program of AEI (DIN2022-012766).

References

1. Daneman R, Prat A (2015) The blood–brain barrier. Cold Spring Harb Perspect Biol 7: a020412. https://doi.org/10.1101/cshperspect.a020412
2. Van Gool AJ, Hendrickson RC (2012) The proteomic toolbox for studying cerebrospinal fluid. Expert Rev Proteomics 9:165–179. https://doi.org/10.1586/epr.12.6
3. de Boer AG, Gaillard PJ (2007) Drug Targeting to the Brain. Annu Rev Pharmacol Toxicol 47:323–355. https://doi.org/10.1146/annurev.pharmtox.47.120505.105237
4. Fung A, Vizcaychipi M, Lloyd D, Wan Y, Ma D (2012) Central nervous system inflammation in disease related conditions: mechanistic prospects. Brain Res 1446:144–155. https://doi.org/10.1016/j.brainres.2012.01.061
5. Eikelenboom P, van Exel E, Hoozemans JJM, Veerhuis R, Rozemuller AJM, van Gool WA (2010) Neuroinflammation—an early event in both the history and pathogenesis of Alzheimer's disease. Neurodegener Dis 7:38–41. https://doi.org/10.1159/000283480
6. Perry VH, Nicoll JAR, Holmes C (2010) Microglia in neurodegenerative disease. Nat Rev Neurol 6:193–201. https://doi.org/10.1038/nrneurol.2010.17
7. Hirsch EC, Hunot S (2009) Neuroinflammation in Parkinson's disease: a target for neuroprotection? Lancet Neurol 8:382–397. https://doi.org/10.1016/S1474-4422(09)70062-6
8. Zeni P, Doepker E, Schulze-Topphoff U, Huewel S, Tenenbaum T, Galla H-J (2007) MMPs contribute to TNF-alpha-induced alteration of the blood-cerebrospinal fluid barrier in vitro. Am J Physiol Cell Physiol 293:C855. https://doi.org/10.1152/ajpcell.00470.2006
9. Tsai MC, Wei CP, Lee DY, Tseng YT, Tsai MD, Shih YL, Lee YH, Chang SF, Leu SJ (2008) Inflammatory mediators of cerebrospinal fluid from patients with spinal cord injury. Surg Neurol 70:19–24. https://doi.org/10.1016/j.surneu.2007.09.033
10. Westermark B (2012) Glioblastoma—a moving target. Ups J Med Sci 117:251–256. https://doi.org/10.3109/03009734.2012.676574
11. Jayaram S, Gupta MK, Polisetty RV, Cho WC, Sirdeshmukh R (2014) Towards developing biomarkers for glioblastoma multiforme: a proteomics view. Expert Rev Proteomics 11:621–639. https://doi.org/10.1586/14789450.2014.939634
12. Wen PY, Kesari S (2008) Malignant gliomas in adults. N Engl J Med 359:492–507. https://doi.org/10.1056/NEJMra0708126
13. Magaña-Maldonado R, Chávez-Cortez EG, Olascoaga-Arellano NK, López-Mejía M, Maldonado-Leal FM, Sotelo J, Pineda B (2016) Immunological evasion in glioblastoma. Biomed Res Int 2016:1–7. https://doi.org/10.1155/2016/7487313
14. Hao C, Parney IF, Roa WH, Turner J, Petruk KC, Ramsay DA (2002) Cytokine and cytokine receptor mRNA expression in human glioblastomas: Evidence of Th1, Th2 and Th3 cytokine dysregulation. Acta Neuropathol 103:171–178. https://doi.org/10.1007/s004010100448
15. Van Meir E, Sawamura Y, Diserens AC, Hamou MF, de Tribolet N (1990) Human glioblastoma cells release interleukin 6 in vivo and in vitro. Cancer Res 50:6683–6688
16. Murphy GM, Bitting L, Majewska A, Schmidt K, Song Y, Wood CR (1995) Expression of interleukin-11 and its encoding mRNA by glioblastoma cells. Neurosci Lett 196:153–156. https://doi.org/10.1016/0304-3940(95)11862-Q
17. Halfter H, Kremerskothen J, Weber J, Hacker-Klom U, Barnekow A, Ringelstein EB, Stögbauer F (1998) Growth inhibition of newly established human glioma cell lines by leukemia inhibitory factor. J Neuro-Oncol 39:1–18
18. Halfter H, Lotfi R, Westermann R, Young P, Bernd Ringelstein E, Stögbauer FT (1998) Inhibition of growth and induction of differentiation of glioma cell lines by oncostatin M (OSM). Growth Factors 15:135–147. https://doi.org/10.3109/08977199809117189

19. Shweiki D, Neeman M, Itin A, Keshet E (1995) Induction of vascular endothelial growth factor expression by hypoxia and by glucose deficiency in multicell spheroids: implications for tumor angiogenesis. Proc Natl Acad Sci USA 92:768–772. https://doi.org/10.1073/pnas.92.3.768
20. Reiss Y, Machein MR, Plate KH (2005) The role of angiopoietins during angiogenesis in gliomas. Brain Pathol 15:311–317. https://doi.org/10.1111/j.1750-3639.2005.tb00116.x
21. Gabrilovich DI, Chen HL, Girgis KR, Cunningham HT, Meny GM, Nadaf S, Kavanaugh D, Carbone DP (1996) Production of vascular endothelial growth factor by human tumors inhibits the functional maturation of dendritic cells. Nat Med 2:1096–1103. https://doi.org/10.1038/nm1096-1096
22. Wilcock DM, Munireddy SK, Rosenthal A, Ugen KE, Gordon MN, Morgan D (2004) Microglial activation facilitates Aβ plaque removal following intracranial anti-Aβ antibody administration. Neurobiol Dis 15:11. https://doi.org/10.1016/j.nbd.2003.09.015
23. Chakrabarty P, Ceballos-Diaz C, Beccard A, Janus C, Dickson D, Golde TE, Das P (2010) IFN-promotes complement expression and attenuates amyloid plaque deposition in amyloid precursor protein transgenic mice. J Immunol 184:5333. https://doi.org/10.4049/jimmunol.0903382
24. Chakrabarty P, Jansen-West K, Beccard A, Ceballos-Diaz C, Levites Y, Verbeeck C, Zubair AC, Dickson D, Golde TE, Das P (2010) Massive gliosis induced by interleukin-6 suppresses A deposition in vivo: evidence against inflammation as a driving force for amyloid deposition. FASEB J 24:548. https://doi.org/10.1096/fj.09-141754
25. Lang FF, Conrad C, Gomez-Manzano C, Yung WKA, Sawaya R, Weinberg JS, Prabhu SS, Rao G, Fuller GN, Aldape KD, Gumin J, Vence LM, Wistuba I, Rodriguez-Canales J, Villalobos PA, Dirven CMF, Tejada S, Valle RD, Alonso MM, Ewald B, Peterkin JJ, Tufaro F, Fueyo J (2018) Phase I study of DNX-2401 (Delta-24-RGD) oncolytic adenovirus: replication and immunotherapeutic effects in recurrent malignant glioma. J Clin Oncol 36:JCO.2017.75.821. https://doi.org/10.1200/JCO.2017.75.8219
26. Foreman PM, Friedman GK, Cassady KA, Markert JM (2017) Oncolytic virotherapy for the treatment of malignant glioma. Neurotherapeutics 14:333–344. https://doi.org/10.1007/s13311-017-0516-0
27. Kleijn A, Kloezeman J, Treffers-Westerlaken E, Fulci G, Leenstra S, Dirven C, Debets R, Lamfers M (2014) The in vivo therapeutic efficacy of the oncolytic adenovirus Delta24-RGD is mediated by tumor-specific immunity. PLoS One 9:e97495. https://doi.org/10.1371/journal.pone.0097495
28. Jiang H, Rivera-Molina Y, Gomez-Manzano C, Clise-Dwyer K, Bover L, Vence LM, Yuan Y, Lang FF, Toniatti C, Hossain MB, Fueyo J (2017) Oncolytic adenovirus and tumor-targeting immune modulatory therapy improve autologous cancer vaccination. Cancer Res 77:3894–3907. https://doi.org/10.1158/0008-5472.CAN-17-0468
29. Nicola NA (1994) Cytokine pleiotropy and redundancy: a view from the receptor. Stem Cells 12 Suppl 1:3–12; discussion 12–4
30. Macron C, Lane L, Núñez Galindo A, Dayon L (2018) Deep dive on the proteome of human cerebrospinal fluid: a valuable data resource for biomarker discovery and missing protein identification. J Proteome Res 17:4113. https://doi.org/10.1021/acs.jproteome.8b00300
31. Cuesta-López L, Escudero-Contreras A, Hanaee Y, Pérez-Sánchez C, Ruiz-Ponce M, Martínez-Moreno JM, Pérez-Pampin E, González A, Plasencia-Rodriguez C, Martínez-Feito A, Balsa A, López-Medina C, Ladehesa-Pineda L, Rojas-Giménez M, Ortega-Castro R, Calvo-Gutiérrez J, López-Pedrera C, Collantes-Estévez E, Arias-de la Rosa I, Barbarroja N (2024) Exploring candidate biomarkers for rheumatoid arthritis through cardiovascular and cardiometabolic serum proteome profiling. Front Immunol 15:1333995. https://doi.org/10.3389/fimmu.2024.1333995
32. Barbarroja N, López-Montilla MD, Cuesta-López L, Pérez-Sánchez C, Ruiz-Ponce M, López-Medina C, Ladehesa-Pineda ML, López-Pedrera C, Escudero-Contreras A, Collantes-Estévez E, Arias-de la Rosa I (2023) Characterization of the inflammatory proteome of synovial fluid from patients with psoriatic arthritis: Potential treatment targets. Front Immunol 14:1133435. https://doi.org/10.3389/fimmu.2023.1133435
33. Arias-de la Rosa I, Ruiz-Ponce M, Cuesta-López L, Pérez-Sánchez C, Leiva-Cepas F, Gahete MD, Navarro P, Ortega R, Cordoba J, Pérez-Pampin E, González A, Lucendo AJ, Collantes-Estévez E, López-Pedrera C, Escudero-Contreras A, Barbarroja N (2023) Clinical features and imune mechanisms directly linked to the altered liver function in patients with rheumatoid arthritis. Eur J Intern

Med 118:49–58. https://doi.org/10.1016/j.ejim.2023.08.002

34. Perez-Sanchez C, Escudero-Contreras A, Cerdó T, Sánchez-Mendoza LM, Llamas-Urbano A, la Rosa IA, Pérez-Rodriguez M, Muñoz-Barrera L, Del Carmen Abalos-Aguilera M, Barbarroja N, Calvo J, Ortega-Castro R, Ruiz-Vilchez D, Moreno JA, Burón MI, González-Reyes JA, Collantes-Estevez E, Lopez-Pedrera C, Villalba JM (2023) Preclinical characterization of pharmacologic NAD+ boosting as a promising therapeutic approach in rheumatoid arthritis. Arthritis Rheum 75(10):1749–1761. https://doi.org/10.1002/art.42528
35. Ruiz-Ponce M, Cuesta-López L, López-Montilla MD, Pérez-Sánchez C, Ortiz-Buitrago P, Barranco A, Gahete MD, Herman-Sánchez N, Lucendo AJ, Navarro P, López-Pedrera C, Escudero-Contreras A, Collantes-Estévez E, López-Medina C, Arias-de la Rosa I, Barbarroja N (2023) Decoding clinical and molecular pathways of liver dysfunction in Psoriatic Arthritis: impact of cumulative methotrexate doses. Biomed Pharmacother 168: 115779. https://doi.org/10.1016/j.biopha.2023.115779
36. Arias de la Rosa I, López-Montilla MD, Román-Rodríguez C, Pérez-Sánchez C, Gómez-García I, López-Medina C, Ladehesa-Pineda ML, Ábalos-Aguilera MDC, Ruiz D, Patiño-Trives AM, Luque-Tévar M, Añón-Oñate I, Pérez-Galán MJ, Guzmán-Ruiz R, Malagón MM, López-Pedrera C, Escudero-Contreras A, Collantes-Estévez E, Barbarroja N (2022) The clinical and molecular cardiometabolic fingerprint of an exploratory psoriatic arthritis cohort is associated with the disease activity and differentially modulated by methotrexate and apremilast. J Intern Med 291(5): 676–693. https://doi.org/10.1111/joim.13447

Chapter 6

Targeted Absolute Quantification of Protein Biomarkers in Cerebrospinal Fluid by FastCAT

Ignacy Rzagalinski, Viditha Rao, Aliona Bogdanova, Lena Hersemann, and Andrej Shevchenko

Abstract

Absolute (molar) quantification of proteins in cerebrospinal fluid (CSF) provides reference values and ranges of their physiological variation for perspective biomarkers across a broad spectrum of neurological disorders. FastCAT (for Fast-track QconCAT) method employs targeted mass spectrometry and multiple recombinant stable-isotope labeled chimeric proteins (CPs) as internal calibrants for the accurate and consistent determination of molar concentrations of dozens of CSF proteins. No purification or external quantification of CPs is required. We demonstrate that FastCAT, powered by parallel reaction monitoring (PRM) detection of multiple quantotypic peptides per target proteins, can quantify the molar concentration of neurological biomarkers from neat CSF over a dynamic range of more than 1000-fold, with high precision (coefficient of variation <20%) and low detection limits (<1 nmol/L) in a single LC-MS/MS analysis.

Key words Absolute quantification, Protein biomarkers, Cerebrospinal fluid, FastCAT, Targeted proteomics

1 Introduction

Cerebrospinal fluid (CSF) plays a pivotal role in maintaining the functionality of central nervous system. Changes in CNS protein and metabolite composition are reflective of a variety of neuropathological conditions [1]. Because of its clinical availability, CSF is widely used for clinical diagnostics [2, 3] and is the preferred biofluid for ongoing biomarker discovery efforts. Biomarker screening typically relies on monitoring fold changes in the protein abundance between cohorts of patients diagnosed with neuropathological condition(s) and a control group. Because of the technical limitations of high throughput screening, the absolute (molar) concentration of candidate proteins and ranges of their variation are not reported, which leads to the limited concordance between

Enrique Santamaría and Joaquín Fernández-Irigoyen (eds.), *Cerebrospinal Fluid (CSF) Proteomics: Methods and Protocols*, Methods in Molecular Biology, vol. 2914, https://doi.org/10.1007/978-1-0716-4462-1_6,

independent studies. This hampers biomarkers validation and their subsequent implementation in clinical practice [4]. While the throughput of liquid chromatography-tandem mass spectrometry (LC-MS/MS) quantification may not be comparable with ELISA [5], its independence from antibodies, high quantification accuracy, analytical transparency, and ease of validation make it appealing for clinical applications [6].

Here, we describe FastCAT (for Fast-track QconCAT) protocol for targeted absolute quantification of protein biomarkers in the human CSF. FastCAT workflow relies on the parallel use of a few relatively short (less than 50 kDa) isotopically labeled chimeric protein (CP) standards that comprise quantotypic (Q-) peptides from target proteins and also reference (R-) peptides for their in-situ molar quantification based on a single reference protein (here we use Pierce BSA standard from Thermo Fisher) (Fig. 1) [7]. Chimeric proteins comprise several Q-peptides per each target protein and FastCAT independently determines their molar

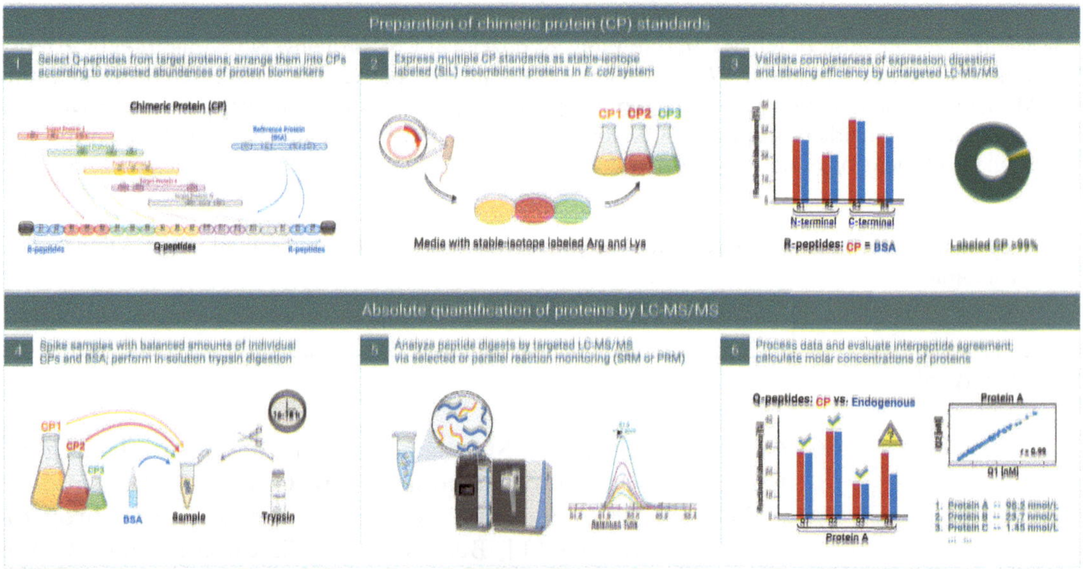

Fig. 1 The FastCAT workflow utilizes multiple chimera proteins (CPs), which include quantotypic (Q-) peptides from target proteins and reference (R-) peptides that link the abundance of CPs to a reference protein, such as BSA. CPs are designed, customized, and produced in an analytical laboratory. FastCAT workflow includes the following steps: 1.CP sequences are designed *in-silico*; 2. The designed DNA is ordered from a synthesis service, cloned into plasmid, and expressed in *E.coli* grown in the medium supplemented with isotopically labeled Arg and Lys. The expressed CP can be used without further purification. 3. Sequences of expressed CPs are validated by trypsin digestion followed by untargeted LC-MS/MS; 4. The appropriate amounts of CPs adjusted to the expected abundance of target proteins and the known amount of reference BSA are spiked into the CSF samples. 5. Samples are digested with trypsin and peptides are analyzed by LC-MS/MS in parallel reaction monitoring (PRM) mode; 6. The exact amounts of spiked CPs are quantified in-situ using R-peptides, which relates them to the known amount of BSA standard. The target proteins are then quantified using isotopically labeled Q-peptides that have the same molar abundance as CP

Table 1
Example of the FastCAT panel targeting 10 neurological biomarkers in CSF [7]

Protein	UniProt accession	Q-peptides	Inter-peptide CV, %
APOE	P02649	SELEEQLTPVAEETR; LGPLVEQGR; QWAGLVEK	<10
APOD	P05090	NILTSNNIDVK; NPNLPPETVDSLK; VLNQELR	<10
AACT	P01011	NLAVSQVVHK; AVLDVFEEGTEASAATAVK; ADLSGITGAR	<15
ZAG	P25311	EIPAWVPFDPAAQITK; WEAEPVYVQR	<15
SCG1	P05060	NYPSLELDK; NYLNYGEEGAPGK; WQQQGDLQDTK	<20
LRG1	P02750	VAAGAFQGLR; GQTLLAVAK	<20
CH3L1	P36222	FPLTNAIK; ILGQQVPYATK; VTIDSSYDIAK; GNQWVGYDDQESVK	<10
CNTN1	Q12860	FIPLIPIPER; ASPFPVYK	<20
VGF	O15240	NSEPQDEGELFQGVDPR; THLGEALAPLSK	<40
NPTX1	Q15818	FQLTFPLR; TPAAETLSQLGQTLQSLK	<10

Proteins are annotated with UniProt accession number; Q-peptides, and inter-peptide coefficients of variation (CVs). Proteins: APOE (apolipoprotein E), APOD (apolipoprotein D), AACT (alpha-1-antichymotrypsin), ZAG (zinc-alpha-2-glycoprotein), SCG1 (secretogranin-1/chromogranin B), LRG1 (leucine-rich alpha-2-glycoprotein 1), CH3L1 (chitinase-3-like protein), CNTN1 (contactin-1), VGF (neurosecretory protein VGF), and NPTX1 (neuronal pentraxin-1)

concentrations in tryptic digests of total CSF. Reliable protein concentration measurements rely on concordant values reported via the simultaneous quantification of multiple Q-peptides as exemplified by 10-protein FastCAT panel (Table 1). In this case study pairwise inter-peptide Pearson correlation coefficients *r* exceeded 0.98, while the coefficient of variation (CV) for all Q-peptides was below 20%. Q-peptides were grouped in CP according to the expected abundance of target proteins and therefore FastCAT allowed us to cover a wider (over 1000-fold) dynamic range of concentrations in a single LC-MS/MS analysis. We propose that the accuracy, precision and low cost (as compared to the same number of isotopically labeled, purified and quantified synthetic peptide standards) of the FastCAT workflow is well-suited for biomarkers validation in accordance with both Tier 1 and 2 guidelines for the targeted MS measurements [8].

2 Materials

Reagents and chemicals were the highest available grade. LC-MS grade solvents (water, acetonitrile), formic acid (FA), and trifluoroacetic acid (TFA) were purchased from Thermo Fisher (Waltham,

MA, USA) or Merck (Darmstadt, Germany). Trypsin/Lys-C proteases mixture (MS grade) was from Promega (Madison, WI, USA); RapiGest detergent from Waters (Eschborn, Germany) and other common chemicals were from Sigma-Aldrich (Munich, Germany).

2.1 Preparation of Chimeric Protein (CP) Standards

1. *E. coli* BL21 (DE3) (*ΔargA ΔlysA*) strain auxotrophic for arginine and lysine [9].
2. Synthetic DNA plasmid (GenScript Biotech).
3. $^{13}C_6$,$^{15}N_4$-L-arginine and $^{13}C_6$-L-lysine (Silantes GmbH).

2.2 In-Solution Digestion of CSF Samples by Trypsin/Lys-C

1. 20 μL of human CSF.
2. 2 mg/mL bovine serum albumin (BSA) standard (Thermo Fisher) (*see* **Note 1**).
3. Sample buffer: 0.2% RapiGest SF surfactant (Waters) in 100 mM ammonium bicarbonate.
4. 100 mM mM 1,4-dithiothreitol (DTT) in 50 mM ammonium bicarbonate.
5. 100 mM iodoacetamide (IAA) in 50 mM ammonium bicarbonate (*see* **Note 2**).
6. 0.1 μg/L trypsin/Lys-C solution in 50 mM ammonium bicarbonate (*see* **Note 3**).
7. 25% TFA in water.
8. Thermostatic mixer.
9. Microcentrifuge with controllable speed.
10. Vacuum concentrator.
11. 1.5 mL low protein binding polypropylene tubes.

2.3 LC-MS/MS

1. Reconstitution buffer: 1% FA in water.
2. LC solvent A: 0.1% FA in water.
3. LC solvent B: 0.1% FA in acetonitrile.
4. Q Exactive HF mass spectrometer equipped with Nanospray Flex ion source (Thermo Fisher).
5. Dionex Ultimate 3000 RSLC (Thermo Fisher) operated in trap-and-elute mode.
6. Acclaim PepMap 100 column (C18, 3 μm, 75 μm × 150 mm), or 50 cm μPAC micropillar HPLC column (both from Thermo Fisher).
7. 100 μm × 2 cm, 5 μm, C18 trap column (Thermo Fisher).

2.4 Data Analysis and Software Tools

1. PeptideCutter [10]—https://web.expasy.org/peptide_cutter/.
2. Protein Blast [11]—https://blast.ncbi.nlm.nih.gov/Blast.cgi?PAGE=Proteins.
3. The neXtProt Peptide Uniqueness Checker [12]—https://www.nextprot.org/tools/peptide-uniqueness-checker.
4. FragPipe software for untargeted proteomics data analysis (University of Michigan) [13].
5. Skyline software for targeted proteomics data analysis (University of Washington) [14].

3 Methods

FastCAT workflow supports targeted absolute quantification of a selection of proteins and is not intended for discovery proteomics. Therefore the list of target proteins should be compiled considering the research goals and (if available) preliminary assessment of their detection feasibility. The important factors to consider are the presumed protein concentration according to previous mass spectrometry or ELISA reports; known isoforms and post-translational modifications and the employed sample preparation, e.g., if the quantification was achieved by direct injection of neat CSF digests or required pre-fractionation or enrichment at the protein or peptide level. In our hands, the practical limit of protein concentration quantifiable in a neat CSF digest was ca. 100 pmol/L, which may serve as a good starting point for detection feasibility assessment.

3.1 Preparation of Chimeric Protein (CP) Standards

1. From each protein of interest, select the candidate Q-peptides using *in-silico* digestion assisting tools (e.g., PeptideCutter [10], Picky [15]) or (preferred) the available proteomics data [16]. Check the Q-peptides uniqueness by Protein Blast [11] or neXtProt Peptide Uniqueness Checker [12]. Use separate (qualitative) experiments to confirm the detection of Q-peptide candidates. Even promising Q-peptide candidates might turn undetectable because of background interference (*see* **Note 4**).
2. Assemble *in-silico* the sequences of Q-peptides into CPs. Since several CPs will be used together in a single analysis (Fig. 1), it is recommended to combine peptides according to an anticipated abundance of target proteins (*see* **Note 5**). Position R-peptides such that at least two R-peptides should be placed at the N- and C-termini of the CP. Note that each CP should

comprise a unique set of R-peptides—they should not be shared between simultaneously used CPs. (Fig. 1, panel 1).

3. DNA sequences encoding individual CPs are codon-optimized for *E. coli* by the GenSmart online tool (https://www.genscript.com/gensmart-free-gene-codon-optimization.html) and their synthesis is ordered from GenScript Biotech company. Sub-cloning of synthetic DNA into plasmids, expression and metabolic labeling of CPs is described in detail (Fig. 1, panel 2) [7, 17].
4. Perform untargeted LC-MS/MS analysis in data-dependent acquisition (DDA) mode of the digests of expressed CP standards and BSA first without spiking them into CSF. Use FragPipe to search against the sequence database (FASTA format) composed of the human and *E. coli* proteomes complemented with BSA and CP sequences (*see* **Note 6**). This is required to validate that full-length CP is correctly expressed and that its metabolic labeling with "heavy" Arg and Lys is complete (>99%). All Q-peptides should be detectable and relative intensities of pairs of labeled R-peptides (originating from CP) and of matching unlabeled peptides (originating from spiked BSA) are similar. (Fig. 1, panel 3). Evaluate the digestion efficiency of CPs by considering the abundance of miscleaved peptides (allow up to 2 miscleaved tryptic sites during database searches). Finally, use DDA data for retention time (RT) scheduling and designing the PRM method.
5. To adjust the optimal amounts (molar ratios) of BSA and CPs, perform a scheduled pilot PRM LC-MS/MS analysis of a pooled CSF sample spiked-in with BSA and with all CP standards. Process LC-MS/MS data with Skyline and confirm that the intensities of peaks of endogenous peptides, Q-peptides, R-peptides and BSA peptides are within the acceptable range and ratios of abundances of peaks of matching pairs of labeled and endogenous (unlabeled) peptides differ less than ten-fold (*see* **Note 7**).

3.2 Enzymatic Digestion of CSF Aliquots

1. Prepare the solutions with molar concentration of the reference protein (BSA) and of CPs adjusted as above (*see* Subheading 3.1) by diluting their stock solutions with the sample buffer. The final total volume of the solution of CPs and BSA should be approximately equal to the starting CSF volume (20 μL). Add BSA and CPs to CSF (*see* **Note 8**).
2. Add DTT to the final concentration of 5 mM and incubate at 56 °C for 30 min at 300 rpm in a thermostatic mixer and then allow samples to cool down to room temperature.
3. Add IAA to the final concentration of 15 mM and incubate at room temperature for 30 min in the dark.

4. Add freshly reconstituted trypsin/Lys-C solution (0.5 μg of enzyme per 20 μL of CSF) and incubate at 37 °C overnight (ca. 16–18 h).
5. Add TFA to a final concentration of 1% and incubate at 37 °C for 30 min.
6. Centrifuge the sample at 13000 rpm for 20 min and transfer the supernatant into a new low protein binding polypropylene tube.
7. Dry down the samples in the vacuum concentrator and store at −20 °C until LC-MS/MS analysis.

3.3 Proteins Quantification by LC-MS/MS

1. Reconstitute CSF digests in 100 μL of reconstitution buffer, centrifuge and transfer to the LC vials (*see* **Note 9**).
2. Place samples in the LC autosampler of Dionex Ultimate 3000 RSLC system.
3. Analyze CSF digests in trap-and-elute mode. Load the samples for 5 min at 7 μL/min using solvent A, and separate peptides using linear 60 min gradient (5–35% solvent B) at 500 nL/min and column temperature of 45 °C (*see* **Note 10**).
4. Analyze the samples by PRM. The acquisition method includes MS1 scan from *m/z* 350 to 1700 with an automatic gain control (AGC) target value of 3×10^6, maximum injection time (IT) of 60 ms and targeted mass resolution $R_{m/z\,=\,200}$ of 60,000. The scheduled PRM method should acquire MS/MS using 5-min retention time (RT) windows; precursor isolation window of 1.0 Th; AGC of 1×10^6; maximum IT of 119 ms; $R_{m/z\,=\,200}$ of 60,000; NCE of 25% (*see* **Note 11**).

3.4 Data Analysis

1. Import the PRM raw files into Skyline software.
2. Manually check the automatic peak selection/integration of all peaks as well as matching retention times for both unlabeled and labeled R- and Q-peptides. Adjust software settings, if necessary.
3. Inspect the fragment ions for their peak purity to exclude interfering background ions. Use the three most consistent XIC traces for peptide quantification (*see* **Note 12**).
4. Use the Document Grid option in Skyline to export the fragment peak areas for both unlabeled and labeled R- and Q-peptides.
5. Use Excel or Phyton script to calculate the exact amount of the individual CPs in each sample by referring the peak areas of labeled (heavy) R-peptides originating from CP to the areas of matching unlabeled (light) R-peptides from BSA:

$$C_{CP} = C_{BSA} \times \frac{S_{RH}}{S_{RL}}$$

where C_{BSA} is the concentration of BSA; S_{RH} is the XIC area of the heavy R-peptide from CP; S_{RL} is the XIC area of the light R-peptide from BSA.

Next, calculate the concentration of target proteins by referring the peak areas of unlabeled (light) Q-peptides originating from endogenous proteins to the areas of matching labeled (heavy) Q-peptides from CP:

$$C_P = C_{CP} \times \frac{S_{QL}}{S_{QH}}$$

where C_P is the concentration of target protein; S_{QL} is the XIC area of the light Q-peptide from endogenous target protein; S_{RL} is the XIC area of the heavy Q-peptide from CP.

6. Evaluate concordance between multiple Q-peptides for each target protein (inter-peptide agreement). Typical inter-peptide coefficients of variation (CVs) are below 20% (Table 1). Use average molar concentrations of concordant Q-peptides to calculate the molar concentration of the target protein (*see* **Note 13**).

4 Notes

1. The NIST-certified BSA standard reference material (SRM 927f) should be used to ascertain accurate reference protein concentration.
2. Prepare iodoacetamide solution freshly before use.
3. Reconstitute enzyme aliquots in 50 mM ammonium bicarbonate shortly before use and keep them on ice.
4. Many clinically relevant proteins are low abundant and are only detectable with one or two peptides. Therefore even seemingly suboptimal sequences should be considered and tested during the method validation [18].
5. Designing CPs sequences with an optimal distribution of Q-peptides may require multiple iterations. Because of moderate cost of CP production, it is advisable to test a few CPs with alternative sequences and use the best-performing sequence for analyzing CSF samples.
6. For the validation of CPs preparation, you can use the IonQuant tool of the FragPipe software with disabled "match between runs (MBR)" and "normalize intensity between runs" options. Use the "combined_ion" output file containing the quantitative information on the precursor level.

7. It is recommended to use condition-specific CSF pools, as protein biomarker concentrations can vary several-fold between control and diseased patient groups.
8. For the most accurate and consistent quantification, the concentration of the spiked-in BSA standard should be close to concentrations CPs. For example, in the 10-protein assay covering a 100-fold range of concentrations from ca. 20 ng/mL for CH3L1 to ca. 2000 ng/mL for APOD (Table 1), the two CPs were used at the amounts of ca. 300 fmol/injection and 30 fmol/injection, while BSA was at 100 fmol/injection.
9. The reconstituted volume should be adjusted according to LC-MS/MS sensitivity. The volumes used in this protocol (20 μL starting CSF, 100 μL reconstitution volume and 5 μL sample injection volumes) resulted in ca. 0.5 μg of total protein loaded on the column (assuming total protein concentration in CSF is close to 0.5 mg/mL).
10. The trap-and-elute chromatography used in this protocol performed online desalting of CSF digests. If using direct injection mode, the samples should be desalted off-line before chromatography.
11. PRM settings should be adjusted according to the number of targeted peptides. Decreasing MS2 resolution by shortening transient acquisition time allows targeting a higher number of peptides; however, it could also compromise the sensitivity and specificity of the analysis. Nonetheless, IT settings should support the collection of 8–10 datapoints per chromatographic peak [19].
12. Using several quantifying fragments per peptide improves selectivity. However, for very low abundant peptides, only fragments detected with the highest signal-to-noise ratio should be used for peptide quantification [20].
13. The concordance between protein concentrations independently calculated from each of Q-peptides is an important indicator of quantification confidence. It can be assessed either by relative (fractional) Q-peptide peak areas (for each sample separately) or by the correlation between different Q-peptides pairs observed in several CSF samples (Fig. 1, panel 6) and expressed as the coefficient of inter-peptide variation (Table 1).

Acknowledgments

Work in the Shevchenko group was supported by Max Planck Society. The authors are grateful to other group members for expert technical support and useful discussions. The figure was assembled using the BioRender tool from BioRender.com.

References

1. Wichmann TO, Damkier HH, Pedersen M (2022) A brief overview of the cerebrospinal fluid system and its implications for brain and spinal cord diseases. Front Hum Neurosci 15: 1–7
2. Zetterberg H, Blennow K (2021) Moving fluid biomarkers for Alzheimer's disease from research tools to routine clinical diagnostics. Mol Neurodegener 16:1–7
3. Hansson O (2021) Biomarkers for neurodegenerative diseases. Nat Med 27:954–963
4. Nakayasu ES, Gritsenko M, Piehowski PD et al (2021) Tutorial: best practices and considerations for mass-spectrometry-based protein biomarker discovery and validation. Nat Protoc 16:3737–3760
5. Ren AH, Diamandis EP, Kulasingam V (2021) Uncovering the depths of the human proteome: antibody-based technologies for ultrasensitive multiplexed protein detection and quantification. Mol Cell Proteomics 20:100–155
6. Teunissen CE, Kimble L, Bayoumy S et al (2023) Methods to discover and validate biofluid-based biomarkers in neurodegenerative dementias. Mol Cell Proteomics 22: 100629
7. Rzagalinski I, Bogdanova A, Raghuraman BK et al (2022) FastCAT accelerates absolute quantification of proteins using multiple short nonpurified chimeric standards. J Proteome Res 21:1408–1417
8. Carr SA, Abbatiello SE, Ackermann BL et al (2014) Targeted peptide measurements in biology and medicine: best practices for mass spectrometry-based assay development using a fit-for-purpose approach. Mol Cell Proteomics 13:907–917
9. Matic I, Jaffray EG, Oxenham SK et al (2011) Absolute SILAC-compatible expression strain allows sumo-2 copy number determination in clinical samples. J Proteome Res 10:4869–4875
10. Gasteiger E, Hoogland C, Gattiker A et al (2005) Protein identification and analysis tools on the ExPASy server. In: The proteomics protocols handbook, vol 112. Humana Press, pp 531–552
11. Altschul SF, Madden TL, Schäffer AA et al (1997) Gapped BLAST and PSI-BLAST: a new generation of protein database search programs. Nucleic Acids Res 25:3389–3402
12. Schaeffer M, Gateau A, Teixeira D et al (2017) The neXtProt peptide uniqueness checker: a tool for the proteomics community. Bioinformatics 33:3471–3472
13. Kong AT, Leprevost FV, Avtonomov DM et al (2017) MSFragger: ultrafast and comprehensive peptide identification in mass spectrometry-based proteomics. Nat Methods 14:513–520
14. Pino LK, Searle BC, Bollinger JG et al (2020) The Skyline ecosystem: informatics for quantitative mass spectrometry proteomics. Mass Spectrom Rev 39:229–244
15. Zauber H, Kirchner M, Selbach M (2018) Picky: a simple online PRM and SRM method designer for targeted proteomics. Nat Methods 15:156–157
16. Guldbrandsen A, Farag Y, Kroksveen AC et al (2017) CSF-PR 2.0: an interactive literature guide to quantitative cerebrospinal fluid mass spectrometry data from neurodegenerative disorders. Mol Cell Proteomics 16:300–309
17. Kumar M, Joseph SR, Augsburg M et al (2018) MS Western, a method of multiplexed absolute protein quantification is a practical alternative to Western blotting. Mol Cell Proteomics 17: 384–396
18. Chiva C, Sabidó E (2017) Peptide selection for targeted protein quantitation. J Proteome Res 16:1376–1380
19. Gallien S, Duriez E, Crone C et al (2012) Targeted proteomic quantification on quadrupole-orbitrap mass spectrometer. Mol Cell Proteomics 11:1709–1723
20. Ruhaak LR, Romijn FPHTM, Begcevic Brkovic I et al (2023) Development of an LC-MRM-MS-based candidate reference measurement procedure for standardization of serum apolipoprotein (a) tests. Clin Chem 69: 251–261

Part III

Post-translational Modifications in Cerebrospinal Fluid

Chapter 7

Cerebrospinal Fluid Phosphoproteome Characterization Using Titanium Dioxide Enrichment and High-Resolution LC-MS/MS

Elena Anaya-Cubero, Adriana Cortés, Mercedes Lachén-Montes, Paz Cartas-Cejudo, Joaquín Fernández-Irigoyen, and Enrique Santamaría

Abstract

Cerebrospinal fluid (CSF) is a valuable source of biomarkers for neurological diseases due to its proximity to the central nervous system. Phosphorylation, a common post-translational modification, plays crucial roles in cellular signaling and is dysregulated in various neurological disorders.

Analysis of the CSF phosphoproteome using titanium dioxide (TiO_2) enrichment coupled with high-resolution liquid chromatography-tandem mass spectrometry (LC-MS/MS) technique unveils a diverse array of phosphorylated proteins and their corresponding phosphorylation sites, shedding light on the intricate regulatory networks within CSF.

The combination of TiO_2 enrichment and high-resolution LC-MS/MS offers a robust and sensitive approach to studying the phosphoproteome, providing valuable insights into cellular signaling dynamics and disease mechanisms.

Key words Mass spectrometry, LC-MS/MS, Proteomics, Phosphoproteome, CSF, Phosphopeptide, Cerebral fluid phosphopeptide enrichment, Titanium dioxide, Label-free

1 Introduction

Post-translational modifications regulate protein structure and function, with protein phosphorylation at serine, threonine, and tyrosine residues being the most common and studied modifications [1]. Phosphorylation modulates numerous regulatory processes, encompassing functions such as facilitating signal transduction, maintaining structural integrity, orchestrating cellular differentiation, regulating proliferation, managing metabolism, and facilitating degradation pathways [2]. Moreover, modifications in the phosphorylation pattern of proteins may induce abnormal cell signaling, contributing to the development or progression of several diseases [3–5].

Enrique Santamaría and Joaquín Fernández-Irigoyen (eds.), *Cerebrospinal Fluid (CSF) Proteomics: Methods and Protocols*, Methods in Molecular Biology, vol. 2914, https://doi.org/10.1007/978-1-0716-4462-1_7,

Large-scale phosphoproteomic studies require efficient and precise enrichment methods as phosphorylation occurs at substoichiometric levels and exhibits a highly dynamic nature. Generally, purification techniques for phosphate groups rely on metals attached to carrier resins as one of the most prominent methods, the Immobilized Metal Ion Affinity Chromatography (IMAC). In these techniques, metals such as Fe^{3+}, Ti^{4+}, Zr^{4+}, and Ga^{3+} are used for phosphopeptide enrichment [6].

Phosphoproteomic analysis using titanium dioxide (TiO_2) enrichment coupled with high-resolution liquid chromatography-tandem mass spectrometry (LC-MS/MS) offers several advantages:

1. *Selective enrichment*, as TiO_2 beads have a high affinity for phosphorylated peptides, allowing for selective enrichment of phosphoproteins from complex biological samples.
2. *Comprehensive Coverage*: The combination of TiO_2 enrichment and high-resolution LC-MS/MS enables the identification and characterization of a large number of phosphorylation events within a single experiment. Moreover, LC-MS/MS offers high sensitivity and specificity for peptide identification, allowing for accurate detection and quantification of phosphorylated peptides. This enables the detection of low-abundance phosphoproteins and precise mapping of phosphorylation sites [7].
3. *Quantitative Analysis*: The use of LC-MS/MS allows for quantitative analysis of phosphorylation events, facilitating the comparison of phosphoproteome under different experimental conditions or disease states. This can reveal dynamic changes in protein phosphorylation and identify potential biomarkers or therapeutic targets.
4. *Label-Free Approach*: TiO_2 enrichment combined with label-free LC-MS/MS workflow eliminates the need for chemical labeling of peptides, reducing sample complexity and experimental variability [8].
5. *Compatibility with Complex Samples*: The TiO_2 enrichment and LC-MS/MS technique is compatible with various sample types, including cell lysates, tissue extracts, and biological fluids such as blood or cerebrospinal fluid (CSF). This versatility enables the study of phosphorylation events in diverse biological contexts [9].

CSF proteome profiling in the realm of neurodegenerative diseases serves to delineate the molecular pathways implicated in disease pathogenesis and uncover potential biomarker candidates [10].

Phosphoproteomics is a branch of proteomics that focuses on the large-scale study of phosphorylation events within proteins. This technique aims to identify, quantify, and characterize phosphorylated proteins and their phosphorylation sites in a given biological sample. Thus, phosphoproteomic analysis is a suitable approximation to identify biomarkers of disease and therapeutic targets in fluid samples such as blood [11] or CSF [12]. Opting for fluid in close proximity to an afflicted organ may enhance the likelihood of identifying a biomarker panel derived from pathological tissue. In this context, CSF, being in intimate contact with the brain, serves as a pertinent reservoir of potentially clinically relevant biomarkers for neurological disorders because of brain inaccessibility [13]. Neurodegenerative diseases involve intricate protein signaling and expression modulation, which vary depending on the analyzed sample. Proteins that decrease in brain tissue may increase in cerebrospinal fluid samples, and vice versa, or show concurrent decrease or increase in both [14]. Consequently, additional studies concentrating on biomarker identification and characterization are needed to better understand and diagnose neurodegenerative diseases.

This chapter describes phosphoproteomic analysis of CSF using titanium dioxide enrichment and high-resolution liquid chromatography-mass spectrometry.

Analyzing the CSF phosphoproteome using TiO_2 enrichment combined with high-resolution liquid chromatography-tandem mass spectrometry is a powerful approach to study phosphorylation events in the central nervous system. Titanium dioxide beads are commonly used for phosphopeptide enrichment due to their high affinity for phosphorylated peptides. The CSF sample is incubated with TiO_2 beads, allowing the phosphopeptides to bind to the beads selectively. After washing away non-phosphorylated peptides, the enriched phosphopeptides are eluted from the beads using high-resolution liquid chromatography, separating peptides based on their hydrophobicity. The separated phosphopeptides are ionized and introduced into the mass spectrometer, where they are fragmented into ions. The resulting spectra are then analyzed to determine the amino acid sequence and the phosphorylation sites within the peptides.

A graphical explanation of the method can be seen in Fig. 1.

2 Materials

2.1 Protein Extraction

1. MARS (Multiple Affinity Removal System) column (Agilent Technologies).
2. MARS buffer A and B (Agilent Technologies).
3. *Lysis buffer*, 8 M urea, 50 mM Dithiothreitol (DTT), 0.1 tablets/mL Phosphatases Inhibitor PhoSTOP™, 1.66 tablets/mL Proteases Inhibitor cOmplete™, 50 mM Ammonium Bicarbonate (ABC).

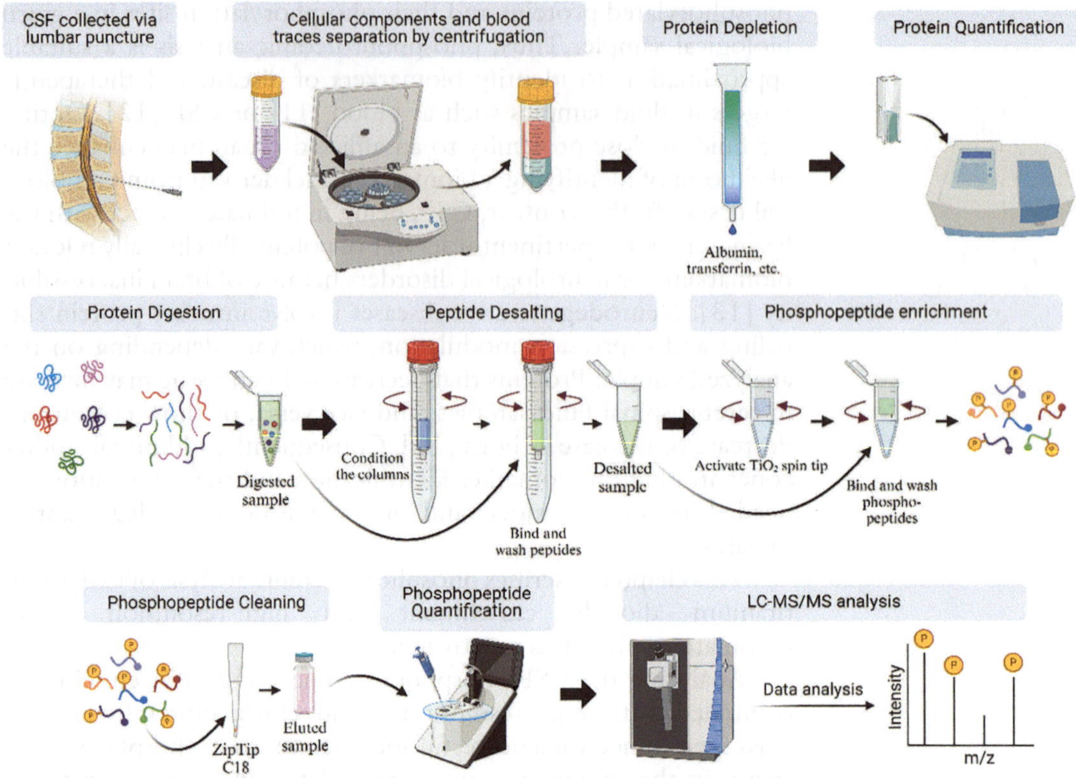

Fig. 1 Schematic illustration of CSF Phosphoproteomic analysis using TiO_2 enrichment and LC-MS/MS

4. Protein standard of bovine serum albumin (BSA) 2 mg/mL.
5. Bradford protein assay dye reagent concentrate.

2.2 Reduction and Alkylation

1. DTT.
2. Iodoacetamide (IAA).
3. ABC 50 mM.

2.3 Digestion

1. Sequencing grade-modified trypsin.
2. Formic acid (FA).

2.4 Peptide Desalting

1. Pierce TM Peptide Desalting Spin Columns (ThermoFisher) containing 20 mg of resin in a 1:1 water/DMSO slurry.
2. Trifluoroacetic acid (TFA) 0.1%.
3. Acetonitrile (ACN) LC-MS Grade 100% and 50%.
4. Water LC-MS Grade.
5. FA 0.1%.

2.5 Phosphopeptide Enrichment

High-Select™ TiO_2 Phosphopeptide Enrichment Kit (ThermoFisher):

1. TiO_2 spin tips.
2. Centrifuge column adaptors.
3. Binding/Equilibration buffer.
4. Wash Buffer.
5. Phosphopeptide Elution Buffer.
6. Water LC-MS Grade.

2.6 Phosphopeptide Cleaning

1. Zip Tip pipette tips p10; C18: Zip Tip with 0.6 μL of C18 resin.
2. Resuspension buffer: 2% ACN, 0,5% FA.
3. Solution A: 100% LC-MS grade water, 0.1% FA.
4. Solution B: 100% ACN, 0.1% FA.
5. Solution C: 50% Solution A, 50% Solution B.

2.7 Other Materials

1. Low Protein Binding Microcentrifuge Tubes, 2.0 mL.
2. Low Protein Binding Microcentrifuge Tubes, 1.5 mL.
3. pH control strips.
4. Ultracentrifuge with adjustable rotor speed up to 100,000× *g*.
5. Microcentrifuge with adjustable rotor speed up to 7000× *g*.
6. Speed-Vacuum centrifuge.

3 Methods

3.1 Sample Preparation

3.1.1 Protein Extraction

1. Collect samples via lumbar or ventricular puncture according to ethically approved protocols (*see* **Note 1**). Place the samples on ice immediately after collection.
2. Centrifuge the CSF samples at 2500× *g* for 10 min at 4 °C to remove cellular components and blood traces (*see* **Note 2**).
3. Deplete the CSF of the most abundant proteins (albumin, transferrin, IgG, IgA, haptoglobin, α1-antitrypsin) using a MARS column or other similar depletion columns (*see* **Note 3**).
4. Concentrate the sample by lyophilization using a speed-vacuum system.

3.1.2 Protein Quantification

1. Prepare samples and BSA standards for protein quantification. BSA standards may range from 0 to 16 μg, diluted from 80 μg/μL BSA stock each time.

2. Dilute the sample between 1/50 and 1/200 with water to a final volume of 600 μL and make six serial dilutions of the sample.
3. Prepare a 1:4 dilution of the Bradford stock. Add 800 μL to each aliquot.
4. Transfer the volume to a spectrophotometer cuvette and measure the absorbance at 595 and 450 nm in the spectrophotometer.
5. Calculate the absorbance ratio 595/450 of both the standard and the sample lines to obtain the protein concentration.

3.1.3 Protein Digestion

Peptide generation for phosphoproteomic LC-MS/MS relies on in-solution digestion, avoiding running a polyacrylamide gel and in-gel digestion. In-solution peptide preparation typically involves two enzymatic steps to convert proteins into peptides.

1. Dilute 400–600 μg of sample to lysis buffer (*see* **Note 4**).
2. Add DTT to a final concentration of 20 mM. Incubate for 30 min at 30 °C with gentle agitation.
3. Add IAA to a final concentration of 30 mM. Incubate for 30 min in the dark at room temperature with gentle agitation.
4. Add ABC 50 mM to reduce urea concentration to 1 M.
5. Check pH is >7.5.
6. Add modified trypsin (enzyme/substrate ratio = 1:20) and incubate for 3–5 h at 37 °C with gentle agitation (*see* **Note 5**).
7. Add modified trypsin (enzyme/substrate ratio = 1:50) and incubate overnight at 37 °C with gentle agitation.
8. Add FA to a final concentration of 1%. Check pH is <6.
9. Dry samples in Speed-Vac.

3.2 Peptide Desalting

The Thermo Scientific™ Pierce™ Peptide Desalting Spin Columns are optimized for desalting 5 μg to 5 mg of peptide samples from protein digests.

3.2.1 Prepare the Samples

1. Resuspend the protein digests with 300 μL of 0.1% TFA.
2. Verify pH of samples is <6 using pH control strips.

3.2.2 Condition the Spin Columns

1. Remove the white plug from the bottom of the column and place it in a 2 mL tube.
2. Centrifuge at 5000× *g* for 1 min to remove buffer and pack the resin. Discard the liquid.
3. Remove the top screw cap and add 300 μL ACN, cap the column and place in the 2 mL tube.

4. Centrifuge at 5000× *g* for 1 min. Discard the ACN and repeat the ACN wash step.
5. Add 300 μL 0.1% TFA, cap the column and place in a new 2 mL tube.
6. Centrifuge at 5000× *g* for 1 min. Discard the TFA and repeat the wash step with 0.1% TFA.

3.2.3 Bind and Wash Samples

1. Add the 300 μL of the resuspended sample with 0.1% TFA to the columns, cap the column and place it in the 2 mL tube.
2. Centrifuge at 3000× *g* for 1 min. Remove the flow-through.
3. Add 300 μL 0.1% TFA, cap the column and place in the 2 mL tube.
4. Centrifuge at 3000× *g* for 1 min. Remove the TFA and repeat the wash step twice to a total of three washes with 0.1% TFA.

3.2.4 Elute Desalted Peptides

1. Add 300 μL 50% ACN, 0.1% TFA, cap the column and place in a new 2 mL Protein Low Bind tube.
2. Centrifuge at 3000× *g* for 1 min. Save the eluate and repeat the elution step with 50% ACN and 0.1% TFA, collecting the eluate into the same Low Bind tube (*see* **Note 6**).
3. Bring to dryness in Speed-Vac.

3.3 High-Select™ TiO_2 Phosphopeptide Enrichment Kit

3.3.1 Prepare the Samples

1. Resuspend the lyophilized peptide samples with 150 μL of Binding/Equilibration buffer. Use a vortex mixer.
2. Verify pH of samples is <3 using pH control strips.

3.3.2 Activation and Calibrating of Columns

1. Place a Centrifuge Column Adaptor in a 2 mL collection tube and insert a TiO_2 Spin Tip into the adaptor.
2. Add 20 μL of Wash buffer and centrifuge at 3000× *g* for 2 min.
3. Add 20 μL of Binding/Equilibration buffer and centrifuge at 3000× *g* for 2 min.
4. Discard the flow-through. Save the microcentrifuge tube for later "Wash column" **step 1** (**step 1**, Subheading **3.3.4**).

3.3.3 Bind Phosphopeptide

1. Transfer the equilibrated TiO_2 Spin Tip and adaptor into a new 2 mL Protein Low Bind microcentrifuge tube.
2. Add the 150 μL of resuspended peptide sample in Binding/Equilibration buffer to the spin tip. Centrifuge at 1000× *g* for 6 min.
3. Reapply the peptide sample from the microcentrifuge tube to the spin tip. Centrifuge at 1000× *g* for 6 min.

4. Bring the flow-through containing the non-modified peptides to dryness in Speed-Vac and store at −80 °C if needed for future experiments.

3.3.4 Wash Column

1. Transfer the TiO_2 Spin Tip and adaptor into the collection tube saved in **step 4** of Subheading 3.3.2.
2. Wash column by adding 20 μL of Binding/Equilibration buffer. Centrifuge at 3000× *g* for 2 min.
3. Wash column by adding 20 μL of Wash buffer. Centrifuge at 3000× *g* for 2 min.
4. Repeat **steps 2** and **3**.
5. Wash column by adding 20 μL of LC-MS grade water. Centrifuge at 3000× *g* for 2 min.

3.3.5 Elute Column

1. Remove excess liquid by blotting the bottom of the spin tip on a clean lab tissue.
2. Place the spin tip and adaptor in a new 1.5 mL Protein Low Bind collection tube and add 50 μL of Elution buffer.
   Centrifuge at 1000× *g* for 5 min.
3. Repeat **step 2**.
4. Bring the eluate to dryness immediately in Speed-Vac to remove the Elution buffer.

3.4 Phosphopeptide Cleaning

3.4.1 ZipTip Cleanup

1. Dilute phosphopeptide-enriched samples in 10 μL of Resuspension buffer (*see* **Note 7**).
2. Attach ZipTip (C18) tip to the appropriate pipette, set to a maximum of 10 μL (*see* **Note 8**).
3. Activate the ZipTip tip by aspirating 10 μL of Solution B into the tip and dispense it to waste. Repeat five times.
4. Equilibrate the ZipTip tip by washing it with 10 μL of Solution A. Repeat five times.
5. Bind the sample to the ZipTip tip. Place the tip in the sample and perform an aspirate-and-dispense cycle of 10 μL 10 times.
6. Wash the ZipTip tip with 10 μL of Solution A. Do three aspirate-and-dispense cycles.
7. Repeat the washing step cycles three times.
8. Elute ZipTip tip with 10 μL of Solution C. Do six aspirate-and-dispense cycles.
9. Dispense 10 μL of eluted samples into a clean microcentrifuge tube using a standard pipette tip.
10. Bring the eluate to dryness in Speed-Vac.
11. Resuspend dried samples in 12 μL of 0,1% FA.

12. Quantify the samples at 205 nm in NanoDrop™ One, with baseline correction at 340 nm before MS/MS experiments.

4 Notes

1. CSF occupies the subarachnoid space and the ventricular system, with a total volume of approximately 135 mL in humans. CSF contains a low number of cells (0–4 cells/μL), low protein concentration, and various salts. Lumbar puncture is the preferred method for CSF collection and is commonly used in neuroproteomics due to its direct contact with the brain. The protein concentration in CSF from the lumbar region is about 15–45 mg/dL, whereas the concentration in cisternal and ventricular collections is lower.
2. Traumatic taps occur in 14–20% of standard lumbar punctures, causing blood contamination in the CSF, affecting CSF proteome and providing incorrect CSF protein biomarker results. To minimize the effect of potential contamination, specific filters can be used to eliminate all membrane and cellular debris, as well as blood traces by centrifugation. Another option is to incorporate a precipitation step using acetone or a methanol/chloroform liquid–liquid partitioning procedure.
3. Since CSF has a low protein concentration, removing the most abundant proteins with depletion columns is advisable to access low-abundant proteins. Other depletion columns, like Top 12 Abundant Protein Depletion Spin Columns (Thermo Scientific), are also an option.
4. Fresh urea solution must be prepared without heating for each application, as the breakdown of urea into ammonium cyanate is accelerated by time and heat. Ammonium cyanate can cause carbamylation of amine groups in proteins. To minimize carbamylation, protein digestion should be carried out at room temperature.
5. Unmodified trypsin can undergo auto-proteolysis, resulting in chymotrypsin-like enzymatic activity. Use modified trypsin to avoid auto-proteolysis.
6. Sample loss can occur due to the procedure (unsuccessful recovery). To overcome it, do not exceed the recommended centrifugation speeds as this can damage the column frit, causing resin leakage and sample loss. In addition, use low protein-binding microcentrifuge tubes to ensure maximum sample recovery.
7. Acetonitrile is volatile and evaporation can occur rapidly. To avoid it, clean samples one at a time, adding the necessary reagents at the start of each process.

8. The resin bed in ZipTip C18 provides a slight backpressure. To mitigate this, do not use the pipettor as a precise volumetric dispenser. To achieve optimal sample uptake and delivery, set the pipettor to 10 μL and securely attach the tip. Depress the plunger to the dead stop and slowly release or dispense the plunger throughout the operation.

Acknowledgments

The Clinical Neuroproteomics Unit is member of the Spanish Olfactory Network—ROE (grant RED2022-134081-T from the Spanish Ministry of Science & Innovation) and is supported by grants PID2023-152593OB-I00 funded by MCIU/AEI/10.13039/501100011033/FEDER, UE to ES and JFI and 0011-1411-2023-000028 (from Government of Navarra- Department of Economic and Business Development-S4) to ES. PC-C is supported by a predoctoral fellowship from the Public University of Navarra (UPNA). ML-M is supported by a postdoctoral fellowship from Miguel Servet Foundation-Navarrabiomed. EA-C is supported by "Programa MRR Investigo 2023" in the framework of the European Union recovery and resilience facility.

References

1. Ardito F, Giuliani M, Perrone D et al (2017) The crucial role of protein phosphorylation in cell signalingand its use as targeted therapy (Review). Int J Mol Med 40:271–280. https://doi.org/10.3892/ijmm.2017.3036
2. Cohen P (2002) The origins of protein phosphorylation. Nat Cell Biol 4:E127. https://doi.org/10.1038/ncb0502-e127
3. Russell CL, Mitra V, Hansson K et al (2016) Comprehensive quantitative profiling of tau and phosphorylated tau peptides in cerebrospinal fluid by mass spectrometry provides new biomarker candidates. J Alzheimers Dis 55: 303–313. https://doi.org/10.3233/JAD-160633
4. Bilbrough T, Piemontese E, Seitz O (2022) Dissecting the role of protein phosphorylation: a chemical biology toolbox. Chem Soc Rev 51: 5691–5730. https://doi.org/10.1039/d1cs00991e
5. Zhang T, Chen D, Lee TH (2020) Phosphorylation signaling in APP processing in Alzheimer's disease. Int J Mol Sci 21. https://doi.org/10.3390/ijms21010209
6. Yang F, Wang C (2020) Profiling of post-translational modifications by chemical and computational proteomics. Chem Commun 56:13506–13519. https://doi.org/10.1039/d0cc05447j
7. Larsen MR, Thingholm TE, Jensen ON et al (2005) Highly selective enrichment of phosphorylated peptides from peptide mixtures using titanium dioxide microcolumns. Mol Cell Proteomics 4:873–886. https://doi.org/10.1074/mcp.T500007-MCP200
8. McNulty DE, Annan RS (2008) Hydrophilic interaction chromatography reduces the complexity of the phosphoproteome and improves global phosphopeptide isolation and detection. Mol Cell Proteomics 7:971–980. https://doi.org/10.1074/mcp.M700543-MCP200
9. Zhou H, Ye M, Dong J et al (2008) Specific phosphopeptide enrichment with immobilized titanium ion affinity chromatography adsorbent for phosphoproteome analysis. J Proteome Res 7:3957–3967. https://doi.org/10.1021/pr800223m
10. del Campo M, Vermunt L, Peeters CFW et al (2023) CSF proteome profiling reveals biomarkers to discriminate dementia with Lewy bodies

from Alzheimer's disease. Nat Commun 14. https://doi.org/10.1038/s41467-023-41122-y

11. Dunphy K, Dowling P, Bazou D, O'Gorman P (2021) Current methods of post-translational modification analysis and their applications in blood cancers. Cancers (Basel) 13. https://doi.org/10.3390/cancers13081930
12. Azevedo R, Jacquemin C, Villain N et al (2022) Mass spectrometry for Neurobiomarker discovery: the relevance of post-translational modifications. Cells 11. https://doi.org/10.3390/cells11081279
13. Lleó A, Cavedo E, Parnetti L et al (2015) Cerebrospinal fluid biomarkers in trials for Alzheimer and Parkinson diseases. Nat Rev Neurol 11:41–55. https://doi.org/10.1038/nrneurol.2014.232
14. Higginbotham L, Ping L, Dammer EB et al (2020) Integrated proteomics reveals brain-based cerebrospinal fluid biomarkers in asymptomatic and symptomatic Alzheimer's disease. Sci Adv 6. https://doi.org/10.1126/sciadv.aaz9360

Chapter 8

Analyzing *N*-Glycans in Cerebrospinal Fluid: A Glycomics Approach

Cristian D. Gutierrez Reyes, Sherifdeen Onigbinde, Joy Solomon, and Yehia Mechref

Abstract

Cerebrospinal fluid (CSF) plays a crucial role in direct interaction with the brain, making it an essential sample for identifying molecular markers that emerge as neurological disorders progress. Notably, changes in protein glycosylation are observed in many neurodegenerative diseases. However, analyzing the glycome profile of CSF is challenging due to low protein concentration. *N*-glycans are chemically modified through permethylation to quantify these changes accurately and analyzed using a highly sensitive liquid chromatography-mass spectrometry (LC-MS) technique. This study outlines a method for preparing samples with optimized conditions for *N*-glycan digestion combined with a sensitive detection system. The approach has proven reliable and consistent in measuring various permethylated *N*-glycans extracted from CSF.

Key words *N*-Glycans, CSF, Permethylation, LC-MS

1 Introduction

Recently, the field of omics sciences has become a key player in studying molecular cell systems, significantly improving our understanding of human diseases [1]. Integrating various omics techniques, when possible, helps reveal the mechanisms involved in the development or progression of diseases. Specifically, the study of *N*-glycomic profiles in human bodily fluids, including cerebrospinal fluid (CSF), has shown essential changes in glycosylation patterns, such as variations in abundance and structural modifications of glycans [2–4].

Despite the more invasive procedure required to collect CSF compared to other fluids, it offers invaluable neurological insights due to its proximity to the brain and spinal cord [5, 6]. Notably, the most significant changes observed in CSF glycosylated proteins are in the abundance of bisecting *N*-acetylglucosamine and core fucose

Enrique Santamaría and Joaquín Fernández-Irigoyen (eds.), *Cerebrospinal Fluid (CSF) Proteomics: Methods and Protocols*, Methods in Molecular Biology, vol. 2914, https://doi.org/10.1007/978-1-0716-4462-1_8,

glycans [7–10]. However, the low protein content in CSF poses a significant challenge for glycomics analysis, as conventional methods require several hundred microliters of fluid for a detailed LC-MS analysis [7, 11].

To address these challenges, glycan molecules are chemically modified to reduce their polarity and enhance ionization efficiency for more effective reverse-phase liquid chromatography-mass spectrometry (RPLC-MS) analysis [12]. Our method employs a sensitive LC-MS technique that begins with only 15 μL of CSF and injects just 5 μL of the prepared sample into the LC-MS system. This process includes a purification step using SPE-C18, followed by the reduction and permethylation of *N*-glycans [13]. Finally, we evaluated our approach against the established Filter-Aided *N*-Glycan Separation (FANGS) method [14, 15].

2 Materials

2.1 Model Glycoproteins

1. Pooled standard CSF (Goden West Bio Solutions, LLC., PN OH1030).

2.2 Solutions for Glycomics Sample Preparation

Unless otherwise specified, all solutions must be made with HPLC-grade water and kept at room temperature.

1. To prepare a 50 mM ammonium bicarbonate (ABC) buffer solution, dissolve 197.64 mg of ABC (with a molecular weight of 79.06) in 50 mL of water. Mix the solution thoroughly using a vortex mixer and store it at room temperature.
2. To prepare 5% acetic acid, pour 50 mL of acetic acid into a 1 L graduated cylinder and add water to the graduated cylinder until the total volume reaches 1 L. Mix the solution thoroughly.
3. To prepare 10 μg/μL ammonium borane complex solution, weigh 2 mg borane ammonia complex and dissolve in 200 μL of water. Mix the solution thoroughly using a vortex mixer and store it at room temperature (*see* **Note 1**).

2.3 Mobile Phase for Separation Using RPLC

1. Mobile Phase A (MPA) consists of 98% water, 2% acetonitrile (ACN), and 0.1% formic acid (FA).
2. Mobile Phase B (MPB) comprises 100% ACN and 0.1% FA (*see* **Note 2**).

3 Methods

The step-by-step process of our methodology is illustrated in Fig. 1.

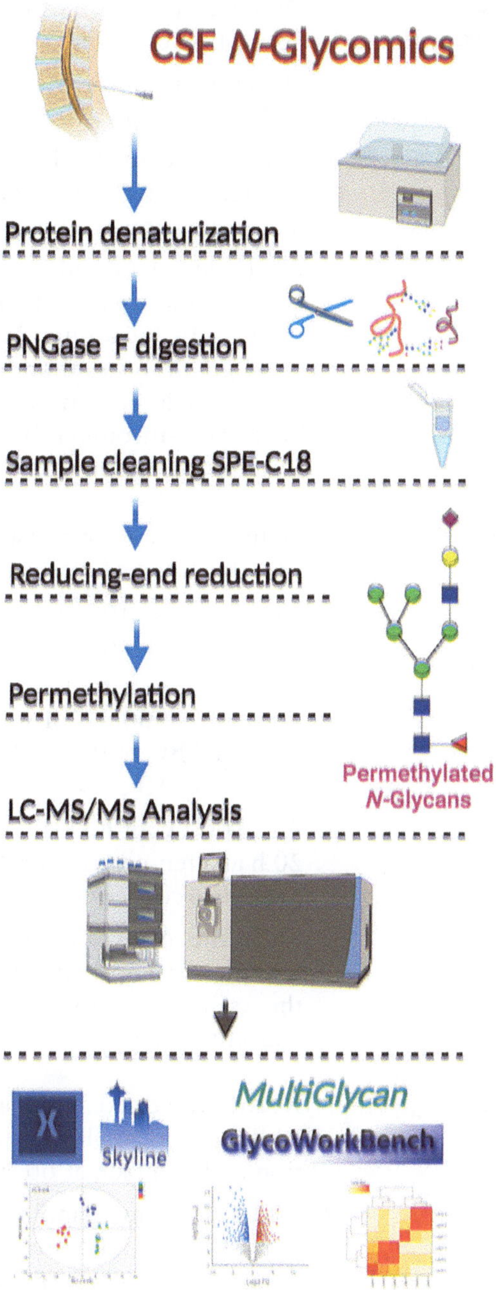

Fig. 1 Analytical workflow CSF *N*-glycomics. A four-digit nomenclature is used, "1-1-1-1," where the numbers correspond to the sugar units of *N*-acetylglucosamine (■), Mannose (●) + Galactose (●), Fucose, and sialic acid (◂ ◆) "GlcNAc, Hex, Fuc, Neu5Ac (sialic acid)"

3.1 N-Glycans Release and SPE-C18 Purification

1. Dilute 15 or 50 μL of CSF to a final volume of 100 μL using 50 mM ABC buffer.
2. Denature the proteins by incubating them in a 90 °C water bath for 15 min.
3. Add 1000 U of PNGase F enzyme to the sample and incubate at a temperature of 37 °C for 20 h to achieve efficient enzymatic digestion. Dry the samples in a vacuum concentrator.
4. Resuspend the dried samples in 300 μL of 5% acetic acid (*see* **Note 3**) and transfer them into the SPE-C18 cartridges. Elute the glycans by adding 300 μL of 5% acetic acid solution and apply gentle vacuum pressure to facilitate passage through the cartridges. Repeat the elution process three times while collecting all flow-through in a new tube. Dry the collected samples in a vacuum concentrator.

3.2 N-Glycan Release and Purification Using FANGS

1. Dilute 15 μL of CSF to a final volume of 100 μL using 50 mM ABC buffer (*see* **Note 4**).
2. Denature the proteins by incubating them in a water bath set to 90 °C for a duration of 15 min.
3. Load the denatured sample onto the membrane filter, equilibrate with 500 μL of water, and wash twice with 100 μL of 50 mM ABC buffer. Finally, resuspend the sample with 80 μL of 50 mM ABC buffer.
4. Add 1000 U of PNGase F enzyme to the samples, incubate for 20 h at a temperature of 37 °C, and then dry the samples using a vacuum concentrator.
5. Elute the *N*-glycans from the filter by centrifugation. Using 100 μL of 50 mM ABC buffer, wash the filter twice and collect the samples. Dry the collected sample in a vacuum concentrator.

3.3 N-Glycan Reduction and Permethylation

1. Reduce the dried *N*-glycans from Subheadings 3.1 and 3.2 by adding 10 μL of 10 μg/μL ammonium borane complex solution and incubate at 60 °C for 1 h (*see* **Note 1**).
2. Add 1 mL of methanol to the sample solution and evaporate to remove residual borane as methyl borate (*see* **Note 5**).
3. Subject the reduced *N*-glycans to solid-phase permethylation. Briefly, resuspend the *N*-glycans by adding 30 μL of DMSO, 1.2 μL of water, and 20 μL of iodomethane (*see* **Note 6**). Place the sample mixture in the microspin columns previously filled with NaOH beads (*see* **Note 7**). Incubate the samples in the dark at room temperature for 25 min. Following the incubation, add another 20 μL of iodomethane and incubate in the dark at room temperature for 15 min.

4. Finally, dry the permethylated *N*-glycans in a vacuum concentrator (*see* **Note 8**) and resuspend in aqueous mobile phase A containing 20% Acetonitrile, and 0.1% formic acid for subsequent LC-MS analysis.

3.4 LC-MS Instrumentation

1. A Dionex UltiMate 3000 nano liquid chromatography system (Thermo Sci., San Jose CA, USA) coupled with a mass spectrometer is employed for *N*-glycan analysis. A reverse phase C18 Acclaim PepMap capillary column (75 μm i.d. × 150 mm, 3 μm, 100 Å) is used for the separation of the permethylated *N*-glycans. The *N*-glycans are purified online using an Acclaim PepMap C18 trapping column (75 μm i.d. × 20 mm, 3 μm, 100 Å) with a flow rate of 3 μL/min and MPA (0.1% formic acid in water) as a loading solution.
2. The analytical gradient is 60 min long operated at a temperature of 55 °C and a chromatographic flow rate of 0.35 μL/min. Start at 20% of MPB (0.1% formic acid in acetonitrile) for the initial 10 min, then increase to 42% MPB. Subsequently, ramp up to 55% MPB in a period of 40 min and increase to 90% MPB over 5 min. Finally, drop the percentage of MPB to 20% in 5 min to pre-equilibrate the system.
3. The mass spectrometer is configured to operate in full-scan mode, scanning a range of 700–1800 *m/z*. The mass tolerance is set to 5 ppm and the mass resolution was 50 K.

3.5 Data Analysis

Analyze the data using Xcalibur software from Thermo Scientific, alongside Skyline® software, to create Extracted Ion Chromatograms (EICs) for all protonated (H^+), sodiated (Na^+), and ammoniated (NH_4^+) molecular forms. Ensure the retention time and MS signals matched within a 5 ppm mass tolerance. Skyline® is also used to measure peak areas, which are normalized before calculating the abundances of identified *N*-glycans.

This analytical strategy compares three methods:

1. The "Benchmark Method," our initial approach analyzing CSF N-glycans with 50 μL of starting material.
2. "Test Method 1," which reduced the starting material to 15 μL.
3. "Test Method 2," based on the previously detailed FANGS protocol.

Figure 2a illustrates the absolute abundance of *N*-glycans across these methods. Figure 2b highlights the top ten most abundant CSF *N*-glycans, while Fig. 2c compares the total *N*-glycan abundance by method. Notably, "Test Method 1," utilizing just 15 μL of CSF, yielded the best results. In contrast, "Test Method 2" (FANGS) shows disproportionate glycan abundance, as

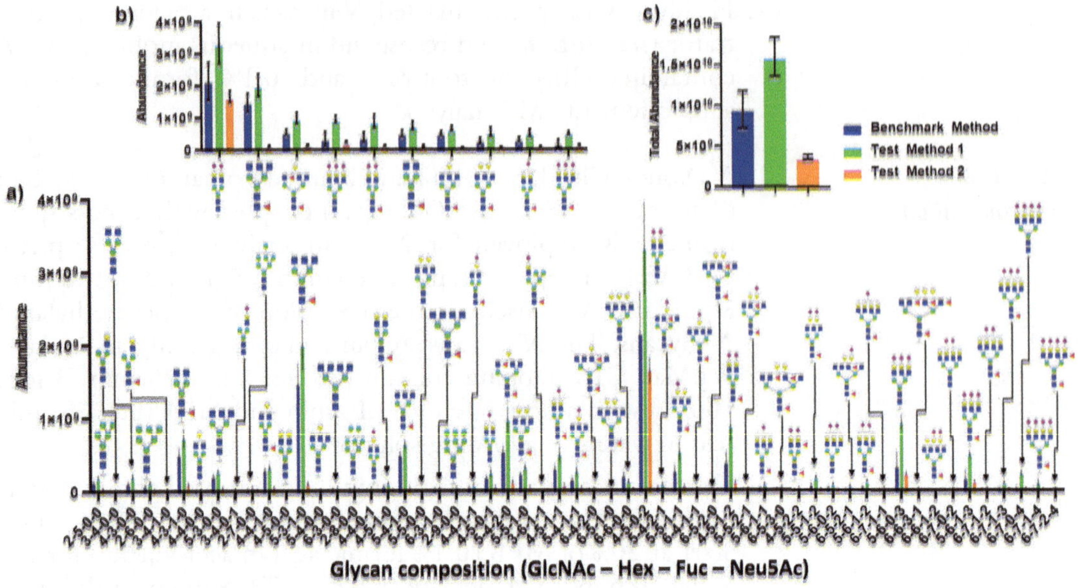

Fig. 2 Quantitative analysis comparing three different methods of *N*-glycan release from CSF: (**a**) absolute abundance of all *N*-glycans detected; (**b**) absolute abundance of top ten most abundant *N*-glycans; (**c**) sum of total *N*-glycan abundance from each method. *N*-Glycan nomenclature as in Fig. 1. (The figure was extracted with permission from the reference [16])

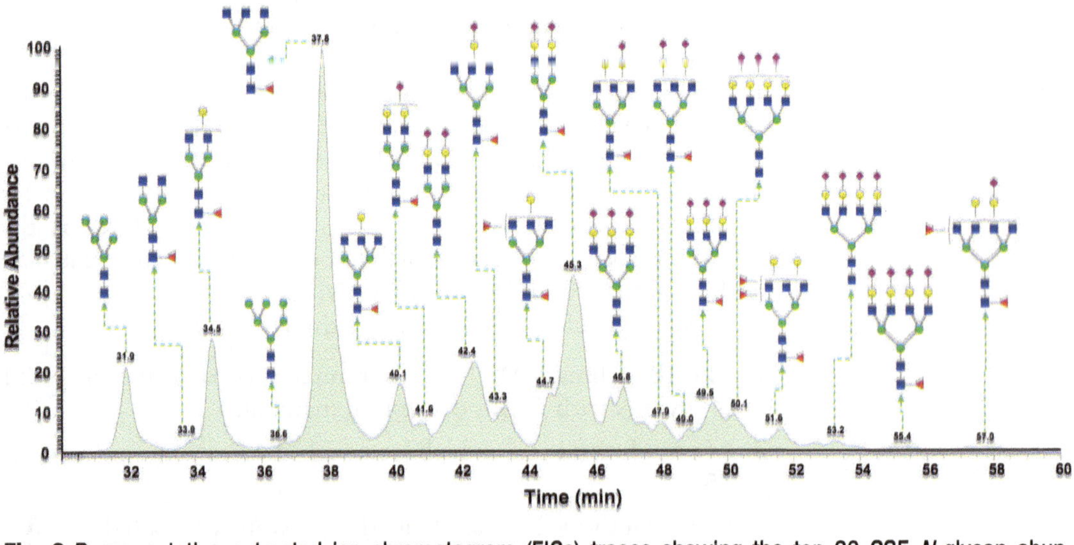

Fig. 3 Representative extracted ion chromatogram (EICs) traces showing the top 20 CSF *N*-glycan abundances. *N*-Glycan nomenclature as in Fig. 1

depicted in Fig. 2b. Figures 3 and 4 provide visual insights, with Fig. 3 showing typical EIC traces of the CSF *N*-glycome, and Fig. 4 presenting mass spectra of a highly abundant CSF *N*-glycan, specified as having a composition of 4-5-0-2 ($HexNAc_4Hex_5NeuAc_2$).

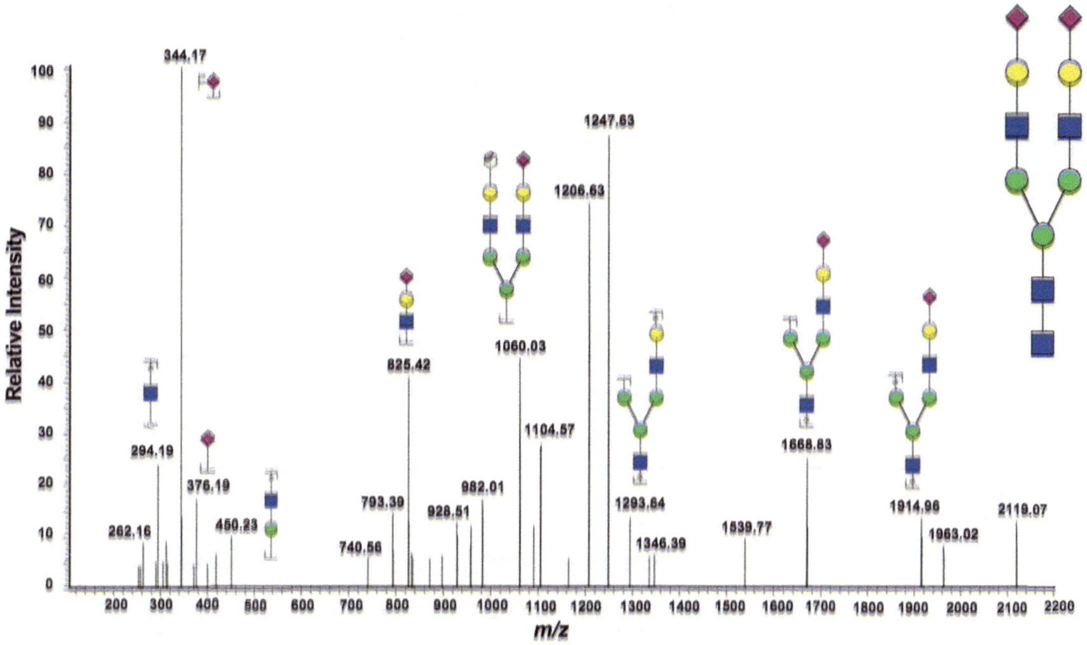

Fig. 4 Representative mass spectra of the *N*-glycan 4-5-0-2 ($HexNAc_4Hex_5Neu5Ac_2$). The spectra facilitate the glycan identification. *N*-Glycan nomenclature as in Fig. 1

4 Notes

1. The 10 μg/μL ammonium borane complex solution should be freshly prepared.
2. Utilize HPLC grade water, ACN, and FA, for mobile phase preparation. The volume of each composition should be added according to the total volume of the mobile phase. After mixing, eliminate gas from the mobile phase. The mobile phase should be kept at room temperature.
3. Wash the SPE-C18 cartridges before use as follows; add 1 mL of methanol, then use vacuum to facilitate the passage of the solvent through the cartridge and discard the solvent, repeat the protocol 3 times. Thereafter, add 1 mL of an aqueous solution with 5% of acetic acid, use vacuum to push the solution through the cartridge and discard the solution. Do not let dry the cartridge before the *N*-glycan purification is complete.
4. The FANGS method is taken and briefly adjusted from Hecht et al. [15].
5. Repeat the addition of methanol till the dried sample appears like a clear, viscous solution. Normally, the mentioned protocol should be repeated between 4 and 6 times.

6. Incorporate the iodomethane into the sample preparation once the microspin columns are packed with the NaOH beads.
7. The NaOH beads are previously resuspended in DMSO. This will facilitate its addition to the microspin columns. After the NaOH beads suspension is introduced into the microspin columns, the DMSO is removed by centrifugation. Finally, the NaOH beads are rinsed by adding 200 μL of DMSO and removed by centrifugation.
8. Dry the sample immediately after the completion of the incubation time. The residual iodomethane undergoes a secondary reaction at room temperature, turning the sample to a yellow color, or sometimes dark yellow.

Acknowledgments

This project is funded by grants from the National Institutes of Health (NIH) grants 1R01GM112490-08, 1R01GM130091-05 and 1U01CA225753-05. Additionally, the work is funded by grants from Robert A. Welch Foundation (No. D-0005), and The CH Foundation.

References

1. Boja ES, Kinsinger CR, Rodriguez H, Srinivas P (2014) Integration of omics sciences to advance biology and medicine. BioMed Central 11:45
2. Reyes CDG, Hakim MA, Atashi M, Goli M, Gautam S, Wang J, Bennett AI, Zhu J, Lubman DM, Mechref Y (2022) LC-MS/MS isomeric profiling of N-Glycans derived from low-abundant serum glycoproteins in mild cognitive impairment patients. Biomol Ther 12(11). https://doi.org/10.3390/biom12111657
3. Dong X, Mondello S, Kobeissy F, Ferri R, Mechref Y (2020) Serum Glycomics profiling of patients with primary restless legs syndrome using LC-MS/MS. J Proteome Res 19(8): 2933–2941. https://doi.org/10.1021/acs.jproteome.9b00549
4. Huang Y, Zhou S, Zhu J, Lubman DM, Mechref Y (2017) LC-MS/MS isomeric profiling of permethylated N-glycans derived from serum haptoglobin of hepatocellular carcinoma (HCC) and cirrhotic patients. Electrophoresis 38(17):2160–2167. https://doi.org/10.1002/elps.201700025
5. Robey TT, Panegyres PK (2019) Cerebrospinal fluid biomarkers in neurodegenerative disorders. Future Neurol 14(1):FNL6
6. Tumani H, Teunissen C, Süssmuth S, Otto M, Ludolph AC, Brettschneider J (2008) Cerebrospinal fluid biomarkers of neurodegeneration in chronic neurological diseases. Expert Rev Mol Diagn 8(4):479–494. https://doi.org/10.1586/14737159.8.4.479
7. Fogli A, Merle C, Roussel V, Schiffmann R, Ughetto S, Theisen M, Boespflug-Tanguy O (2012) CSF N-glycan profiles to investigate biomarkers in brain developmental disorders: application to leukodystrophies related to eIF2B mutations. PLoS One 7(8):e42688
8. Regan P, McClean PL, Smyth T, Doherty M (2019) Early stage glycosylation biomarkers in Alzheimer's disease. Medicines 6(3):92
9. Kizuka Y, Nakano M, Kitazume S, Saito T, Saido TC, Taniguchi N (2016) Bisecting GlcNAc modification stabilizes BACE1 protein under oxidative stress conditions. Biochem J 473(1):21–30
10. Akasaka-Manya K, Manya H, Sakurai Y, Wojczyk BS, Spitalnik SL, Endo T (2008) Increased bisecting and core-fucosylated N-glycans on mutant human amyloid precursor proteins. Glycoconj J 25(8):775–786. https://doi.org/10.1007/s10719-008-9140-x
11. Goyallon A, Cholet S, Chapelle M, Junot C, Fenaille F (2015) Evaluation of a combined

glycomics and glycoproteomics approach for studying the major glycoproteins present in biofluids: application to cerebrospinal fluid. Rapid Commun Mass Spectrom 29(6): 461–473

12. Gutierrez Reyes CD, Jiang P, Donohoo K, Atashi M, Mechref YS (2021) Glycomics and glycoproteomics: approaches to address isomeric separation of glycans and glycopeptides. J Sep Sci 44(1):403–425
13. Li Q, Xie Y, Wong M, Barboza M, Lebrilla CB (2020) Comprehensive structural glycomic characterization of the glycocalyxes of cells and tissues. Nat Protoc 15:1–37
14. Abdul Rahman S, Bergström E, Watson CJ, Wilson KM, Ashford DA, Thomas JR, Ungar D, Thomas-Oates JE (2014) Filter-aided N-glycan separation (FANGS): a convenient sample preparation method for mass spectrometric N-glycan profiling. J Proteome Res 13(3):1167–1176. https://doi.org/10.1021/pr401043r
15. Hecht ES, McCord JP, Muddiman DC (2016) A quantitative Glycomics and proteomics combined purification strategy. J Vis Exp 109. https://doi.org/10.3791/53735
16. Cho BG, Gutierrez Reyes CD, Mechref Y (2021) N-Glycomics of cerebrospinal fluid: method comparison. Molecules 26(6). https://doi.org/10.3390/molecules26061712

Chapter 9

Application of Isotope Dilution Gas Chromatography-Mass Spectrometry Using Selected Ion-Monitoring to Evaluate Nonenzymatic Protein Damage in Cerebrospinal Fluid

Natalia Mota-Martorell, Mariona Jové Font, Anna Fernàndez-Bernal, Idoia Company-Marín, Alba Juanes-Casado, and Reinald Pamplona

Abstract

Mass spectrometry (MS)-based approach for the detection and quantification of nonenzymatic posttranslational modifications (PTMs) is a powerful technology to measure the dynamics of proteome damage with high throughput and deep coverage. Here, we describe a gas chromatography-mass spectrometry with a selected ion-monitoring mode (SIM-GC/MS) method for protein damage profiling of biological and clinical samples. This approach can be carried out on materials ranging from isolated proteins to tissue/biofluid samples. In particular, the SIM-GC/MS method is applied to detect and quantify diverse protein damage markers in the human cerebrospinal fluid (CSF) proteome. The protocol includes protein extraction, digestion, derivatization, PTMs detection and quantification by GC/MS, followed by computational data processing. Different protein damage markers of oxidation, glycoxidation, lipoxidation and succination can be confidently detected and quantified simultaneously. This protocol provides a general protein nonenzymatic modification measurement tool, enabling the dissection of protein and proteome modifications in a particular biofluid, the CSF.

Key words Cerebrospinal fluid, Gas chromatography-mass spectrometry, Glycoxidation, Lipoxidation, Oxidation, Proteome, Selected ion-monitoring, Succination, Trifluoroacetic acid methyl esters

1 Introduction

A posttranslational modification (PTM) is a chemical-biological modification that takes place to one or more amino acids on a protein after that protein has been translated by a ribosome [1]. In contrast to enzymatic PTMs, which allow cells to expand their protein function and coordinate their signaling networks, nonenzymatic PTMs include chemical modifications mediated by reactive compounds on proteins, which is a central event in endogenous cell chemical damage. Examples of nonenzymatic PTMs include glycation, glyco- and lipoxidation, nitrosylation,

Enrique Santamaría and Joaquín Fernández-Irigoyen (eds.), *Cerebrospinal Fluid (CSF) Proteomics: Methods and Protocols*, Methods in Molecular Biology, vol. 2914, https://doi.org/10.1007/978-1-0716-4462-1_9,

sulphonation, sulphenation, sulphydration, succination and oxidation [2–4]. As a rule, all amino acids in any protein are susceptible to being nonenzymatically modified by a broad diversity of endogenous reactive species, such as free radicals and carbonyl compounds. Nevertheless, commonly modified residues are lysine (Lys), arginine (Arg), cysteine (Cys), methionine (Met), and tyrosine (Tyr) [5, 6]. Nonenzymatic modifications of amino acid side chains in proteins are mostly irreversible and usually a hallmark of structural damage and loss-of-function. However, amino acid modification may also have a protective effect and may serve regulatory or structural purposes [7]. Nonenzymatic PTMs have been associated with a wide spectrum of physiological and pathological conditions such as aging [8–10], diabetes, and neurodegenerative diseases, among others [2, 11], providing useful biomarkers of these processes.

MS-based techniques—specifically gas chromatography-MS (GC/MS)—have been applied to the detection and quantification of the steady-state level of protein damage at the subcellular, cellular, tissue, and organ levels, including biofluids such as plasma and cerebrospinal fluid (CSF) [8, 9, 12]. CSF is a useful biofluid for analyzing brain metabolism and is potentially a source of biomarkers associated with human disease and disease targets. The CSF serves as a dynamic system delivering metabolites to neuronal cells, distributing a broad spectrum of compounds throughout the central nervous system, and collecting waste products of cellular metabolism. As a result, CSF becomes a complex system with a wide compositional diversity, even though its low protein concentration (35 vs 7000 mg/dL, for CSF and plasma, respectively) and simple metabolome compared to plasma [13–16]. In addition to water, metal ions or salts, multi-omics characterization of CSF has annotated to date 250 miRNAs, 3400 proteins, 900 peptides and 440 metabolites (including neurotransmitters, amino acids, carbohydrates, lipid species, and carbonyl compounds, among others). However, there is a scarcity of studies on nonenzymatic modifications to proteins/peptides in the CSF and their physiological and pathological implications.

In this chapter, we describe a protocol for the direct detection and quantification of different specific markers of nonenzymatic protein damage—the specific protein carbonyls glutamic semialdehyde (GSA) and aminoadipic semialdehyde (AASA), glycoxidation (carboxyethyl-lysine, CEL), lipoxidation (malondialdehyde-lysine, MDAL), and mixed glyco- and lipoxidation (carboxymethyl-lysine (CML) and carboxymethyl-cysteine (CMC))—, as well as the marker of protein succination 2-S-(2-succino)cysteine (2-SC), by GC–MS. In this approach, protein damage markers are determined as TFAME (trifluoroacetic acid methyl esters) derivatives in protein samples previously delipidated and hydrolyzed in acidic conditions.

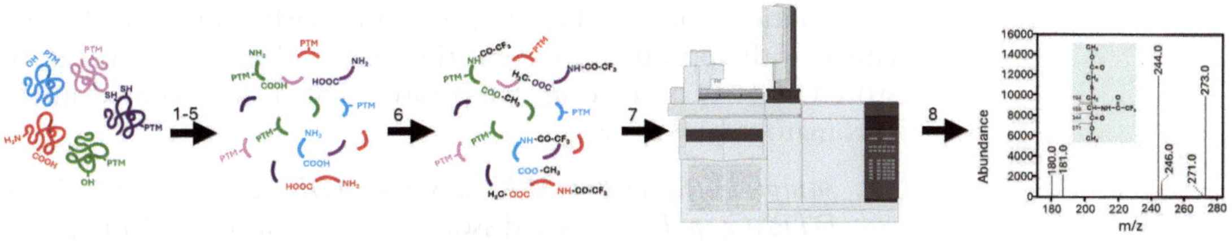

Fig. 1 General workflow for the analysis of nonenzymatically modified proteins in CSF. (**1**) Sample delipidation with methanol/chloroform (2:1, v/v). Chloroform facilitates the separation of lipids from polar compounds. The chloroform phase is discarded. (**2**) Protein precipitation with trichloroacetic acid (TCA). TCA is a strong acid that precipitates proteins out of the solution. This process separates proteins from soluble contaminants like salts, nucleic acids and metabolites. (**3**) Protein reduction with borate buffer. Borate buffer maintains a stable alkaline pH (8–9), preserving protein structural integrity. (**4**) Protein re-precipitation of stabilized proteins with TCA. (**5**) Acidic protein hydrolysis with HCl. (**6**) Amino acids are submitted to a double derivatization to form volatile and stable trifluoroacetyl methyl esters (TFAMEs) for GC/MS analysis. (**7**) Injection of TFAMES in a GC/MS system. (**8**) Detection of PTMs. Electron impact mass spectrum of CMC derivatized as the TFAME is shown

Amino acid analysis after complete digestion allows the determination of particular nonenzymatic modifications and their absolute concentrations (determined by the use of isotope-labeled standards together with the parent species). Care must be taken in sample handling to prevent artefactual modifications, and also during protein hydrolysis, because some markers are labile. GC–MS analyses have many advantages including high specificity, high sensitivity, and the capacity to detect many different modifications and parent species concurrently. Later, quantification can be made by internal and external standardization using standard curves created from mixtures of deuterated and nondeuterated standards. Quantification relative to non-modified amino acids, and preferably against added heavy-isotope-labelled materials, is recommended to overcome potential artifacts arising from sample handling and preparation. Analyses are carried out with a GC coupled to a MS in selected ion-monitoring (SIM) mode (SIM-GC/MS). Finally, the amount of product is expressed as mili- or micromoles of the damage marker per mol of lysine. Figure 1 shows the general workflow for the analysis of nonenzymatically modified proteins in CSF.

2 Materials

Sample contamination is the most critical point (*see* **Note 1**). Prepare all solutions using miliQ water and analytical-grade reagents. Harmful organic solvents and inorganic reagents are used. Work in the hood (▲) every time you need to add toxic solvents or reagents and use positive displacement pipettes for organic solvents (●).

Solutions must be kept in previously washed glass bottles and closed with appropriate solvent-tight caps. Most solutions can be stored at 4 °C or at room temperature (on a shelf located in the chromatography zone).

1. *Sample buffer (180 mM KCl, 5 mM MOPS, 2 mM EDTA, 1 mM DTPAC; pH = 7.4):* dissolve 3.36 g of KCl, 260 mg of 3-(N-morpholino)propanesulfonic acid (MOPS), 186 mg of ethylenediaminetetraacetic acid (EDTA), and 98 mg diethylenetriaminepentaacetic acid (DTPAC) in 100 mL of water. Adjust pH to 7.4 and add water to a final volume of 250 mL. This solution can be stored at 4 °C for extended periods.
2. *BHT 1 mM:* dissolve 55.09 mg dibutylhydroxytoluene (BHT) in 250 mL of water. This solution can be stored at room temperature for extended periods.
3. *Sample buffer containing 1 μM BHT:* add 1 μL of BHT 1 mM to each mL of sample buffer. This solution must be prepared freshly and maintained at 4 °C.
4. *TCA 10%:* dissolve 10 g trichloroacetic acid (TCA) in 100 mL of water (▲). This solution can be stored at 4 °C for extended periods.
5. *TCA 20%:* dissolve 20 g TCA in 10 mL of water (▲). This solution can be stored at 4 °C for extended periods.
6. *HCl 6 N:* add HCl 37% to water in a ratio 1:1 (v/v) (▲). Never add water over HCl. This solution can be stored at room temperature for extended periods.
7. *Borax buffer 0.2 M (pH = 9.2):* Heat 150 mL of water. Add 19.07 g of Borax ($Na_2B_4O_7$, mw = 381.4 g/mL) and mix. Let the solution cool down to room temperature. Adjust pH to 9.2 and add water to a final volume of 250 mL. This solution can be stored at room temperature for extended periods. If crystals are formed, heat the solution (<100 °C) prior usage.
8. *Borate solution:* dissolve 0.56 g of sodium borohydride ($NaBH_4$) in 30 mL of pre-heated borax buffer 0.2 M (in case crystals are formed at the bottom of the containing bottle). This solution is effervescent and must be prepared freshly in an Erlenmeyer to be immediately added to samples.
9. *Methanolic acid solution:* measure 20 mL of methanol (MeOH) and transfer the volume into an Erlenmeyer (▲). Lower the protective glass of the hood. Carefully and slowly add 2.14 mL of acetyl chloride (●) while facing the Erlenmeyer opening to the end of the hood and rotating it gently in circular motions. It's an exothermic solution and as HCl anhydrous is generated during the addition of the reagent, it is possible for the solution to pop up. To prevent it, rotate the Erlenmeyer continuously to allow the newly generated HCl anhydrous to disperse along the

Table 1
Concentration of stock, intermediate, and working solutions (WS) of ISD and IS. All solutions are prepared in water and can be stored at −80 °C for long periods

ISTD	Stock	Units	Intermediate	Units	WS	Units
d8-Lys/Lys	120	mmol/μL	120	μmol/μL	120	nmol/μL
d5-HAVA/HAVA	30	μmol/μL	30	nmol/μL	30	pmol/μL
d4-HACA/HACA	23	μmol/μL	23	nmol/μL	23	pmol/μL
d-CMC/CMC	8	μmol/μL	8	nmol/μL	8	pmol/μL
d2-2-SC/2-SC	14	μmol/μL	14	nmol/μL	14	pmol/μL
d4-CML/CML	267	μmol/μL	267	nmol/μL	267	pmol/μL
d4-CEL/CEL	192	μmol/μL	192	nmol/μL	192	pmol/μL
d8-MDAL/MDAL	55	μmol/μL	55	nmol/μL	55	pmol/μL

vessel. Prepare a minimum solution volume of 20 mL. The solution must be prepared freshly to be immediately added to samples.

10. *Deuterated internal standards (ISD):* Included ISD were prepared as described previously: [$^2H_5$]5-hydroxy-2-aminovaleric acid (*d5*-HAVA, stable derivative for GSA) and [$^2H_4$]6-hydroxy-2-aminocaproic acid (*d4*-HACA, stable derivative for AASA) [17], U–$^{13}C_3{}^{15}N$-CMC (*d*-CMC) [18], [$^2H_4$]CML (*d4*-CML) [19], [$^2H_4$]CEL (*d4*-CEL) [20] and [$^2H_8$]MDAL (*d8*-MDAL) [21]. [$^2H_2$]SC (d2-2-SC) was synthesized by PolyPeptide Group AG (Baar, Switzerland). [$^2H_8$]Lys (*d8*-Lys) was purchased from CDN Isotopes. Currently, some of these compounds can be purchased from Iris Biotech GmbH (Marktredwitz, Germany). For dilution preparation, *see* **Note 2**. Standard stock, intermediate and working solutions are prepared in water and can be stored at −80 °C for long periods (for details, *see* Table 1). ISD mix (Table 2) is prepared freshly in water. Final working solutions have been optimized and prepared to use the minimum amount of ISD in each sample and ensure the maximum response. According to the equipment, adjustments might be necessary.
11. *Internal standards (IS):* Included ISs were CMC, CML, CEL and SC, purchased from Iris Biotech GmbH (Marktredwitz, Germany). Lys was purchased from Sigma (Madrid, Spain). 5-hydroxy-2-aminovaleric acid (HAVA, stable derivative for GSA), and 6-hydroxy-2-aminocaproic acid (HACA, stable derivative for AASA) were prepared as described previously [17]. MDAL was prepared as described previously [21]. For dilution preparation, *see* **Note 2**. Standard stock, intermediate

Table 2
Volumes for the preparation of ISD mix. Volumes are calculated for each sample. A total of 21 μL of ISD mix is added to each sample. ISD mix must be prepared freshly

ISTD mix	pmol	WS	Units
*d*8-Lys	360,000	3	μL
d5-HAVA	90	3	μL
d4-HACA	70	3	μL
d-CMC	25	3	μL
d2-2-SC	43	3	μL
d4-CML	800	3	μL
d4-CEL	576	3	μL
*d*8-MDAL	166	3	μL
Volume per sample		**21**	**μL**

Table 3
Volumes for the preparation of IS calibration curve. IS mix must be prepared freshly. When preparing a calibration curve, a Cal 0 (no IS) and double blank (no IS or ISD) must be included. Besides, ISD mix must be added to each calibration point

	Cal 1			Cal 2			Cal 3			Cal 4		
	Quant	Units	μL WS	Quant	Units	μL WS	Quant	Units	μL WS	Quant	Units	μL WS
Lys	360	nmol	3	540	nmol	4.5	720	nmol	6	1080	nmol	9
HAVA	90	pmol	3	135	pmol	4.5	180	pmol	6	270	pmol	9
HACA	70	pmol	3	105	pmol	4.5	140	pmol	6	210	pmol	9
CMC	25	pmol	3	37.5	pmol	4.5	50	pmol	6	75	pmol	9
2-SC	43	pmol	3	64.5	pmol	4.5	86	pmol	6	129	pmol	9
CML	800	pmol	3	1200	pmol	4.5	1600	pmol	6	2400	pmol	9
CEL	576	pmol	3	864	pmol	4.5	1152	pmol	6	1728	pmol	9
MDAL	166	pmol	3	249	pmol	4.5	332	pmol	6	498	pmol	9

and working solutions are prepared in water and can be stored at −80 °C for long periods (for details *see* Table 1). The calibration standard solution (Table 3) is prepared freshly in water. ISD mix is added to each calibration standard solution.

3 Methods

3.1 Experimental Design and Development

Every sample batch includes blank processed samples, quality control (QC), and NIST (*see* **Notes 3** and **4**). Blank processed samples (500 μL of sample buffer containing 1 μM BHT) serve to control the signal baseline that typically arises from solvents and reagents. QCs monitor the consistency, stability, and robustness of experimental procedures and data acquisition within and across experiments. NIST is a commercial human plasma sample that allows for experiment comparability across laboratories.

The protocol includes four main steps: (1) sample pre-processing: sample aliquot; (2) sample processing (delipidation, protein precipitation, reduction, re-precipitation and hydrolysis); (3) derivatization; and (4) injection of TFAMEs in a GC-MS system. The protocol (pre-processing, processing and derivatization) is accomplished in 3 days (*see* **Note 5**).

(!) Work in the hood (▲) every time you need to add toxic solvents or reagents and use positive displacement pipettes when needed (●).

3.2 Sample Aliquot (Day 1, d1)

After obtaining CSF, samples were centrifuged, and protein concentration was determined. Then samples were stored at −80 °C until analysis.

Prepare before start:

– Sample buffer containing 1 μM BHT (maintained on ice).

Detailed procedures:

1. Defrost the tubes containing CSF samples on ice (▲) (*see* **Note 6**).
2. In a 7-mL glass tube, mix 300 μL of CSF with 200 μL sample buffer containing 1 μM BHT (▲) (total sample volume: 500 μL).

   (!) Steady-state levels of PTMs are low. To detect them, a minimum of 100 μg of protein is necessary. Considering that CSF protein concentration is around 35 mg/dL, 300 μL of CSF will ensure 100 μg of protein.
3. Close the 7-mL glass tube with Teflon caps and vortex the tubes (▲).
4. Transfer the samples on ice (▲).

3.3 Sample Delipidation and Protein Isolation (d1)

In general, sample delipidation is accomplished by preforming 2 extractions with methanol/chloroform ($MeOH/CH_3Cl$) 1: 2 (v/v) (Fig. 2). When analyzing tissues with high content of lipids (e.g., brain or adipose tissue) 3 extractions with $MeOH/CH_3Cl$ 1: 2 might be necessary.

Prepare before start:

- Turn on the centrifuge and let it cool down to 4 °C.

Detailed procedures:

1. Add 500 μL of cold MeOH (●) while keeping the samples on ice (▲).
2. Vortex and place the samples on ice (▲).
3. Add 1000 μL of cold CH_3Cl (●) while keeping the samples on ice to perform the first lipid extraction (▲).
4. Vortex and place the samples on ice (▲).
5. Centrifuge at 4400 rpm at 4 °C for 15 min. Centrifugation at low temperatures will facilitate the separation of an upper polar layer (MeOH) and a lower non-polar layer (CH_3Cl). An intermediate thick and white layer (formed by proteins) will be formed.
6. Place the samples on ice (▲). Low temperatures facilitate (and maintain) the separation between the two organic layers.
7. Discard the lower non-polar layer (CH_3Cl). The upper polar layer (MeOH) and intermediate thick and white layer must be maintained on the 7-mL glass tube and kept on ice (▲).

   (!) For the extraction of the lower non-polar layer (CH_3Cl), we recommend using a glass Pasteur pipette coupled with a rubber bulb (Fig. 2). Slide the Pasteur pipette along the

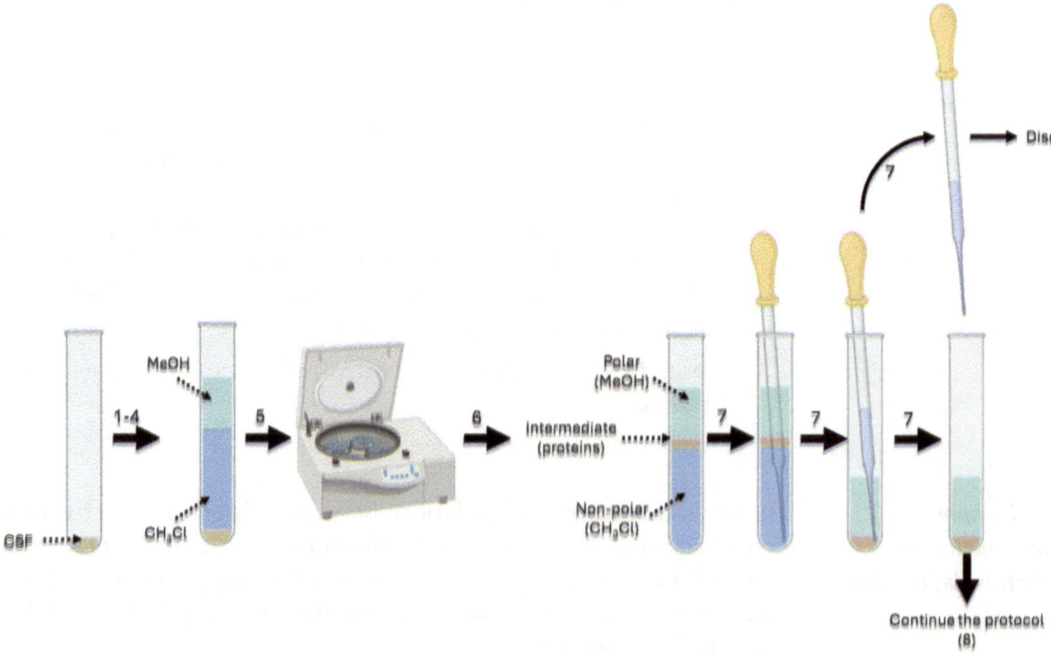

Fig. 2 Detailed procedures of sample delipidation with MeOH:CH_3Cl 1:2 (v/v)

inner wall of the glass tube while pressing the bulb. During this process, expel air to generate bubbles when reaching the upper and intermediate phases. Once the tip of the Pasteur pipette is in the lower phase, carefully release the rubber bulb to allow the liquid to be drawn up. Once the desired volume is aspirated, remove the pipette and transfer the volume to the desired container. When removing the pipette, do not apply pressure to the rubber bulb to avoid expelling the aspirated sample.

8. Add 1000 μL of cold CH_3Cl (●) while keeping the samples on ice to perform the second lipid extraction (▲).
9. Repeat **steps 4** to **8** to perform a second lipid extraction.

3.4 Protein Precipitation and Reduction (d1)

Detailed procedures:

1. For protein precipitation, add 1 mL of cold TCA 20% (●) and vortex vigorously (▲). Vortex as necessary to mix TCA with the sample, promoting uniform protein precipitation, and remove protein adhered to the glass tube.
2. Place the samples on ice (▲).
3. Centrifuge at 4400 rpm at 4 °C for 15 min.
4. Place the samples on ice (▲).
5. Decant supernatant and discard it (▲).

   (!) The pellet could dislodge from the glass. To prevent it, rotate the glass tube gently and slowly while decanting the supernatant so that the pellet adheres to the wall of the tube.
6. Leave tubes facing downward to finish drying by capillarity (▲).
7. Add a drop of 1-hexanol (serves as an anti-foaming agent) (▲).
8. Add 1 mL of freshly prepared borate solution (●) and cover the tube quickly with Teflon caps (▲).

   (!) The solution becomes effervescent. Ensure that the appropriate volume is collected during pipetting, that the samples are properly covered with borate solution, and that tubes are properly closed (*see* **Note 7**).
9. Store the tubes in darkness overnight at room temperature to allow for protein reduction (to prevent artefactual protein modifications).

3.5 Protein Re-precipitation (d2)

Prepare before start:

- Turn on the centrifuge and let it cool down to 4 °C.
- Defrost ISD working solution on ice.

Detailed procedures:

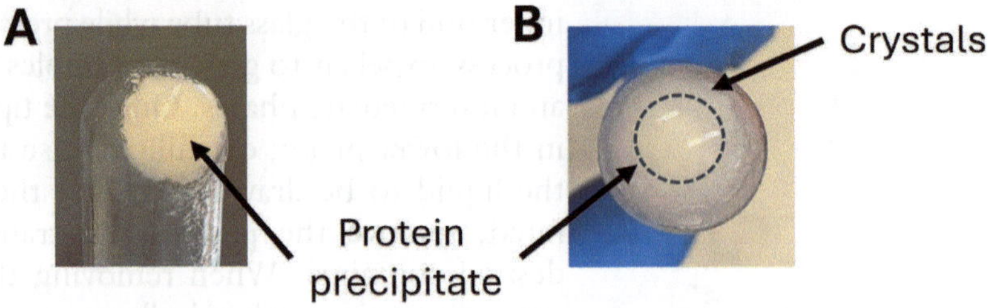

Fig. 3 Crystal formation after the addition of TCA. (**a**) Protein precipitate without crystals. (**b**) Protein precipitate with crystals

1. Place the samples on ice (▲).
2. Add 1 mL of cold TCA 20% (●▲).

   (!) After leaving the tubes overnight, a decompression should be noticeable when opening the glass tube Teflon caps.
3. Vortex vigorously to remove protein stacked to glass tube (▲) (*See* **Note 8**).
4. Place the samples on ice (▲).
5. Centrifuge at 4400 rpm at 4 °C for 30 min.
6. Place the samples on ice (▲).
7. Decant supernatant and discard it (▲).

   (!) The pellet could dislodge from the glass. To prevent it, rotate the glass tube gently and slowly while decanting the supernatant so that the pellet adheres to the wall of the tube.
8. If there are no crystals in the pellet (Fig. 3a), proceed to the following step. If there are crystals in the pellet (Fig. 3b), place the samples on ice and re-precipitate the proteins by adding 1 mL of cold TCA 10%. Repeat **steps 3** to **8**.
9. Leave tubes facing downward to ensure TCA solution elimination.
10. Add ISD using a properly calibrated pipette (*see* Table 2).
11. Vortex to ensure that ISD mixes properly with the sample pellet.

    (!) Ensure that ISD doesn't get stuck in the glass tube walls and mixes properly with the sample pellet.
12. From now on, samples don't need to be placed on ice.

3.6 Protein Hydrolysis (d2)

Prepare before start:

- Turn on the SpeedVac refrigerator and allow it to cool down to 4 °C.
- Turn on the block and allow it to heat up to 155 °C.
- IS curve solutions (for details, *see* Table 3).

Detailed procedures:

1. Transfer the appropriate IS curve solution to 7-mL glass vials (for details, *see* Table 3). From now on, both samples and IS curves will be processed simultaneously.
2. For protein hydrolysis, add 1 mL of HCl 6 N (●) and vortex (▲).
3. Incubate the tubes in a block at 155 °C for 30 min (preferably 32 min to allow the tubes to equilibrate to the temperature).

   (!) Ensure that the tubes are properly capped, as elevated temperatures may lead to solution evaporation.
4. After incubation, allow the tubes to stand on the bench for 5 min to equilibrate to room temperature.

   (!) Handle the tube by the cap, not by the glass (it is very hot, and it could burn your fingers).
5. Evaporate to dryness in the SpeedVac. The time to complete evaporation varies according to the evaporation program. It takes 3 h to evaporate to dryness at 65 °C and 100 mTorr.

   (!) Ensure that samples are completely evaporated. Humidity impedes sample derivatization and leads to sample loss.
6. If necessary, cap the tubes properly and store them at 4 °C to proceed the following day.

3.7 Amino Acid Derivatization (d3)

Prepare before start:

1. Turn on the block and allow it to heat up to 65 °C.

Detailed procedures:

1. Add 1 mL of freshly prepared methanolic acid solution (●▲).
2. Vortex and ensure that the solution mixes properly with the sample (▲).
3. Incubate the tubes in the block at 65 °C for 30 min (preferably 32 min to allow the tubes to equilibrate to the temperature).

   (!) Ensure that the tubes are properly capped, as elevated temperatures may lead to solution evaporation.

   (!) After incubation, allow the tubes to stand on the bench for 5 min to equilibrate to room temperature. Handle the tube by the cap, not by the glass (it is very hot, and it could burn your fingers).
4. Evaporate to dryness under nitrogen stream beam (approximately 30 min) (▲).
5. Add 1 mL of pure trifluoroacetic acid anhydrous (TFAA) to form TFAMEs (●▲).

   (!) Ensure that TFAA mixes properly with the sample pellet.
6. Vortex samples (▲).

7. Incubate for 60 min at room temperature (▲).
8. Evaporate to dryness under nitrogen stream beam (approximately 10 min) (▲).
9. Re-dissolve the TFAME hydrolysates with 80 μL of dichloromethane (CH_2Cl_2) (▲).
10. Vortex (▲).
11. Collect the solution quickly and transfer it to appropriated vials containing inserts to perform GC/MS analysis.
12. Close the tube with the appropriated caps (▲) (*see* **Note 9**).
13. Inject the sample immediately in a GC/MS system for analysis (*see* **Note 10**).

3.8 Sample Injection in the GC/MS System

GC-MS analysis was performed in an Intuvo 9000 GC equipped with an HP-5MS UI capillary column coupled to a 5977-mass selective detector (MSD) (Agilent Technologies, Barcelona, Spain). The temperature of the injection port was held at 250 °C. Four microliters of sample were injected in a splitless mode and underwent the following temperature program: 5 min at 110 °C, then 2 °C/min at 150 °C, then 5 °C/min at 240 °C, followed by 15 °C/min at 300 °C, and finally maintained at 300 °C for 5 min. Total runtime was 52 min followed by 3 min post run at 110 °C.

Analytes were detected by selected ion-monitoring GC–MS. MSD temperature was set at 250 °C; transfer line was set at 310 °C. Helium was used as a carrier gas. The ions used were Lys and *d*8-Lys, *m/z* 180 and 187, respectively; HAVA acid and *d*5–HACA, *m/z* 280 and 285, respectively; HACA and *d*4–HACA, *m/z* 294 and 298, respectively; CEL and *d*4-CEL, *m/z* 379 and 383, respectively; MDAL and *d*8-MDAL, *m/z* 474 and 482, respectively; CML and *d*4-CML, *m/z* 392 and 396, respectively; CMC and U–$^{13}C_3{}^{15}N$-CMC, *m/z* 271 and 275, respectively, and 2-SC and *d*2–2-SC, *m/z* 284 and 286, respectively. Ion response in human CSF is reported in Fig. 4.

3.9 Calculation of Analyte Content

Quantification was performed using standard curves from mixtures of deuterated (ISD) and nondeuterated standards (IS) for each marker. *See* Table 3 for standard curve generation. The amount of each marker can be expressed as the ratio of micromoles of GSA, AASA, CEL, MDAL, CML, CMC, and 2-SC per mol of Lys.

4 Notes

1. To prevent sample contamination, we recommend setting a place in the laboratory dedicated only to process the samples and to store all the needed material (e.g., glass material and pipettes). It is highly recommended that material is exclusively

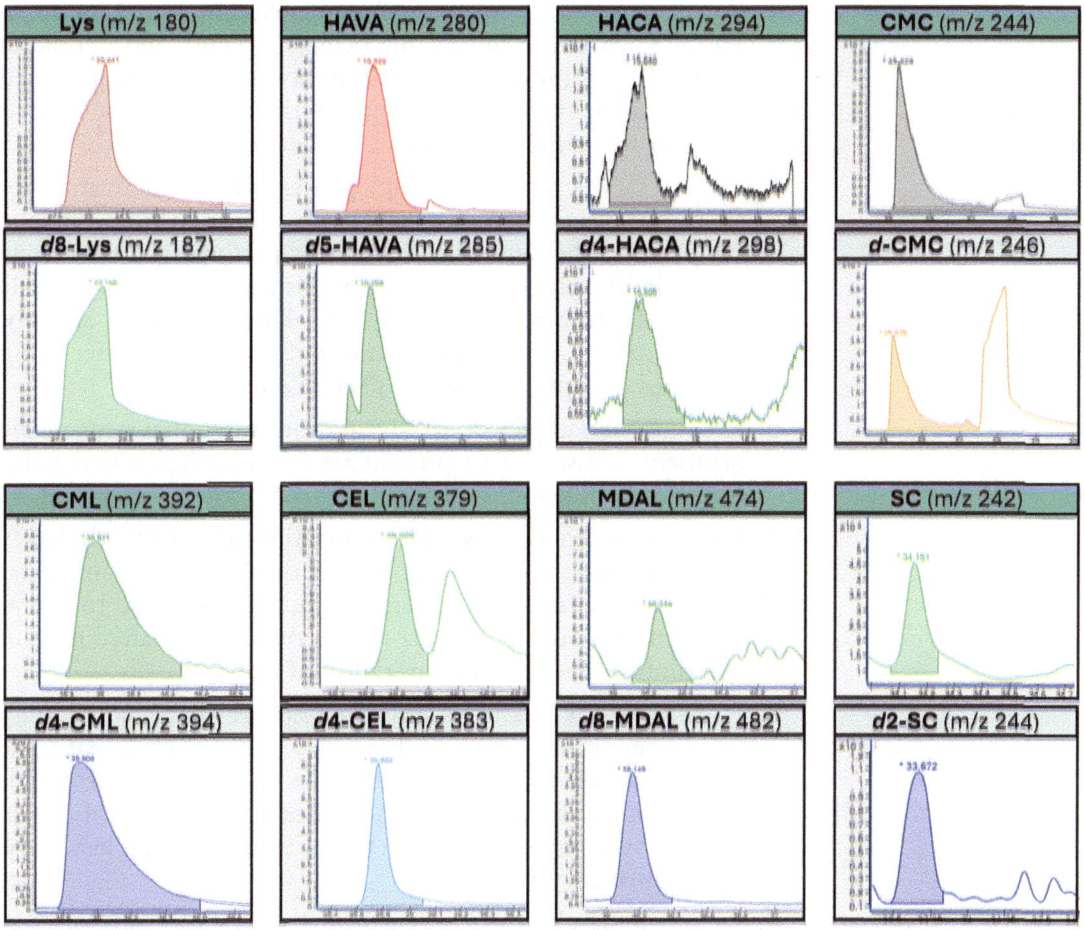

Fig. 4 Detected biomarkers of protein nonenzymatic modification in CSF and their respective ISD. ISD peak area corresponds to the amounts described in Table 2

used for this analytical technique. Good laboratory practices include cleaning the glass materials before and after use with appropriate solvents (never use soaps) and checking reagents for contamination or crystal formation before use.

2. Pipetting is crucial to ensure the correct amounts of standards in a STD mix and the same STD mix in every sample. To ensure this, we recommend the use of properly calibrated pipettes dedicated exclusively to these procedures.
3. QCs appropriate selection is a critical step. Typically, we opted for the use of tissue samples of an experimental model with a known concentration of protein damage markers (e.g., adult male rat liver). Besides the lesser content of proteins, plasma samples would also be a good choice to be used as QCs. The most important is to previously characterize the average content of protein modification markers.

4. For every project, we recommend including a minimum of 10 QCs across the run (at the beginning, end, and middle of the run) to monitor equipment stability. Also include processed blanks (at the beginning, end and in the middle of the run) to monitor for sample carryover.
5. Ideally, sample pre-processing, processing and derivatization should be performed in 3 days. However, depending on the sample size, the derivatization can be performed immediately after sample processing to accomplish the protocol in only 2 days. However, we recommend taking your time and not rushing, as samples should be injected immediately after being derivatized to prevent loss of TFAMEs stability.
6. Keeping the samples on ice throughout the initial steps of the protocol is essential to prevent its modification and maintenance of sample integrity.
7. Using appropriate caps and closing the tubes properly is key when adding effervescent solutions or incubating samples at high temperatures. Missclosing the tubes might lead to alterations in the protocol (e.g., no decompression after protein overnight reduction or evaporation of reagents).
8. Homogeneous mixing is essential. Ensure that when adding new reagents to sample tubes, they mix properly with the sample. When adding small volumes (e.g., ISTD mix), it is usual that the small volumes get stuck on the glass tube wall. To prevent it, try to pipette on top of the pellet or, if necessary, vortex it or spin it down.
9. We recommend the use of polypropylene (PTFE) caps for vials.
10. TFAMEs must be injected immediately after derivatization. We recommend not to process and derivatize more samples than the maximum number of samples that can be injected in a 24 h in GC/MS system (considering blank, QC and NIST samples, and standard curve). When many samples are injected, TFAMEs can be stored at 4 °C until injection. Capping the samples properly is essential, as sample evaporation leads to their loss.

Acknowledgments

This work was supported by the Spanish Ministry of Science, Innovation, and Universities (grant PID2023-152233OB-I00, co-funded by the European Regional Development Fund, “A way to build Europe”), and the Generalitat of Catalonia: Agency for Management of University and Research Grants (2021SGR00990) to R.P.

References

1. Walsh C (2006) Posttranslational modifications of proteins. Roberts and Company, Englewood
2. Thorpe SR, Baynes JW (2003) Maillard reaction products in tissue proteins: new products and new perspectives. Amino Acids 25:275–281. https://doi.org/10.1007/s00726-003-0017-9
3. Couvertier SM, Zhou Y, Weerapana E (2014) Chemical-proteomic strategies to investigate cysteine posttranslational modifications. Biochim Biophys Acta Proteins Proteom 1844: 2315–2330. https://doi.org/10.1016/j.bbapap.2014.09.024
4. Golubev A, Hanson AD, Gladyshev VN (2017) Non-enzymatic molecular damage as a prototypic driver of aging. J Biol Chem 292: 6029–6038. https://doi.org/10.1074/jbc.R116.751164
5. Dean RT, Fu S, Stocker R, Davies MJ (1997) Biochemistry and pathology of radical-mediated protein oxidation. Biochem J 324: 1–18. https://doi.org/10.1042/bj3240001
6. Pamplona R, Costantini D (2011) Molecular and structural antioxidant defenses against oxidative stress in animals. Am J Phys Regul Integr Comp Phys 301:R843–R863. https://doi.org/10.1152/ajpregu.00034.2011
7. Schindeldecker M, Moosmann B (2024) Cysteine is the only universally affected and disfavored proteomic amino acid under oxidative conditions in animals. Antioxidants 13:267. https://doi.org/10.3390/antiox13030267
8. Jové M, Mota-Martorell N, Pradas I et al (2020) The advanced lipoxidation end-product malondialdehyde-lysine in aging and longevity. Antioxidants 9:1132. https://doi.org/10.3390/antiox9111132
9. Jové M, Pradas I, Mota-Martorell N et al (2020) Succination of protein thiols in human brain aging. Front Aging Neurosci 12. https://doi.org/10.3389/fnagi.2020.00052
10. Gladyshev VN, Kritchevsky SB, Clarke SG et al (2021) Molecular damage in aging. Nat Aging 1:1096–1106. https://doi.org/10.1038/s43587-021-00150-3
11. Jové M, Mota-Martorell N, Torres P et al (2021) The causal role of lipoxidative damage in mitochondrial bioenergetic dysfunction linked to Alzheimer's disease pathology. Life 11:388. https://doi.org/10.3390/life11050388
12. Dakterzada F, Jové M, Cantero JL et al (2023) Plasma and cerebrospinal fluid nonenzymatic protein damage is sustained in Alzheimer's disease. Redox Biol 64:102772. https://doi.org/10.1016/j.redox.2023.102772
13. Yuan X, Desiderio DM (2005) Proteomics analysis of human cerebrospinal fluid. J Chromatogr B 815:179–189. https://doi.org/10.1016/j.jchromb.2004.06.044
14. Hühmer AF, Biringer RG, Amato H et al (2006) Protein analysis in human cerebrospinal fluid: physiological aspects, current progress and future challenges. Dis Markers 22:3–26. https://doi.org/10.1155/2006/158797
15. Wishart DS, Lewis MJ, Morrissey JA et al (2008) The human cerebrospinal fluid metabolome. J Chromatogr B 871:164–173. https://doi.org/10.1016/j.jchromb.2008.05.001
16. Hooshmand K, Xu J, Simonsen AH et al (2024) Human cerebrospinal fluid sample preparation and annotation for integrated lipidomics and metabolomics profiling studies. Mol Neurobiol 61:2021–2032. https://doi.org/10.1007/s12035-023-03666-4
17. Requena JR, Chao C-C, Levine RL, Stadtman ER (2001) Glutamic and aminoadipic semialdehydes are the main carbonyl products of metal-catalyzed oxidation of proteins. Proc Natl Acad Sci 98:69–74. https://doi.org/10.1073/pnas.98.1.69
18. Wang Y (2002) Modification of cysteine residues in proteins: identification and quantitation of sulfhydryl AGEs. M.S. Thesis, Univeristiy of South Carolina
19. Knecht KJ, Dunn JA, McFarland KF et al (1991) Effect of diabetes and aging on carboxymethyllysine levels in human urine. Diabetes 40:190–196. https://doi.org/10.2337/diab.40.2.190
20. Ahmed MU, Frye EB, Degenhardt Thorsten P et al (1997) N-epsilon-(Carboxyethyl)lysine, a product of the chemical modification of proteins by methylglyoxal, increases with age in human lens proteins. Biochem J 324:565–570. https://doi.org/10.1042/bj3240565
21. Requena JR, Fu MX, Ahmed MU et al (1997) Quantification of malondialdehyde and 4-hydroxynonenal adducts to lysine residues in native and oxidized human low-density lipoprotein. Biochem J 322:317–325. https://doi.org/10.1042/bj3220317

Chapter 10

Phosphoproteomics of Cerebrospinal Fluid-Derived Extracellular Vesicles Applied to Primary Central Nervous System Lymphoma

Yuanyuan Deng, Anton Iliuk, W. Andy Tao, and Zhuoying Xie

Abstract

Cerebrospinal fluid (CSF)-derived extracellular vesicles (EVs) are significant media for the cell-cell interaction in the central nervous system (CNS). Thus, characterization and analysis of CSF EVs can provide a pile of information for the discovery of molecular biomarkers of CNS diseases. Furthermore, phosphorylation of proteins plays a vital role in the regulation of protein function and cell signal transduction mechanism, and dysfunction of phosphorylation is always associated with the occurrence and progression of disease. Aiming to conduct mass spectrometry (MS)-based phosphoproteomic analysis of CSF EVs, we describe here a complete workflow to isolate EVs from CSF and profile their phosphoproteome, employing EVtrap-based EV isolation, spin-tip PolyMAC-Ti phosphopeptide enrichment and label-free quantitation of phosphoproteome. The approach provides a powerful CSF EV phosphoproteome analysis pipeline that can contribute to the discovery and identification of EV-based molecular markers.

Key words Cerebrospinal fluid, Extracellular vesicles, Phosphoproteomics, EVtrap PolyMAC, Label-free quantitation, PCNSL

1 Introduction

The diagnosis of tumors in the central nervous system (CNS) and brain always involves stereotactic biopsy of brain lesions, which may lead to a significant risk of morbidity, especially for tumors of deep location or multiple lesions. Given that, liquid biopsy becomes an attractive field for developing novel tools for the diagnosis and monitoring of CNS diseases and brain tumors. Compared to tissue biopsy, the detection of body fluids is less invasive and much more convenient. For CNS diseases and brain tumors, cerebrospinal fluid (CSF) is the most appropriate body fluid to be the resource for biomarker discovery. Though the obtainment of CSF needs a more invasive method than blood, there are still multiple advantages to choosing CSF as the preferred sample. CSF circulates around the

Enrique Santamaría and Joaquín Fernández-Irigoyen (eds.), *Cerebrospinal Fluid (CSF) Proteomics: Methods and Protocols*, Methods in Molecular Biology, vol. 2914, https://doi.org/10.1007/978-1-0716-4462-1_10,

CNS and is closer to the tumor, which means greater possibilities to gain tumor-associated data. Studies on glioma [1], pilocytic astrocytoma [2], and primary diffuse large B-cell lymphoma of the CNS [3] have shown the feasibility of CSF liquid biopsy in the early diagnosis and progress monitoring for CNS diseases.

In the field of liquid biopsy, extracellular vesicles (EVs) have won their place as an abundant resource for potential biomarkers [4–6]. EVs package DNAs, RNAs, and proteins within their lipid bilayer membrane and contribute to intercellular communication between different cell types by delivering biological information. The cargo in EVs varies due to the change of cell origin and biological process, which further strengthens the potential of EVs as diagnostic and prognostic biomarkers. In the central nervous system, EVs can facilitate the tumor-microenvironment interaction and enhance the process of tumor development [7–9]. Studies have demonstrated that CSF-derived EVs can be a promising tool as a biomarker in CNS pathology [10–13], including mass spectrometry (MS)-based proteomic analysis on CSF EVs.

MS-based proteomics is a high-sensitivity and high throughput technique to detect and analyze multiple proteins, but there is only a handful of research on phosphoproteome analysis of CSF EVs [12, 13]. Phosphorylation, one of the most extensive post-translational modifications of proteins, plays a vital role in the regulation of many physiological processes and the dysfunction of phosphorylation can lead to the genesis of diseases [14–17]. However, for many reasons, work on phosphoproteomic profiling of CSF-derived EVs is a great lack. On the one hand, methods for EV isolation from CSF need improvement in efficiency as the volume of CSF obtained from the individual patient and also the number of EVs enriched from CSF are both limited. On the other hand, phosphoproteins account for a small fraction of all proteins, so their detection can be easily affected by many other relatively abundant proteins.

Here, we describe a strategy and workflow (see Fig. 1 for the overall workflow) to isolate EVs from CSF, enrich phosphopeptides, and carry out downstream LC-MS analysis. For EV isolation, we employed a chemical affinity magnetic beads-based method called extracellular vesicles total recovery and purification (EVtrap). EVtrap is highly compatible with downstream LC-MS/MS detection and provides high purity and high recovery rate of EVs enough for downstream phosphoproteome analysis [18, 19]. For phosphopeptide enrichment, we used a commercial Spin-Tip PolyMAC-Ti Phosphopeptide Enrichment Kit (Tymora Analytical), which allows low amounts of starting materials from CSF EVs, and this step can also be replaced by other methods. Detailed procedures of the EVtrap method, EV lysis, protein digestion, Stage-Tip desalting, phosphopeptide enrichment, and LC-MS settings are recorded in this chapter. By employing such a protocol, over 19,000 peptides

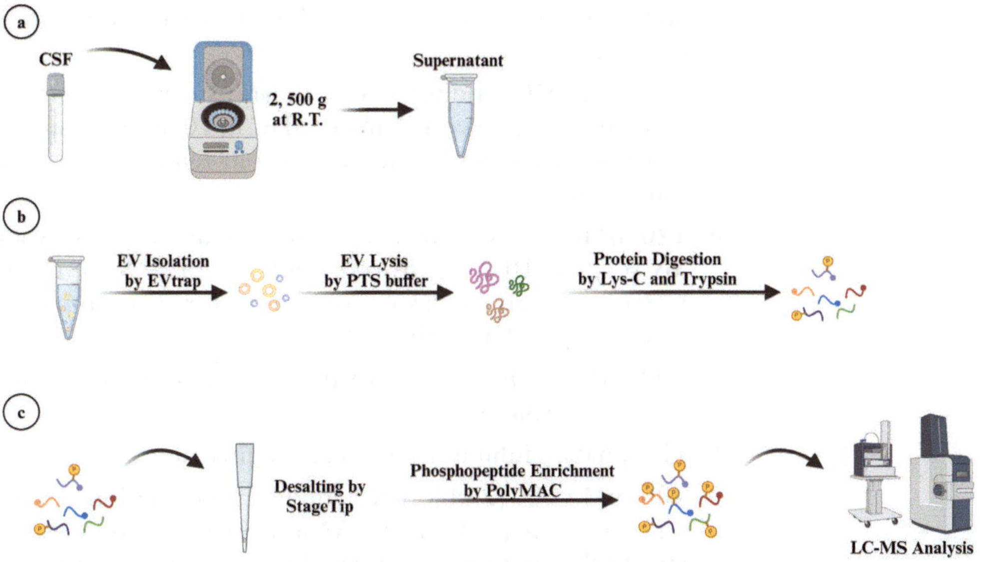

Fig. 1 Overview of the Workflow to isolate CSF EVs and extract phosphopeptides for LC-MS

representing 1,841 proteins can be identified from just 30 of CSF and up to 3,000 phosphopeptides representing over 1,000 phosphoproteins can be identified from about 2 mL CSF [13]. We also carried out label-free quantification analysis on CSF EVs phosphoproteome of primary central nervous system lymphoma (PCNSL) patients and non-PCNSL controls based on our workflow. PCNSL is a rare and highly aggressive DLBCL confined in CNS. From the perspective of CSF EVs phosphoproteomic, multiple reported phosphoproteins that related to PCNSL, including osteopontin (OSTP), myristoylated alanine-rich C-kinase substrate (MARCS), nucleophosmin (NPM) and vimentin (VIME), can be found up-regulated in the PCNSL group. These results demonstrated the feasibility of EVtrap-based analytical strategy in CSF EV phosphoproteomic analysis for CNS marker development.

2 Materials

1. EVtrap beads (Tymora Analytical).
2. EVtrap CSF dilution buffer, consisting of 0.01% Triton X-100 and 0.01% NP-40 in Phosphate buffered saline.
3. EVtrap elution buffer, 100 mM TEA (Triethylamine) in ultrapure water (*see* **Note 1**).
4. 500 mM Tris-HCl stock buffer, pH 8.5. To prepare 20 mL, dissolve 1.576 g of trizma hydrochloride (Tris-HCl) in 18 mL of ultrapure water, then adjust pH to 8.5 with 28% ammonia solution (NH_4OH). Bring to a final volume of 20 mL by

adding ultrapure water. Stock buffer can be stored at room temperature.

5. 120 mM SDC (sodium deoxycholate) stock solution. To prepare 10 mL, dissolve 497 mg of SDC in 10 mL ultrapure water. Stock solution can be stored at room temperature for up to 6 months.
6. 120 mM SLS (N-Lauroylsarcosine sodium salt) stock solution. To prepare 10 mL, dissolve 352 mg of SLS in 10 mL of ultrapure water. Stock solution can be stored at room temperature for up to 6 months.
7. 500 mM TCEP (Tris(2-carboxyethyl)phosphine hydrochloride) stock solution.
8. Phosphatase inhibitor cocktail 2 (Sigma-Aldrich).
9. 400 mM CAA (2-chloroacetamide) stock solution. To prepare 1 mL, dissolve 37.4 mg of CAA in 1 mL of ultrapure water. The stock solution must be prepared fresh and used immediately.
10. PTS lysis buffer. To prepare 500 μL of PTS lysis solution, add the following stock solutions in that order below to 285 μL of ultrapure water immediately before use. Vortex after each added ingredient.

    50 μL 500 mM Tris–HCl stock solution, pH 8.5.

    50 μL 120 mM SDC (sodium deoxycholate) stock solution.

    50 μL 120 mM SLS (sodium lauroyl sarcosinate) stock solution.

    10 μL 500 mM TCEP stock solution.

    5 μL Phosphatase inhibitor cocktail 2.

    50 μL 400 mM CAA (chloroacetamide) stock solution.
11. 50 mM TEAB (triethylammonium bicarbonate). To prepare 1 mL of 50 mM TEAB, dilute 50 μL 1 M TEAB stock (Sigma-Aldrich) and bring to a final volume of 1 mL by adding ultrapure water.
12. Lysyl endopeptidase (Lyc-C; Wako).
13. Trypsin from bovine pancreas (Sigma-Aldrich).
14. Trifluoroacetic acid (TFA).
15. Ethyl acetate (Sigma-Aldrich).
16. CDS Empore SDB-XC extraction disks.
17. Acetonitrile (ACN).
18. Spin-Tip PolyMAC-Ti Phosphopeptide Enrichment Kit (titanium-based; Tymora Analytical).
19. Phosphopeptide enrichment loading buffer, consisting of 100 mM glycolic acid, 50% (vol/vol) ACN, 1% (vol/vol) TFA in ultrapure water.

20. Phosphopeptide enrichment wash buffer 1, consisting of 25 mM glycolic acid, 80% (vol/vol) ACN, and 0.2% (vol/vol) TFA in ultrapure water.
21. Phosphopeptide enrichment wash buffer 2, consisting of 80% (vol/vol) ACN in ultrapure water.
22. Phosphopeptide enrichment elution buffer, consisting of 400 mM ammonium hydroxide solution (NH_4OH), 50% (vol/vol) ACN in ultrapure water.
23. Formate (FA).
24. nanoElute LC buffer A, consisting of 0.1% FA in ultrapure water.
25. nanoElute LC buffer B, consisting of 0.1% FA in ACN.
26. Water bath.
27. Microcentrifuge tubes.
28. Pipette tips (10 μL, 200 μL, 1 mL).
29. Refrigerated bench-top centrifuge.
30. Micro high-speed centrifuge.
31. Vortex mixer.
32. Thermo shaker incubator.
33. Centrifugal vacuum concentrator.
34. LC instrument.
35. Mass spectrometer.
36. Software for analyzing raw proteomics data.

3 Methods

3.1 CSF Sample Preparation

The CSF samples used in this protocol were obtained from Huashan Hospital affiliated with Fudan University (the Huashan Hospital Institutional Review Board (HIRB), approved No. 2018767), Shanghai, China, and all patients properly consented before samples were collected.

1. Thaw frozen CSF samples (~ 3 mL starting volume is suitable for phosphoproteomics analysis) in a water bath at 37°C.
2. Centrifuge CSF at 2500 g for 10 min at room temperature to remove apoptotic bodies and cell debris (*see* **Note 2**).
3. Collect the supernatant carefully (*see* **Note 3**).

3.2 EV Isolation and Lysis

1. Dilute CSF fourfold with the EVtrap CSF dilution buffer.
2. Add EVtrap beads to the diluted solution (6 μL EVtrap beads per 100 μL original volume of CSF) and incubate end-over-end for 1 h.

3. Separate the beads by a magnetic separator and remove the supernatant. EVs are bound onto the EVtrap beads.
4. Wash the beads with the EVtrap CSF dilution buffer by inverting them 10–20 times and remove the washing solution using a magnetic separator.
5. Wash the beads with 1XPBS by inverting them 10–20 times and remove the washing solution using a magnetic separator.
6. Wash the beads with 1XPBS for 2 more times.
7. Add 50–400 μL of the EVtrap elution buffer to the beads (the elution volume should be enough to fully resuspend the beads and allow for efficient interaction) and incubate for 10 min by vigorous shaking.
8. Collect the eluted sample (this contains EVs) and dry the solution completely in a vacuum centrifuge (*see* **Note 4**).
9. Prepare fresh PTS lysis buffer.
10. Add lysis buffer to the EV pellet and resuspend it by vortexing (*see* **Note 5**).
11. Incubate at 95 °C for 5–10 min in a water bath and ensure the sample is in the dark.
12. Cool the sample to room temperature and dilute it fivefold with 50 mM TEAB.

3.3 Protein Digestion and Surfactant Removal

1. Add Lys-C into the diluted sample with a 1:100 ratio of Lys-C to protein w/w and incubate at 37 °C for 3 h.
2. Add trypsin into the sample with a 1:50 ratio of trypsin to protein w/w and digest overnight at 37 °C.
3. Add 10% TFA to acidify the sample to pH <3 (make a final TFA concentration at ~1%) and vortex.
4. Add ethyl acetate equal to the volume of the sample and shake vigorously for 2 min.
5. Centrifuge the sample at 15,000 g for 3 min and carefully remove the organic phase on the top. Remember not to disturb the interface membrane (*see* **Note 6**).
6. Repeat **steps 4** and **5** for one more time.
7. Dry the sample using a vacuum centrifuge (*see* **Note** 7).

3.4 StageTip Desalting

In-house-made plugged StageTips by packing CDS Empore SDB-XC extraction disks into pipettes were used in this protocol. Other alternatives (such as C18 solid-phase extraction (SPE) cartridges) are available; however, buffer solutions may need to be changed as appropriate.

1. Condition the StageTip by passing the solution of 0.1% TFA and 80% ACN through the disk with slow centrifugation (*see* **Note 8**).
2. Equilibrate the StageTip by passing the solution of 0.1% TFA through the disk with slow centrifugation.
3. Dissolve the dried sample in 0.1% TFA and load onto the StageTip disk by slow centrifugation for 5–10 min.
4. Wash the StageTip by passing 0.1% TFA through the disk with slow centrifugation.
5. Transfer the tip with the disk into a new microcentrifuge tube.
6. Elute peptides from the disk by adding the solution of 0.1% TFA and 80% ACN. Centrifuge slowly and collect the flow-through.
7. Repeat elution **step 6** one more time.
8. Take 1–5% of the total elution solution into another tube and dry it completely with a vacuum centrifuge for total proteomics analysis. Dry the rest of the elution solution using a vacuum centrifuge for the following phosphopeptide enrichment.

3.5 Phosphopeptide Enrichment

Spin-Tip PolyMAC-Ti Phosphopeptide Enrichment Kit was used in this protocol and alternative phosphopeptide enrichment methods can be used (IMAC, TiO2, among others); however, buffers may vary from the ones outlined in this protocol.

1. Take out 50 μL PolyMAC beads for each sample and change the original solvent of PolyMAC beads into an equal volume of phosphopeptide enrichment loading buffer (*see* **Note 9**).
2. Dissolve the dried desalting sample with ~150 μL phosphopeptide enrichment loading buffer and add the PolyMAC beads into the solution.
3. Incubate the sample in a shaker at room temperature for 15 min.
4. Load the beads onto the tip with a frit and centrifuge at 60 g for 2 min, followed by 130 g for 1 min. The centrifugation duration needs to be adjusted to make the solution pass through.
5. Wash the beads in the tip by adding 200 μL phosphopeptide enrichment loading buffer to the tip and centrifuge as described in **step 4**.
6. Wash the beads in the tip by adding 200 μL phosphopeptide enrichment wash buffer 1 to the tip and centrifuge as described in **step 4**.
7. Wash the beads in the tip by adding 200 μL phosphopeptide enrichment wash buffer 2 to the tip and centrifuge as described in **step 4**.

8. Transfer the tip with beads to a new microcentrifuge tube.
9. Elute the phosphopeptides from the beads by adding 50 μL of phosphopeptide enrichment elution buffer to the tip and centrifuge as described in **step 4**.
10. Repeat **step 9**. Combine the two eluates and dry the sample using a vacuum centrifuge.

3.6 LC-MS/MS Analysis

1. Dissolve the dried sample of peptides after desalting (for total proteome analysis) and phosphopeptides after enrichment (for phosphoproteome analysis), each in 10 μL of nanoElute LC buffer A.
2. Inject 2 μL for the total proteome sample and 8 μL for the phosphopeptide sample into the nanoElute LC system in this protocol. In this protocol, peptides were separated on a 25 cm in-house packed column (360 μm OD × 75 μm inner diameter (ID)) containing C18 resin (1.9 μm, 100 Å; Michrom Bioresources) at a flow rate of 300 nL/min with a linear 120 min gradient (2–22% solvent B for 90 min, 22–37% B for 10 min, 37–80% B for 10 min, and 80% B for 10 min) for proteome analysis and a linear 90 min gradient (2–22% solvent B for 67.5 min, 22–37% B for 7.5 min, 37–80% B for 7.5 min, and 80% B for 7.5 min) for phosphoproteome analysis.
3. Analyze each sample by mass spectrometry. We used a timsTOF pro (Bruker) in this protocol. The mass spectrometer was operated in the data-dependent mode for all the experiments described here, which consists of a full scan MS (from m/z 100 to 1700 with the intensity of 1.5 X 106 at the "target value" resolution of 40,000) along with 10 PASEF MS/MS frames. The quadrupole isolation width was set to 2 Th for m/z < 700 and 3 Th for m/z > 700. Collision energy was set stepwise, changing within a TIMS elution ramp: 52 eV for 0–19%, 47 eV for 19–38%, 42 eV for 38–57%, 37 eV for 57–76%, and 32 eV for the remainder.
4. The raw file of MS analysis can now be searched by appropriate software. We use PEAKS Studio X+ software (Bioinformatics Solutions Inc.) in this protocol. The raw files were searched directly against the Uniprot database downloaded in November 2019 with no redundant entries. MS1 precursor mass tolerance was set at 15 ppm initially and the final was set at 6 ppm. MS2 tolerance was set at 0.6 Da. Modification criteria included a static modification of carbamidomethylation on cysteines (+57.0214 Da) and variable modifications of acetylation (+42.011 Da) at the N terminus of proteins, oxidation (+15.9949 Da) on methionine residues and an extra modification of phosphorylation (+79.996 Da) on serine, threonine, or tyrosine residues for phosphoproteome analysis. Search was

performed with semi-specific trypsin/P digestion, allowing a maximum of three missed cleavages on the peptides. The false discovery rates (FDRs) for peptides, proteins, and phosphosites were all set at 1%. All protein and peptide identifications were grouped with no redundant entries.

4 Notes

1. It's recommended to prepare the elution buffer fresh. Triethylamine is a light-sensitive reagent, which is susceptible to the reaction of light and can lead to mass changes.
2. It's better to use fresh CSF and immediately centrifuge at 2500 g, room temperature, to remove cell debris and larger apoptotic bodies. Then collect the supernatant and store at −80°C until use.
3. It's critical for discarding cell debris and apoptotic bodies. Ensure not to disturb the bottom 10–30 μL of CSF and leave it behind when collecting the supernatant.
4. To accelerate the drying process, freeze the sample at −80 °C for 5–10 min and then put the sample into the vacuum centrifuge.
5. The volume of PTS lysis buffer can be adjusted from 50 to 100 μL according to the protein mass and 100 μL of the solution is enough for most EV samples.
6. The intermediate may contain a proportion of peptides. A small volume of ethyl acetate phase can be kept and does not disturb the two-phase interface.
7. It's not necessary to dry the sample completely after discarding the ethyl acetate phase, but no less than 80% should be removed to ensure an optimal solution environment for desalting.
8. Do not plug the disk too tightly in pipettes to ensure the use of low centrifugal speed for StageTip. High speed may destroy the structure in the disk and thus cause the loss of peptides.
9. The stock solution environment of PolyMAC beads can affect the solution state for phosphopeptide and bead binding. It's recommended to exchange the original solution for phosphopeptide enrichment loading buffer.

References

1. Miller AM, Shah RH, Pentsova EI, Pourmaleki M, Briggs S, Distefano N, Zheng Y, Skakodub A, Mehta SA, Campos C, Hsieh WY, Selcuklu SD, Ling L, Meng F, Jing X, Samoila A, Bale TA, Tsui DWY, Grommes C, Viale A, Souweidane MM, Tabar V, Brennan CW, Reiner AS, Rosenblum M, Panageas KS, DeAngelis LM, Young RJ, Berger MF, Mellinghoff IK (2019) Tracking tumor evolution in glioma through liquid biopsies of cerebrospinal fluid. Nature 565(7741):654–658. https://doi.org/10.1038/s41586-019-0882-3
2. Behling F, Schittenhelm J (2019) Oncogenic BRAF alterations and their role in brain tumors. Cancers (Basel) 11(6). https://doi.org/10.3390/cancers11060794
3. Baraniskin A, Kuhnhenn J, Schlegel U, Chan A, Deckert M, Gold R, Maghnouj A, Zollner H, Reinacher-Schick A, Schmiegel W, Hahn SA, Schroers R (2011) Identification of microRNAs in the cerebrospinal fluid as markers for primary diffuse large B-cell lymphoma of the central nervous system. Blood 117(11): 3140–3146. https://doi.org/10.1182/blood-2010-09-308684
4. Guo Z, Tian C, Shi Y, Song XR, Yin W, Tao QQ, Liu J, Peng GP, Wu ZY, Wang YJ, Zhang ZX, Zhang J (2024) Blood-based CNS regionally and neuronally enriched extracellular vesicles carrying pTau217 for Alzheimer's disease diagnosis and differential diagnosis. Acta Neuropathol Commun 12(1):38. https://doi.org/10.1186/s40478-024-01727-w
5. Cano A, Ettcheto M, Bernuz M, Puerta R, Esteban de Antonio E, Sanchez-Lopez E, Souto EB, Camins A, Marti M, Pividori MI, Boada M, Ruiz A (2023) Extracellular vesicles, the emerging mirrors of brain physiopathology. Int J Biol Sci 19(3):721–743. https://doi.org/10.7150/ijbs.79063
6. Morad G, Moses MA (2019) Brainwashed by extracellular vesicles: the role of extracellular vesicles in primary and metastatic brain tumour microenvironment. J Extracell Vesicles 8(1): 1627164. https://doi.org/10.1080/20013078.2019.1627164
7. Russo MN, Whaley LA, Norton ES, Zarco N, Guerrero-Cazares H (2023) Extracellular vesicles in the glioblastoma microenvironment: a diagnostic and therapeutic perspective. Mol Asp Med 91:101167. https://doi.org/10.1016/j.mam.2022.101167
8. Yekula A, Yekula A, Muralidharan K, Kang K, Carter BS, Balaj L (2019) Extracellular vesicles in glioblastoma tumor microenvironment. Front Immunol 10:3137. https://doi.org/10.3389/fimmu.2019.03137
9. Osti D, Del Bene M, Rappa G, Santos M, Matafora V, Richichi C, Faletti S, Beznoussenko GV, Mironov A, Bachi A, Fornasari L, Bongetta D, Gaetani P, DiMeco F, Lorico A, Pelicci G (2019) Clinical significance of extracellular vesicles in plasma from glioblastoma patients. Clin Cancer Res 25(1):266–276. https://doi.org/10.1158/1078-0432.CCR-18-1941
10. Hirschberg Y, Valle-Tamayo N, Dols-Icardo O, Engelborghs S, Buelens B, Vandenbroucke RE, Vermeiren Y, Boonen K, Mertens I (2023) Proteomic comparison between non-purified cerebrospinal fluid and cerebrospinal fluid-derived extracellular vesicles from patients with Alzheimer's, Parkinson's and Lewy body dementia. J Extracell Vesicles 12(12):e12383. https://doi.org/10.1002/jev2.12383
11. Whitehead CA, Luwor RB, Morokoff AP, Kaye AH, Stylli SS (2017) Cancer exosomes in cerebrospinal fluid. Transl Cancer Res 6(S8): S1352–S1370. https://doi.org/10.21037/tcr.2017.08.31
12. Sun J, Li Q, Ding Y, Wei D, Hadisurya M, Luo Z, Gu Z, Chen B, Tao WA (2023) Profiling phosphoproteome landscape in circulating extracellular vesicles from microliters of biofluids through functionally tunable paramagnetic separation. Angew Chem Int Ed Engl 62(29):e202305668. https://doi.org/10.1002/anie.202305668
13. Deng Y, Li Q, Sun J, Ma L, Ding Y, Cai Y, Iliuk A, Chen B, Xie Z, Tao WA (2023) Phosphoproteome analysis of cerebrospinal fluid extracellular vesicles in primary central nervous system lymphoma. Analyst 148(15): 3594–3602. https://doi.org/10.1039/d3an00670k
14. Chen IH, Xue L, Hsu CC, Paez JS, Pan L, Andaluz H, Wendt MK, Iliuk AB, Zhu JK, Tao WA (2017) Phosphoproteins in extracellular vesicles as candidate markers for breast cancer. Proc Natl Acad Sci USA 114(12): 3175–3180. https://doi.org/10.1073/pnas.1618088114
15. Humphrey SJ, James DE, Mann M (2015) Protein phosphorylation: a major switch mechanism for metabolic regulation. Trends

Endocrinol Metab 26(12):676–687. https://doi.org/10.1016/j.tem.2015.09.013

16. Hao BB, Ma K, Xu JY, Fan RF, Zhao WS, Jia XL, Zhai LH, Lee S, Xie D, Tan MJ (2024) Proteomics analysis of histone deacetylase inhibitor-resistant solid tumors reveals resistant signatures and potential drug combinations. Acta Pharmacol Sin 45(6):1305–1315. https://doi.org/10.1038/s41401-024-01236-5
17. Fricke AL, Muhlhauser WWD, Reimann L, Zimmermann JP, Reichenbach C, Knapp B, Peikert CD, Heberle AM, Faessler E, Schauble S, Hahn U, Thedieck K, Radziwill G, Warscheid B (2023) Phosphoproteomics profiling defines a target landscape of the basophilic protein kinases AKT, S6K, and RSK in skeletal myotubes. J Proteome Res 22(3):768–789. https://doi.org/10.1021/acs.jproteome.2c00505
18. Wu X, Li L, Iliuk A, Tao WA (2018) Highly efficient phosphoproteome capture and analysis from urinary extracellular vesicles. J Proteome Res 17(9):3308–3316. https://doi.org/10.1021/acs.jproteome.8b00459
19. Iliuk A, Wu X, Li L, Sun J, Hadisurya M, Boris RS, Tao WA (2020) Plasma-derived extracellular vesicle phosphoproteomics through chemical affinity purification. J Proteome Res 19(7):2563–2574. https://doi.org/10.1021/acs.jproteome.0c00151

Part IV

Shotgun Cerebrospinal Fluid Proteomics

Chapter 11

High-Throughput Human Cerebrospinal Fluid Proteome Analysis with Direct Data-Independent Acquisition (dDIA)

Ramón Díaz-Peña, Pol Andrés-Benito, Erica Peral, Raúl Domínguez, Mónica Povedano, Enrique Santamaría, and Joaquín Fernández-Irigoyen

Abstract

Cerebrospinal fluid (CSF) is a low-risk, rapid, and mid-invasive sampling for diagnosis, prognosis, and treatment of neurological pathologies. The CSF liquid biopsies disponibility and sampling homogeneity foster the research for biomarker discovery for neurological disorders and pathologies and, importantly, enable extensive population studies. Liquid chromatography-mass spectrometry (LC-MS) proteomics is a powerful tool for biomarker discovery. Proteomics large studies provide more robust and reliable results, offering exceptional FDR control and outlier identification, allowing high-precision results. Consequently, a robust and reliable pipeline proteomics methodology must be required to analyze hundreds of samples, from sample preparation to data analysis. Here, we describe a detailed workflow for analyzing human CSF samples for large studies by direct data-independent (dDIA).

Key words dDIA, Cerebrospinal fluid, Proteomics, Mass-spectrometry

1 Introduction

LC-MS proteomics has revealed itself as a source of biomarkers in different neurological pathologies [1–3]. Cerebrospinal fluid (CSF), obtained by lumbar puncture (LP), is the preferred liquid biopsy for biomarker discovery of neurological disorders and pathologies. Mid invasiveness and its composition, a snapshot of intracranial milieu, overcomes the perils of solid biopsies and their heterogeneity. To maximize the success of analyzing CSF samples, one must consider sample obtention and managing sample preparation, chromatograph-MS instrumentation, and MS method and data analysis.

Obtaining and managing CSF samples are meticulous processes crucial for ensuring accurate diagnostic results, research aims, and patient safety. Sample preparation precision is crucial for the fate of the analysis. It must be higher than the instrumental one by

Enrique Santamaría and Joaquín Fernández-Irigoyen (eds.), *Cerebrospinal Fluid (CSF) Proteomics: Methods and Protocols*, Methods in Molecular Biology, vol. 2914, https://doi.org/10.1007/978-1-0716-4462-1_11,

definition. Simplicity, a few steps protocol, and minimal sample manipulation should be the goal of an ideal protocol. Due to the complexity of CSF composition (i.e., proteins, saccharides, lipids, vitamins, salts), sample preparation before MS analysis must be accurate. Some of its components could inhibit protease enzymes (ex., Trypsin, Lys-C) in a bottom-up experiment, and others affect chromatography or even suppress MS signal (ex., phospholipids). Suspension trapping (S-Trap) sample preparation was our choice to solve it. Using SDS as a protein denaturant, S-Trap sample preparation columns-plates allow the removal of polar and non-polar contaminants and in-filter protein digestion to finally obtain cleaned-up peptides ready for MS with no need for additional contaminant removal steps. The S-Trap protocol is also faster than Filter-Aided Sample Preparation (FASP), which was previously found to be an alternative to in-solution digestion to remove sample contaminants prior to digestion [4–6]. The S-Trap sample preparation outperforms traditional ones (i.e., in-solution digestion, FASP), eliminating contaminants, enhancing protease digestion activity, reducing sample processing times, and preserving instruments.

Chromatograph proper selection for high-throughput LC-MS analysis is decisive. We decided to use the Evosep One instrument (Evosep) to shorten run times since the chromatograph overlaps sample loading time over gradient run time with exceptional chromatographic robustness. A remarkable feature of the instrument is that the sample is loaded in a C18-tip (Evotip), acting as a trap column, adding a clean-up step. This is enhanced as the acetonitrile gradient percentage increases only up to 35%, impeding the most hydrophobic contaminants eluting to the LC-MS system. Besides, Evosep One performs a low-pressure gradient with minimal wear and tear on mechanical components, avoiding mechanical interventions interrupting large studies. Altogether, it ensures a significant column lifetime, reliable chromatography, minimal cross-contamination, and reduced time analysis, all required in large studies.

One of the challenges of CSF analysis is the dynamic range of sample protein concentration. Historically, data-dependent acquisition (DDA)—for label-free experiments—has been the usual MS approach until recently. As a precursor ion abundance-based quantitation method, the stochastic selection of the most abundant ions discriminates against low-abundant CFS ones, hindering peptide identification. To surpass the inherent limitation of DDA, we used a direct Data-Independent Acquisition (directDIA). In DIA, all precursor peptides are selected to be fragmented independently of their abundance. MS scan survey is divided into 20–30 Th windows on all precursors belonging to a particular window are fragmented at once. The result is a complex MS2 spectrum containing all fragments from different precursors. A powerful search engine software then achieves the peptide assignation (and quantitation) from those fragments (and quantitation) and protein inference. This is possible after two decades of developing robust chromatography, faster high-resolution MS, and search engine software [7]. Former DIA experiments consisted of generating high-resolution spectral libraries with DDA narrow precursor windows (to avoid precursor interference) to aid the DIA spectra match. Spectral libraries are typically obtained from the offline fractionation of a pool of samples participating in the analysis and posterior acquisition of these fractions in DDA or acquiring a certain number of representative samples in DDA. Either fractionation or acquisition of several samples in DDA is time-consuming and hardly feasible if one imagines a study comprising hundreds of samples. A more recent alternative to library-based DIA has been developed to bypass spectral library generation: direct DIA (dDIA). In dDIA, acquired data in a high-resolution MS (in our case, Exploris 480) is first processed by the specific part of the search engine, which generates pseudo-DDA data (spectral library) for further DIA peptide assignment analysis. Spectronaut (Biognosys) and DIA NN (free licensed) are two examples of software capable of performing such analysis with similar results to library-based DIA assisted by deep learning machine [8–10].

In summary, our workflow includes S-Trap sample preparation, LC-MS analysis with Evosep One chromatograph, and a high-resolution MS to perform a dDIA experiment for proteome label-free quantitation, making the standardization of large studies possible.

2 Materials

2.1 CSF Sample Obtention and Processing

1. Spinal needle of 22 gauge (22G).
2. Polypropylene or glass sterile tubes.
3. Centrifuge adapted for selected tubes that spin at 3000 rpm for 15 min at room temperature.
4. All solutions must be prepared using HPLC MS-grade reagents and freshly.

2.2 S-TRAP™ Digestion (up to 100 ug of Protein)

1. S-Trap™ micro spin column (1–100 μg protein loading capacity) or S-Trap™ 96-well plate (100–250 μg protein loading capacity).
2. Denaturing buffer: 10% SDS, 100 mM TEAB pH 8.5 (triethylammonium bicarbonate).

   120 mM TCEP Bond-Breaker™ solution (neutral pH) in 100 mM TEAB (*see* **Note 1**).

3. Cysteine-blocking buffer: 500 mM MMTS (methyl methanethiosulfonate) in isopropanol.
4. 27.5% ortho-phosphoric for S-Trap™ micro spin column. Dilute ortho-phosphoric acid 85% with water.
5. Binding buffer: 90% methanol, 100 m TEAB (triethylammonium bicarbonate).
6. Trypsin solution: dissolve 2 μg of MS-grade trypsin (Promega) in 20 μL of 100 mM TEAB.
7. Elution solution 1: 50 mM TEAB.
8. Elution solution 3: 50% acetonitrile (ACN), 50 mM TEAB pH 8.5.

2.3 Desalting Proteolytic/Tryptic Peptides After Digestion Using Oasis HLB Cartridges

1. Oasis HLB 96-well μElution Plate, 2 mg Sorbent per Well, 30 μm (Waters).
2. Extraction Plate Manifold for Oasis 96-Well Plates (Waters).
3. ACN.
4. TFA 0.1% in water.
5. 70% ACN in water.

2.4 Chromatography and Mass Spectrometry

1. Mobile Phase A: 0.1% formic acid in water.
2. Mobile Phase B: 0.1% formic acid in ACN.
3. Liquid chromatography (Evosep):
   (a) Evosep One LC.
   (b) Analytical column. Performance Evosep column E1112 (C18, 1.9 μm beads, 15 cm × 75 μm).
   (c) E1111 Silica emitter (10 μm) (*see* **Note 2**).
4. Mass Spectrometer: Exploris-480 or other high-resolution mass spectrometry instrument.

3 Methods

3.1 CSF Sample Obtention

Lumbar puncture (LP) is the procedure used to access the subarachnoid space in the lumbar region. It is used to obtain CSF, measure CSF pressure, or administer intrathecal treatment. As an invasive procedure, several items must be achieved and/or taken into consideration.

1. *Patient preparation*
   1.1 Review the patient's medical history and confirm there are no contraindications to performing the LP procedure. Consider the following scenarios:
      (a) If the procedure is not urgent, cranial imaging (CT scan or MRI) should be available and must be no more than 6 months old.
      (b) A blood cell count, including platelet count, must be available from the previous 6 months, with the platelet count above 50,000 × 10^6/L.
      (c) The use of anticoagulants or antiaggregants should be assessed prior to the LP and suspended with enough time.
   1.2 Explain the procedure to the patient, ensuring they understand the risks and benefits, and then obtain their informed consent (IC).
   1.3 Position the patient for the LP, either in a lateral decubitus position (lying on their side) or in a sitting position with their back arched forward. This positioning helps to open up the intervertebral spaces for easier needle insertion (*see* **Notes 3** and **4**).
2. *Site selection:* It is the most frequently used technique to access this space, because the spinal cord ends in the low thoracic (T11-T12) or upper lumbar (L1-L2) region in most people, so no direct injury in the central nervous system can be made. The preferred site for LP is between the L3-L4 vertebrae, but it can also be done in an upper or lower space.
3. *Aseptic technique:* CSF is a sterile liquid, so the procedure should be done in asepsis conditions. The skin over the chosen puncture site is thoroughly cleaned with an antiseptic solution, such as chlorhexidine or iodine, to reduce the risk of infection. Sterile drapes are used to maintain a sterile field around the puncture site.
4. *Local anesthesia:* A local anesthetic, typically lidocaine, is administered to numb the puncture site. This is done to minimize discomfort during the procedure (*see* **Note 5**).
5. *Needle insertion:* A spinal needle with a stylet is carefully inserted into the subarachnoid space. The use of 22 gauge (22G) or smaller needle is recommended, although larger needles can also be used (*see* **Note 6**). The physician will advance the needle slowly, and a characteristic "pop" sensation indicates the needle has passed through the dura mater into the subarachnoid space. Inform the patient prior to performing each action, especially when trespassing the skin with the needle, and encourage the patient to report if they note any discomfort and check repeatedly how the patient is feeling.

6. *CSF collection:* When subarachnoid space is reached, CSF should be collected in polypropylene or glass sterile tubes. Once the stylet is removed, CSF begins to drip out of the needle hub. CSF is collected in sterile tubes, typically three to four tubes, for various analyses (e.g., microbiology, chemistry, cell count, and cytology) (*see* **Notes** 7 and **8**).
7. *Needle removal:* After CSF is collected, be sure to put the needle hub again before removing the needle. Removing the needle is recommended when there is negative pressure in the abdomen (ex. during exhalation). After LP, 15 min of absolute rest is mandatory and relative rest should be suggested for the patient for 24 h, although not mandatory. Inform the patient about possible secondary effects and alarm symptoms and actions to take for each one.

3.2 CSF Sample Processing

1. Each tube, with a CSF volume of 1.0–1.5 mL, must be immediately labeled with the patient's information, date, and time of collection. Accurate labeling is critical to avoid sample mix-ups.
2. The CSF samples should be transported to the laboratory as quickly as possible. Some tests require immediate processing, while others can wait, but prompt transport helps preserve the integrity of the sample.
3. Samples must be proceeded immediately or temporally stored at appropriate temperatures at 4 °C and processed as soon as possible. Do not leave the sample at 4 °C for more than 12 h.
4. For proteomics analysis, CSF must be centrifuged at 3000 rpm for 15 min at room temperature. The supernatant must be collected and aliquoted in volumes of 220 μL and stored at − 80 °C until use.

By following these steps meticulously, healthcare providers ensure the safe and effective collection and management of CSF samples, which are critical for diagnosing various neurological conditions and research purposes.

3.3 S-TRAP™ Microspin Column Digestion

1. Mix 50 μL of CSF with 50 μL denaturing buffer in a 1.5 mL microcentrifuge tube (*see* **Note 9**).
2. Add 4.4 μL of TCEP 120 mM and incubate at 55 °C for 15 min in a thermoblock with gentle mixing (final concentration 5 mM).
3. Add 4.4 μL of 500 mM cysteine-blocking reagent MMTS (final concentration 20 mM) and incubate for 10 min at room temperature in a thermoblock with gentle mixing.
4. Add 11 μL of 27.5% ortho-phosphoric acid solution (final concentration 2.5%) and mix (*see* **Note 10**).

5. Add 719 μL (dilution 1:7 vol/vol) of binding buffer. Colloidal protein particulate is instantaneously formed in this step.
6. Place the S-Trap filter in a 2-mL tube for flow through. Load 200 μL of the whole sample (colloidal and non-colloidal part) into the column and centrifuge at 4000× *g* 30 s. Repeat the step as many times as required to filter the sample volume totality (*see* **Note 11**).
7. At this point, protein is trapped by the matrix of the spin column.
8. Add 150 μL of binding buffer to wash the spin column. Repeat three times (4000 rcf, 1 min.).
9. Place the filter onto a clean 1.5-mL tube for protein digestion. Add *n* μg of MS-grade trypsin in 20 μL of a 50 mM TEAB solution into the top of the column in a ratio 1:10 (w/w) and allow digestion at 47 °C for 1 h.
10. After 1 h, add 50 μL more of Elution solution 1 (50 mM TEAB) —to avoid the sample drying out, and return the plate to the incubator for another 1 h.
11. Elute proteolytic peptides:
    (a) Add 40 μL of elution solution 1 to the S-Trap and centrifuge (4000 rcf, 1 min.).
    (b) Add 40 μL of elution solution 2 to the S-Trap and centrifuge (4000 rcf, 1 min.).
12. Pool eluted peptides and dry them down in a vacuum concentrator system. After resuspension with a suitable solution, the sample is ready for reverse-phase chromatography. Additional C18 clean-up may be required with samples with high contaminant content.

3.4 S-Trap™ 96-Well Plate Digestion

1. Mix 25 μL of CSF with 25 μL denaturing buffer in a 1.5 mL microcentrifuge tube (final volume should not exceed 50 μL).
2. Add 2.2 μL of TCEP 120 mM and incubate at 95 °C for 10 min in a thermoblock with gentle mixing (final concentration 5 mM). Let cool down the sample.
3. Add 2.2 μL of 500 mM cysteine-blocking reagent MMTS (final concentration 20 mM) and incubate for 10 min at room temperature in a thermoblock with gentle mixing.
4. Add 5.5 μL of 27.5% ortho-phosphoric acid solution (final concentration 2.5%) and mix.
5. Add 360 μL (dilution 1:7 vol/vol) of binding buffer.
6. Place *S-Trap™ 96-well plate* in a 2-mL plate for waste flow through. Load the sample into the S-Trap™ 96-well plate and centrifuge at 2000 rcf 2 min. Since all wells exhibit flow behavior, a vacuum manifold may also be used, applying 5 Hg (2.5 Psi) of negative pressure to filter and wash the sample.

7. Add 200 μL of binding buffer to wash S-Trap plate. Repeat three times (2000 rcf, 2 min).
8. Place *S-Trap™ 96-well plate* in a new clean 2-mL plate to recover the peptides. Add n μg of MS-grade trypsin in 125 μL of a 50 mM TEAB solution into the top of the column in a ratio 1:10 (w/w). Make sure the digestion solution completely covers the protein capture matrix (*see* **Note 12**).
9. Loosely cover the digestion plate to limit evaporative loss and allow digestion at 47 °C for 2 h. Do not seal the plate) as pressure buildup will force the solution prematurely through the trapping matrix.
10. After 1 h, add 70 μL of Elution solution 1 (50 mM TEAB)—to avoid the sample drying out—and return the plate to the incubator for another 1 h.
11. Remove the plate from the incubator, add 75 μL of Elution solution 1, and centrifuge the plate at 2000 rcf min or until all solution has passed through.
12. Elute proteolytic peptides:
    (a) Add 80 μL of elution solution 1 to the S-Trap and centrifuge (2000 rcf, 2′).
    (b) Add 80 μL of elution solution 2 to the S-Trap and centrifuge (2000 rcf, 2′).
13. Pool eluted peptides and dry them down in a vacuum concentrator system. After resuspension with a suitable solution, the sample is ready for reverse-phase chromatography. Additional C18 clean-up may be required (e.g., Oasis HLB Cartridges) with samples with high contaminant content.

3.5 Desalting Tryptic Peptides Using Oasis HLB Cartridges (Additional Clean-up)

1. Resuspend dry-down tryptic peptides from S-Trap™ digestion with 500 μL 0.1% TFA.
2. Place Oasis HLB 96-well μElution Plate onto the extraction plate manifold with collection-wash plate inside (*see* **Note 13**).
3. Add 200 μL of ACN to conditionate HLB plate.
4. Add 200 μL of TFA 0.1% to equilibrate HLB plate.
5. Load the acidified tryptic peptides onto the HLB plate.
6. Wash the bound peptides with 800 μL of TFA 0.1%.
7. Wash the bound peptides with 200 μL of water to remove salts and buffers.
8. Replace collection plate with the collection-elution plate to recover the peptides and elute peptides, adding 50 μL of 70% ACN.
9. Dry down eluted peptides in a vacuum concentrator system and kept at −80 °C.

3.6 LC-MS/MS Analysis

1. Dried-down peptide samples are reconstituted with 2% ACN-0.1% FA and quantified by NanoDrop™ spectrophometer (ThermoFisher) before LC-MS/MS analysis.
2. 100 ng per sample is analyzed by reverse-phase LC-MS/MS using an Evosep LC coupled to an Exploris 480 mass spectrometer (Thermo).
3. Peptides are resolved using a performance Evosep column E1112 (C18, 1.9 μm beads, 15 cm x75 μm) using SP20 whisper method (58 min gradient, 40 °C).
4. Sample data is acquired in a data-independent mode (DIA) with a full MS scan (scan range: 400 to 900 m/z; resolution: 12000; maximum injection time: 50 ms; normalized AGC target: 300%) and 50 periodical MS/MS segments, applying 10 Th isolation windows (0.5 Th overlap; resolution: 15000; maximum injection time: 22 ms; normalized AGC target: 100%).
5. Peptides are fragmented using a normalized HCD collision energy of 30%, acquiring data in profile and centroid mode for full MS scan and MS/MS, respectively (*see* **Note 14**).

3.7 Identification and Quantification

DIA data files are analyzed using Spectronaut (Biognosys) by directDIA analysis (dDIA). Sample raw data is first processed by the Pulsar Spectronaut search engine generating a pseudo-spectral library, searching MSMS spectra against Uniprot-Swissprot Homo Sapiens database (MS1/MS2 calibration and main search tolerance are set to dynamic). Maximum precursor ion charge is set to 4, and fragment selection to intensity-based. Carbamidomethyl (C) is selected as a fixed modification, and Oxidation (M), Acetyl (Protein N-term), Deamidation (N), and Gln-> pyro-Glu as variable modifications (3 maximum modifications per peptide). Enzyme is set to trypsin in a specific mode (two missed cleavages maximum). The target-decoy-based false discovery rate (FDR) filter for PSMs, peptide, and protein groups is set to 1%.

Once the Pulsar search is performed, the Spectronaut DIA search engine algorithm starts the DIA search using the default settings and filtering the precursors and protein groups by a 1% Qvalue.

4 Notes

1. Use TCEP Bond-Breaker™ solution with neutral pH to avoid pH bias in the reducing buffer. Solutions are acidified by powder TCEP.
2. Several applications/solutions have been developed by the Evosep team to balance deep proteome analysis, number of samples analyzed a day and system robustness. Higher sample

analysis per day has a detrimental effect on deep proteome analysis, requiring a robust system when injecting a large (i.e., thousands) number of samples. Researchers should evaluate the scope of their study to select the appropriate Evosep solution. In this protocol, we selected the performance E1112 column, which allows the analysis of 20 or 40 a day (SP20 and SP40 method, respectively), being the best Evosep column in terms of chromatographic features.

3. Ensure the vertebral column is as straight as possible to facilitate access to the vertebral canal.
4. For right-handed physicians, the left lateral decubitus position is recommended. For left-handed physicians, the right lateral decubitus position is recommended.
5. The use of local anesthetics is recommended but not mandatory.
6. If more than 5 mL needs to be extracted, no smaller than 22G needles should be used.
7. Collect a tube for cell count in every LP, even if the assessment of the CSF is not the objective of the LP. At least 1 mL.
8. Collect a tube for microorganism culture in every LP, even if the assessment of the CSF is not the objective of the LP. At least 1.5 mL.
9. Sample protein concentration must be checked to ensure sample quality, and calculation of trypsin quantity for tryptic digestion.
10. Adding ortho-phosphoric acid is essential to trap the proteins, denaturing proteins completely. The pH has to be ≤ 1. Add additional phosphoric acid to reach pH ≤ 1 if required.
11. Rotate the S-Trap micro units 180 degrees between the centrifugations of binding and wash steps. This is especially important when using a fixed-angle rotor because the spin column does not experience homogenous flow. Additional wash (es) may be performed and should be performed if contamination is observed by mass spectrometry.
12. Do not apply less than 10 μg of trypsin per well for effective digestion.
13. Apply 5 Hg of vacuum to filter through the well matrix plate.
14. The MS method described here, in combination with the 20SP whisper method gradient, offers a 2.15 s cycle time, allowing 6–7 quantification points per peak, a reasonable number for label-free quantitation. For other Evosep methods in whose gradient and extension peptide peak lengths are different, it is required to adjust the MS settings to obtain suitable points per peak.

Acknowledgments

The Clinical Neuroproteomics Unit is member of the Spanish Olfactory Network—ROE (grant RED2022-134081-T from the Spanish Ministry of Science & Innovation) and is supported by grants PID2023-152593OB-I00 funded by MCIU/AEI/10.13039/501100011033 / FEDER, UE to ES and JFI and 0011-1411-2023-000028 (from Government of Navarra—Department of Economic and Business Development-S4). PAB is funded by the Carlos III Health Institute and co-financed by the European Union through grant CD23/00097.

References

1. Schmid D, Warnken U, Latzer P, Hoffmann DC, Roth J, Kutschmann S, Jaschonek H, Rübmann P, Foltyn M, Vollmuth P, Winkler F, Seliger C, Felix M, Sahm F, Haas J, Reuss D, Bendszus M, Wildemann B, von Deimling A, Wick W, Kessler T (2021) Diagnostic biomarkers from proteomic characterization of cerebrospinal fluid in patients with brain malignancies. J Neurochem 158(2): 522–538. https://doi.org/10.1111/jnc.15350
2. Hinsinger G, Du Trieu De Terdonck L, Urbach S, Salvetat N, Rival M, Galoppin M, Ripoll C, Cezar R, Laurent-Chabalier S, Demattei C, Agherbi H, Castelnovo G, Lehmann S, Rigau V, Marin P, Thouvenot E (2024) CD138 as a specific CSF biomarker of multiple sclerosis. Neurol Neuroimmunol Neuroinflamm 11(3):e200230. https://doi.org/10.1212/NXI.0000000000200230
3. van Zalm PW, Ahmed S, Fatou B, Schreiber R, Barnaby O, Boxer A, Zetterberg H, Steen JA, Steen H (2023) Meta-analysis of published cerebrospinal fluid proteomics data identifies and validates metabolic enzyme panel as Alzheimer's disease biomarkers. Cell Rep Med 4(4):101005. https://doi.org/10.1016/j.xcrm.2023.101005
4. Zougman A, Selby PJ, Banks RE (2014) Suspension trapping (STrap) sample preparation method for bottom-up proteomics analysis. Proteomics 14(9):1006. https://doi.org/10.1002/pmic.201300553
5. Haile Mariam M, Eguez RV, Singh H, Bekele S, Ameni G, Pieper R, Yu Y (2018) S-Trap, an ultrafast sample-preparation approach for shotgun. Proteomics 17(9): 2917–2924. https://doi.org/10.1021/acs.jproteome.8b00505
6. Bang G, Lee H, Kim H, Han EH, Park YH, Kim JY (2022) Comparison of protein characterization using in solution and S-Trap digestion methods for proteomics. Biochem Biophys Res Commun 22(589):197–203. https://doi.org/10.1016/j.bbrc.2021.12.026
7. Kitata RB, Yang JC, Chen YJ (2023) Advances in data-independent acquisition mass spectrometry towards comprehensive digital proteome landscape. Mass Spectrom Rev 42(6): 2324–2348. https://doi.org/10.1002/mas.21781
8. Demichev V, Messner CB, Vernardis SI, Lilley KS, Ralser M (2020) DIA-NN: neural networks and interference correction enable deep proteome coverage in high throughput. Nat Methods 17(1):41–44. https://doi.org/10.1038/s41592-019-0638-x
9. Bekker-Jensen DB, Bernhardt OM, Hogrebe A, Martinez-Val A, Verbeke L, Gandhi T, Kelstrup CD, Reiter L, Olsen JV (2020) Rapid and site-specific deep phosphoproteome profiling by data-independent acquisition without the need for spectral libraries. Nat Commun 11(1):787. https://doi.org/10.1038/s41467-020-14609-1
10. Zhang F, Ge W, Huang L, Li D, Liu L, Dong Z, Xu L, Ding X, Zhang C, Sun YAJ, Gao J, Guo TA (2023) A comparative analysis of data analysis tools for data-independent acquisition mass spectrometry. Mol Cell Proteomics 22(9):100623. https://doi.org/10.1016/j.mcpro.2023.100623

Chapter 12

Olink® Explore for High-Throughput Protein Biomarker Discovery in Cerebrospinal Fluid

Mireia Boluda-Navarro

Abstract

The Olink® Explore platform enables high-throughput protein biomarker discovery through Proximity Extension Assay (PEA) technology combined with Next Generation Sequencing (NGS) on Illumina instruments. This approach allows for the simultaneous measurement of thousands of human plasma proteins with minimal sample volumes. The Explore 3072 library offers approximately 3000 protein assays, while the smaller Explore 384-plex panels cater to more targeted studies. The platform excels in detecting low-abundance proteins, such as cytokines and chemokines, and is particularly effective for challenging sample types like cerebrospinal fluid (CSF), where protein content is typically low. In this chapter, we emphasize critical dry-lab considerations, including CSF handling, study design, sample size determination, instrumentation requirements, and post-experiment data management. Proper planning and execution of these factors are essential for optimizing performance and ensuring reliable outcomes when using Olink®'s platform.

Key words Proteomics, Olink, PEA, Proximity Extension Assay, Antibodies, NGS

1 Introduction

Traditional proteomic techniques faced limitations in quality, sample consumption, and the ability to measure a broad range of proteins simultaneously (dynamic range). Olink® Explore technology overcomes these challenges by utilizing Proximity Extension Assay (PEA) [1, 2], a novel approach coupled with next-generation sequencing (NGS) readout on Illumina instruments. This enables the measurement of thousands of human proteins in multiplex assays using miniscule sample amounts.

PEA employs antibodies (Abs) labeled with DNA oligonucleotides to quantify protein levels (see Fig. 1). These labeled Abs bind to their specific target proteins in a solution. The design ensures that when two specific Abs bind to their target protein, their complementary oligonucleotides come in proximity and hybridize to form a unique DNA template that is then amplified during a

Enrique Santamaría and Joaquín Fernández-Irigoyen (eds.), *Cerebrospinal Fluid (CSF) Proteomics: Methods and Protocols*, Methods in Molecular Biology, vol. 2914, https://doi.org/10.1007/978-1-0716-4462-1_12,

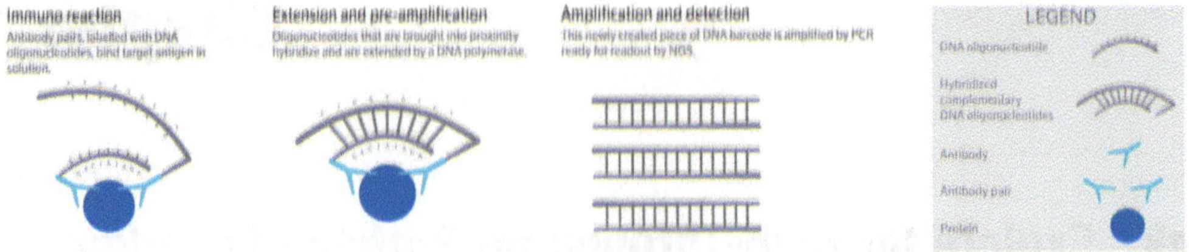

Fig. 1 Olink® PEA technology. PEA is comprised of three core steps: (1) Immune reaction: labeled antibodies pair to the targeted protein. (2) Extension and pre-amplification: oligonucleotides linked to the antibodies hybridize if in close proximity. Then a first PCR extends the hybridized oligos, creating unique protein barcodes. (3) Amplification and detection: a second PCR involves sample indexing the PCR1 extension products with 96 unique sample indexes and further amplification for NGS readout

PCR. This double-stranded oligonucleotide acts as a barcode for the specific protein-sample combination. Only specific binding generates a signal, as non-specific binding will not result in oligonucleotide hybridization. The resulting DNA amplicons are then analyzed using NGS, and protein concentrations are calculated using dedicated software and reported in relative concentration units (NPX).

Olink® offers significant advantages in terms of specificity and sensitivity compared to traditional proteomic techniques like mass spectrometry. The PEA technology's dual-recognition, DNA--coupled readout ensures exceptional specificity even in highly multiplexed assays (e.g., 99.7% of assays exhibit no cross-reactivity). In addition, while traditional methods struggle with proteins below ~10 ng/mL concentrations, Olink® boasts a detection limit (LoD) as low as 0.2 pg/mL and a limit of quantification (LoQ) of 0.4 pg/mL. This translates to a wider dynamic range, allowing the detection of proteins across a vast spectrum (from femtograms/mL to milligrams/mL) within a single assay.

Cytokines and chemokines, crucial components of cellular signaling networks, are present at exceptionally low concentrations in biofluids, making their measurement difficult. Cerebrospinal fluid (CSF) is commonly used for biomarker assessment due to its reliability in reflecting brain activity. However, CSF analysis presents challenges due to its low protein content. The remarkable dynamic range obtained with Olink® has opened doors for analyzing alternative sample matrices like CSF [3].

Additionally, these performance yields are achieved with a minimal sample volume requirement of ≤6 μL per sample. This makes Olink® a powerful tool for researchers working with limited sample volumes, such as those obtained from CSF collections [4–7].

2 Materials

2.1 Reagents

2.1.1 Olink Explore 384 Reagent Kit

This kit consists of one of the Olink Explore 384 panels (Inflammation, Oncology, Cardiometabolic, Neurology, Inflammation II, Oncology II, Cardiometabolic II or Neurology II).

Each kit contains reagents for up to 90 user samples and 6 Olink controls and is intended for the preparation of one single panel library. Each kit comprises the following reagents:

1. Olink® Explore Negative Control.
2. Olink® Explore PCR 1 Solution (x2).
3. Olink® Explore PCR 1 Enzyme.
4. Olink® Explore PCR 2 Solution.
5. Olink® Explore PCR 2 Enzyme.
6. Olink® Explore Index Plate 1.
7. Olink® Target96 / Explore Sample Diluent.
8. Olink® Explore Incubation Solution.
9. Olink® Explore PCR 1 Enhancer (x2).
10. Olink® Explore Plate Control.

2.1.2 Sample Controls

Sample controls are provided by the user and must be plasma, regardless of the sample matrix used in the experiment. Each plate should include duplicate samples to evaluate run-to-run and plate-to-plate variability. These samples are essential for calculating inter-assay and intra-assay coefficients of variation (CVs) and troubleshooting. With Sample Controls included, the system can process up to 88 user samples simultaneously.

2.1.3 Fungible

1. Twin.tec 384-well skirted PCR plate.
2. 96-well PCR plate, preferably with full skirt (*see* **Note 1**).
3. Filter pipette tips.
4. Adhesive PCR films. Make sure the films can be used in wide temperature ranges including −80 °C to safeguard samples during transport or storage (*see* **Note 2**).

   Marker pen.

2.2 Instrumentation

Olink® follows a semi-automated workflow through several instrument combinations, which can be divided into pre-PCR and post-PCR instruments.

2.2.1 Pre-PCR

1. Dragonfly®—SPT Labtech.
2. Mosquito®—SPT Labtech. †
3. F.A.S.T.™—Formulatrix®. †

2.2.2 Post-PCR

1. Microlab STAR—Hamilton. †
2. epMotion®—Eppendorf. †
3. F.A.S.T.™—Formulatrix®. †

† Note: either of these instruments can be used.

2.2.3 Other Instruments

1. Either a 2100 Bioanalyzer (Agilent) or a 4200 TapeStation (Agilent).
2. 384-well thermocycler.
3. Vortex (able to reach 3000 rpm).
4. 96-well plate centrifuge.
5. Pipettes (0.5–10 μL).

3 Methods

3.1 CSF Sample Handling

This section outlines the essential steps for collecting, preparing, and normalizing CSF samples for proteomic analysis. Following standardized protocols is crucial to ensure reliable and reproducible data.

3.1.1 CSF Collection and Preparation

- *Collection:* CSF should be collected using established clinical guidelines to minimize pre-analytical variability. Refer to [8] for detailed considerations on standardized collection, including:
  - Volume of withdrawal.
  - Location of puncture.
  - Use of polypropylene collection tubes.
  - Time of day of withdrawal.
- *Storage and processing:*
  - Centrifuge samples at ≥500× *g* for 10 minutes to remove cells and debris.
  - Freshly collected protein from CSF has limited stability at room temperature. Protein degradation can be reduced by keeping samples cold on ice during handling and transportation.
  - Samples must be delivered on dry ice to minimize the degradation of proteins.
  - The long-term storage temperature for samples should be −80 °C or lower.
  - Samples should be aliquoted on small polypropylene tubes from the same withdrawal to reduce multiple freeze-thaw

cycles that affect protein concentration, especially those represented on a smaller scale. *See* **Note 3** for further understanding of sample usability that has undergone several freeze-thaw cycles.

- *Blood contamination:* including samples with apparent blood contamination is discouraged due to:
  - Blood-derived proteases can lead to protein degradation.
  - Heme can interfere with downstream applications like PCR.
  - Blood proteins can introduce confounding factors in CSF proteome analysis.

3.1.2 CSF Sample Normalization

Standardized protocols for CSF samples are crucial to ensure statistically robust results. Increasing sample size enhances statistical power, but this can be undermined by pre-analytical variability. Sample normalization is a critical step to mitigate this variability and isolate the effects of biological events on protein concentrations.

Addressing Pre-analytical Variability

Sample normalization refers to adjusting sample volume or concentration before data acquisition (pre-acquisition) and/or adjusting acquired signals afterward (post-acquisition) to ensure comparable total protein signals across samples. Here, we focus on pre-acquisition normalization, with post-acquisition normalization methods discussed in Subheading 3.4.

Due to the lack of a universal reference compound that reflects stability concentration in CSF [9], pre-acquisition normalization relies on a standardized volume withdrawal protocol.

Total protein levels can vary significantly between CSF samples, potentially masking true biological differences in individual proteins. Normalizing to a set volume minimizes the impact of sample quantity variation on protein quantification. This ensures that biologically relevant differences, such as elevated cytokine levels in patients with certain diseases, are accurately detected. Adhering to established protocols, like the volume withdrawal method described by Teunissen et al. (2009) [8], is crucial for achieving this normalization.

3.2 Study Design

3.2.1 General Considerations

Study design should consider all procedures involved, from initial planning through sample collection and analysis, to minimize factors that distort or bias results. Researchers should contemplate the following questions to ensure a streamlined experimental path:

- What is the study objective?
- What are the study groups?
- How many samples are needed?

- What is the sample type?
- What are the pre-analytical variations?
- How can data be normalized?
- What is the appropriate statistical analysis?

3.2.2 Selecting the Optimal Olink® Explore Platform for Your Research

The Olink® Explore platform offers two options for protein analysis: Explore 3072 and Explore 384-plex panels. Choosing the right platform depends on your research goals and sample availability.

- *Olink® Explore 3072.* This comprehensive platform offers a protein library of approximately 3000 human proteins. It is ideal for exploratory proteomics and multiomics studies where a broad protein landscape is desired. Explore 3072 covers metabolic pathways in various areas like cardiology, oncology, neurology, and immunology.
- *Olink® Explore 384-plex panels.* This option provides eight focused panels, each measuring 384 proteins relevant to specific research areas like cardiometabolic diseases, inflammation, oncology, and neurology. These panels are suitable for researchers with more defined protein targets.

Both Explore options offer advantages for analyzing the less abundant proteome while requiring minimal biological sample volume. However, a crucial factor to consider when choosing between them is the power of the study desired. *See* **Note 4**.

The protocols for Olink® Explore can be accessed on the Olink® website under the "Knowledge" section by navigating to "Documents" and applying the filter for "Manual & Instruction." Additionally, refer to **Notes 5** and **6** for guidance on experiment handling to prevent potential downstream issues.

3.2.3 Study Design Scenarios

There are four main study design approaches:

- *Small study single plate.* This simple yet powerful approach typically involves 44 cases and 44 controls analyzed on a single 96-well plate.
- *Large study multi-plate.* These studies are performed with a large sample size at a single timepoint. The larger sample size maximizes statistical power, making it ideal for detecting even subtle effects (low effect size).
- *Continuous study multi-batch.* These are for batch studies in which samples are prepared at different time points. It is useful for exploring non-standardized samples (e.g., not serum/plasma) where a pilot experiment is performed to guide further experimental analysis and design.

- *Longitudinal study multi-timepoint.* Designed to study multiple timepoints per patient. Ideal for studying processes or responses over an extended period.

3.2.4 Power Analysis

Power refers to the probability of correctly detecting a true difference between groups. Higher power reduces the risk of Type II errors (false negative result). Typically, a power level of 0.8 (80%) is employed in research studies. Power provides a framework for determining the minimum sample size required to achieve a desired level of sensitivity, reducing false-negative results. The main factors affecting power analysis include:

- *Sample size.* The sample size is the number of samples in each group included in the study and it is often the variable we want to estimate. However, in cases where it is not possible to increase the number of samples, a power analysis can be used to determine the smallest effect size at which the proposed study will have a reasonably high power. Increasing the sample size will always increase the power of a study.
- *Spread.* Spread of data, commonly measured as the standard deviation, describes how similar observed data points are and must be estimated in order to perform a power calculation. If the standard deviation is high for a variable, a larger sample size is needed to reach a certain power compared to a variable with a low standard deviation. A study may have multiple sources of variation, and the spread may therefore be hard to estimate. If no good estimate is available, it may still be worth performing a power analysis with both low and high variability levels to see what difference it makes to the calculated sample size or power.
- *Effect size (ES):* ES reflects the magnitude of the expected difference between study groups. It is a standardized measure calculated based on the difference between means divided by the pooled standard deviations of two groups. A larger ES indicates less overlap between groups, requiring a smaller sample size to detect the difference. Conversely, a smaller ES needs a larger sample size to achieve the same level of power. A common method for calculating effect size is Cohen's d, which is obtained by dividing the difference between the means of two samples by their pooled standard deviations.

$$\text{Cohen's } d = (\text{Mean}_2 - \text{Mean}_1)/\text{Standard Deviation}_{\text{pooled}}$$

An effect size can be standardized, and it contains information about the variability in the measurements. Cohen's d values of 0.2, 0.5, and 0.8 represent small, medium, and large effect sizes, respectively. Because Cohen's d is the difference in means relative to the pooled variance, regardless of sample size, it doesn't change with the sample size.

- *Significance level (α)*. The significance level is the probability of a false positive result and is commonly set to 0.05 (*p*-value) if one statistical test is carried out.

Of the four variables discussed (spread, effect size, sample size and significance level), effect size has by far the biggest influence on power, but sample size is most often the only factor we can control to affect the power of the study.

Since Olink® technology was fully evaluated on serum and plasma samples, for CSF samples, pilot studies should be considered during the experiment design. *See* **Note 7**.

3.2.5 Olink Insight: a Statistical Analysis Tool

Olink® Explore experiments benefit from Olink Insight, a free statistical online tool that incorporates a "Study size calculator" (*see* **Note 8**). This tool helps you determine the optimal sample size based on desired power and ES, without needing data upload. The utilization of this tool can vary based on the availability of samples:

- *Planning prior to data collection:* determine the sample size needed, setting your desired power and effect size parameters. Use these parameters to calculate the necessary number of samples for the specific number of groups to be compared.
- *Using pre-collected samples:* perform a power calculation to evaluate statistical significance using an existing group of samples. If your groups are unequal, use the smallest sample size among your groups.
- *Using previous knowledge of effect size:* utilize prior research or pilot study data to inform sample size calculations. Incorporate the known ES parameter and an estimate of population dispersion to accurately calculate the required sample size.

The software then calculates the minimum sample size per group to achieve your desired power level (Fig. 2).

3.2.6 Sample Selection Considerations: Sample Groups

When selecting sample groups, several crucial factors must be considered to ensure the validity and reliability of the study. The following considerations are essential:

- *Define controls and cases*: clearly distinguish between control groups and case groups. This step is fundamental for comparative analysis.
- *Selection of control group samples*: while control groups often comprise healthy individuals, this choice may not always align with the study's objectives. The selection of controls should be tailored to the specific research question.
- *Demographic variables*: consideration of ethnicity, age, and sex differences is vital as these factors can introduce bias into the study results. Careful matching or stratification may be required to mitigate such biases.

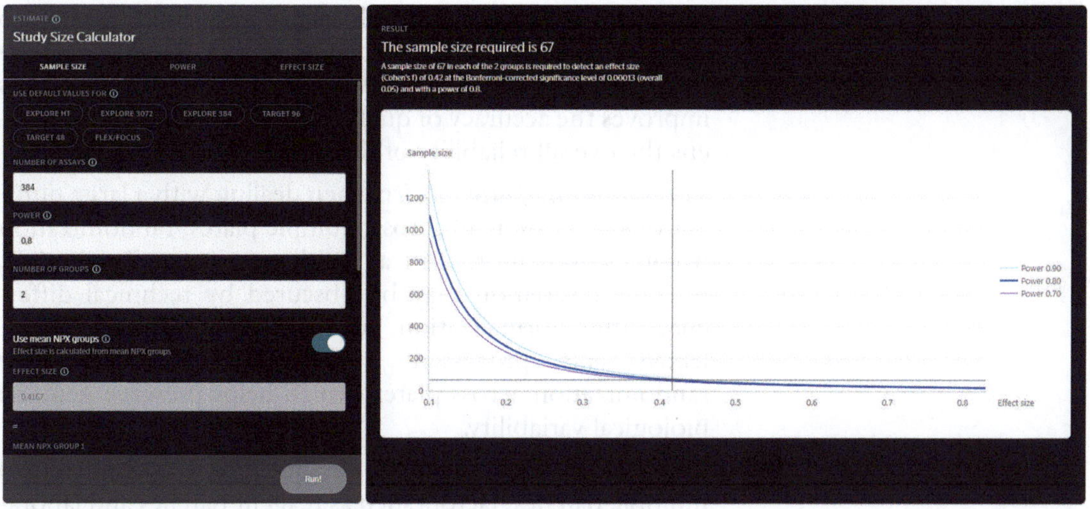

Fig. 2 Study size calculator snapshot from Olink® Insight platform

- *Drug treatment and pathophysiological parameters*: consider any drug treatments or underlying pathophysiological conditions that may affect the study outcomes. This ensures that the groups are comparable and that confounding variables are controlled.

These considerations aid in identifying which parameters need to be controlled to establish comparable groups for analysis.

3.2.7 Avoiding/Minimizing Experimental Variables

Minimizing experimental variability is crucial for accurately identifying true biological differences between study groups. The following practices are recommended:

1. *Pre-analytical variability.*

Reduce technical variability to enhance the detection of genuine biological differences. This includes using the same collection procedure for each sample, ensuring uniformity in the number of freeze/thaw cycles for all samples and maintaining even storage conditions for all samples.

Ensure all samples originate from the same matrix. Comparing samples from different matrices can introduce interference or bias due to matrix effects and varying dilution needs.

2. *Randomization.*

   Randomization plays a crucial role in any Olink® experiment to eliminate systematic bias and ensure that samples are representatively distributed across experimental conditions. It is essential to incorporate randomization into every study design, with specific strategies depending on the experiment's scale:

- *Single-plate studies*: randomly assigning sample groups to wells helps eliminate any inherent row or column associations, which might otherwise skew the results. This improves the accuracy of quality control (QC) and strengthens the overall reliability of the data.
- *Large or multi-plate studies*: when dealing with a large number of samples spread across multiple plates, randomization becomes even more critical. Without proper distribution, biological variation may be obscured by technical differences after normalization. Ideally, samples should be collected and processed simultaneously, allowing for randomization across plates, which helps preserve genuine biological variability.
- *Continuous, multi-batch studies*: for studies conducted in multiple batches, factors such as reagent batches and laboratory conditions can affect reproducibility. To account for this, it is recommended to include bridging samples, which are run in every batch of the study. These samples serve as a consistent reference across batches, allowing normalization and correction for any batch-to-batch variability. For details on selecting bridging samples, refer to **Note 9**.
- *Longitudinal studies*: in longitudinal studies, the goal is to track changes in protein expression over time within the same individual, as well as to compare time points across different groups. To minimize technical variation, all time points for a given subject should ideally be placed on the same plate. Furthermore, randomizing samples both within and between plates helps reduce bias and ensure robust results.

Olink® offers an R package to facilitate randomization. Refer to **Note 10** for further information.

3. *Pre-dilution strategies.*

A critical concept in protein analysis is the high-dose hook effect. This occurs when there is an excess of antigen relative to the antibody probes, leading to falsely lowered protein values (Fig. 3). If not addressed, this can result in the misinterpretation of results. Each analyte has a specific hook threshold, which is reported in pg/mL.

Each Olink panel is optimized for the expected physiological and pathological ranges found in plasma and serum for the 92 biomarker assays included. Dilution factors are set with plasma and serum in mind, so when running other sample matrices, these dilutions may need adjustment depending on the expected protein concentrations in the specific matrix. *See* **Note 7** for specific recommendations for CSF samples.

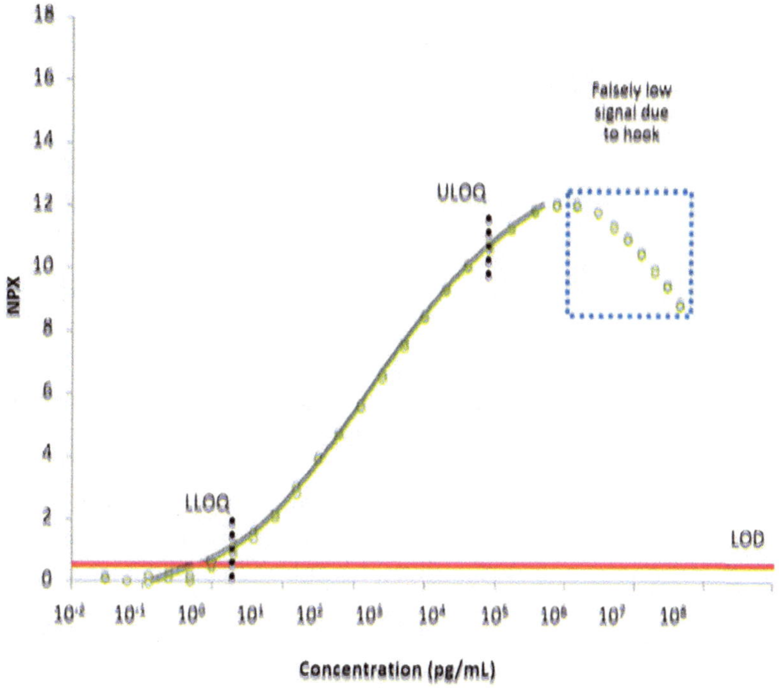

Fig. 3 High-dose hook effect representation

3.3 Olink's Quality Control System for Multiplex Biomarker Panels

Olink has developed an integrated quality control (QC) system, incorporating both internal and external controls, for its multiplex biomarker panels. This system enables comprehensive monitoring of assay performance and sample quality throughout each step of the Olink protocol.

3.3.1 Quality Controls

Internal Controls

Olink's internal controls are added to each well and sample at the same concentration, ensuring consistent signals across the entire plate. These controls are designed to monitor different stages of the assay and include:

- *Immuno control*: a protein targeting a non-human antigen and monitors all steps starting from the immune reaction.
- *Extension control*: an antibody linked to two matched oligonucleotides that monitors the extension and readout phases. This control is used for data normalization across samples.
- *Amplification control*: a synthetic double-stranded template used to track the amplification and readout steps.

These internal controls are critical for assessing both individual sample quality and overall plate performance.

Control levels for each sample are compared to the plate median, and if deviations exceed ±0.3 NPX, a QC warning is generated by the NPX processing software. Excessive variation in

Fig. 4 Sample plate layout. Sample controls in yellow, negative controls in red and plate controls in green

these controls can result in the rejection of the entire plate, signaling potential issues such as pipetting errors or pre-analytical sample variations, including matrix effects.

External Controls

Each 96-sample plate contains eight dedicated control samples located in the last column, known as the "Strip control" (Fig. 4). These external controls include:

- *Sample controls (SCs)*: a pooled plasma sample provided by the user, tested in duplicate. These controls are used to evaluate precision, enabling the calculation of the coefficient of variation (CV) across different plates/panels. For more details on how SCs aid in troubleshooting, refer to **Note 11**.
- *Negative controls*: negative controls are run in triplicate and consist of buffer treated as normal samples. They monitor background noise resulting from the proximity of DNA tags without protein binding. Hence, negative controls are used to calculate the limit of detection (LOD) of each assay in every plate.
- *Plate controls (PC)*: a pooled plasma sample run in triplicate, designed to produce high signals across all assays, serving as a positive control and ensuring reagent quality. The median signal from the triplicates is also used to normalize the assays, helping to compensate for potential variation both within and between runs.

3.3.2 Limit of Detection

As with any molecular technique, Olink® assays have background noise due to interactions between free probes. The LOD for each protein is defined as three standard deviations above this background, reported in pg/mL.

Although the NPX™ Explore HT & 3072 and NPX™ Explore CLI HT & 3072 software do not automatically report LOD, it can be calculated using the Olink Analyze R package (*see* **Note 12**).

LOD is crucial for technical evaluations such as CV calculations, which should be based on data above the LOD threshold.

However, for downstream statistical analyses, including data below the LOD is recommended. This is because proteins below the LOD may still provide valuable insights, particularly in biomarker discovery, where differential expression between groups can highlight strong candidates, even if LOD filtering would exclude such data.

3.3.3 Measuring Ranges

The analytical measuring range, or dynamic range, is defined by the lower limit of quantification (LLOQ) and the upper limit of quantification (ULOQ). These limits are calculated based on relative error, ensuring that all assays fall within this range through sample dilution.

3.3.4 Precision

Repeatability (intra-assay variation) and reproducibility (inter-assay variation) are assessed by calculating the CV from the sample controls.

During Olink's validation process, normal CV ranges were established using six individual samples across seven separate runs for analytes with detectable signals in serum and plasma. However, for alternative matrices such as CSF, panels containing proteins that are less abundant may present higher CVs than anticipated.

Given that Olink's limits are based on plasma and serum, it is advisable to incorporate an additional strategy when designing studies with alternative matrices. For guidance on this, refer to **Note 7**, which discusses the considerations for sample dilution.

3.4 Protein Concentration Calculation: Normalized Protein Expression (NPX)

Olink® Explore panels use Normalized Protein Expression (NPX), a relative quantification unit of protein concentration expressed on a log2 scale.

3.4.1 NPX Calculation

NPX is calculated from raw data adjusted against the Extension Control per sample to reduce technical variation introduced in the extension step. This yields the intermediate NPX (ExtNPX) value, calculated as:

$$\text{ExtNPX } i, j = \log 2\left(\frac{\text{counts(sample } j, \text{assay } i)}{\text{counts(ExtCtr } j)}\right)$$

Here, i refers to the assay and j refers to the sample.

3.4.2 NPX Normalization

When analyzing protein concentration changes, it is essential to control for non-biological variability. Normalizing samples before conducting the analysis helps isolate the biological signal, ensuring that any observed metabolite concentration changes reflect true biological events rather than technical artifacts.

3.4.3 NPX Normalization for Single-Plate Studies

After NPX calculation, an additional normalization step is required to reduce inter-assay variability. ExtNPX values are normalized against the plate controls for each assay and plate.

$$\text{NPX } i,j = \text{ExtNPX}i,j - \text{median}(\text{ExtNPX}(\text{PC } i))$$

where i refers to the assay, j refers to the sample and PC to plate controls.

Single-plate normalization can also be performed using "Intensity normalization," as outlined below, provided the samples have been properly randomized. Refer to **Note 13** for guidance on selecting the most appropriate approach for CSF samples.

3.4.4 NPX Normalization for Multi-plate Studies

For multi-plate studies, PC normalization should be replaced with either intensity normalization or reference sample normalization. When applied correctly, both strategies increase the power in a given study by reducing technical variation.

Intensity Normalization

For randomized studies, intensity normalization should be applied to increase statistical power and reduce technical variation between plates and projects. This method adjusts for differences by setting the median level of all assays to the same value across plates (Fig. 5).

The assumption behind intensity normalization is that, on average, there is no expected difference between the median signal for an assay on one plate compared to another as long as samples have been previously randomized. If any such difference is seen between plates, it can be construed as technical bias and be safely removed, resulting in more comparable values. Intensity normalization can be summarized in the following formula:

$$\begin{aligned}&\text{IntNorm NPX}i,j\\&\quad = (\text{ExtNPX}i,j - \text{plate median}(\text{NPX allsamples } i))\end{aligned}$$

This step excludes external controls and helps standardize protein levels across large datasets, making results comparable. NPX™ Explore software can perform intensity normalization automatically.

Reference Sample Normalization

For non-randomized studies (e.g., continuous, multi-batch studies), using reference sample on each plate can be used for normalization. This method involves adjusting data from different assay plates to ensure consistency across experiments. This is done by selecting a reference plate and calculating the differences between overlapping samples on that plate and others. A normalization factor is then derived for each plate and assay based on these

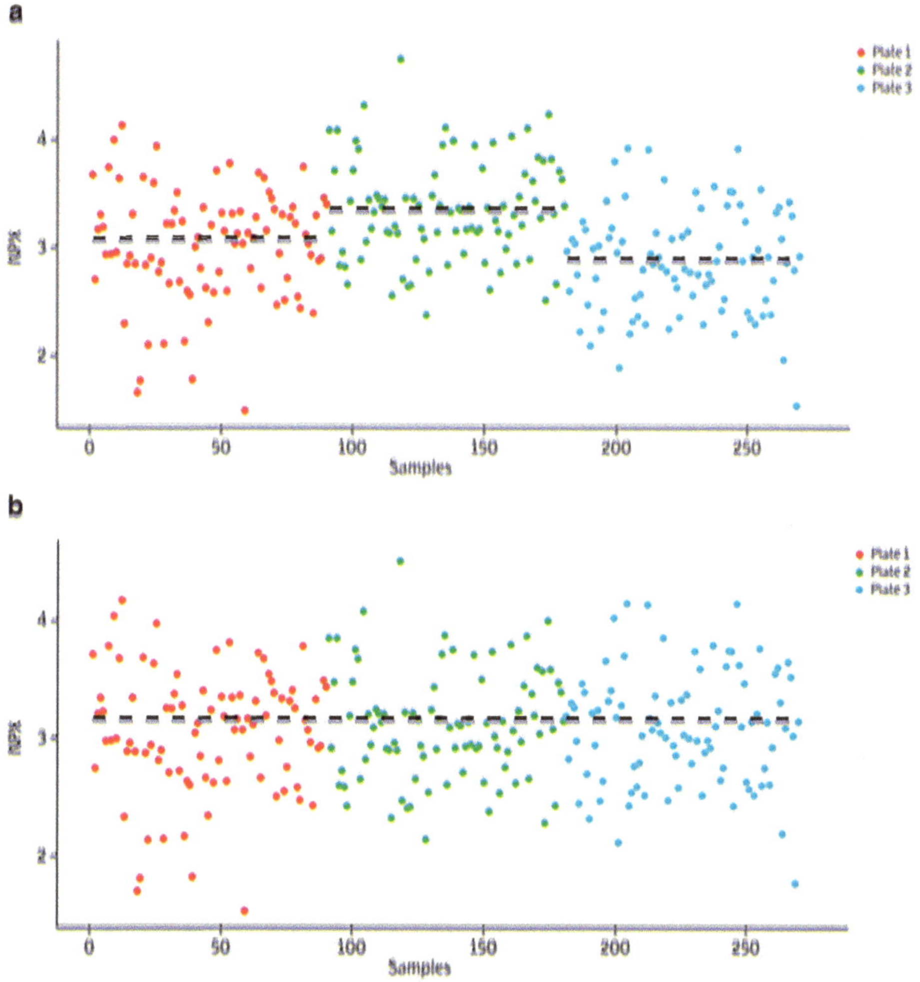

Fig. 5 NPX across three plates for each protein before (**a**) and after (**b**) intensity normalization. The analysis shows that after intensity normalization the median NPX is leveled

differences. Finally, the values for each assay and plate are adjusted using this factor, aligning them to the reference plate, which minimizes technical variation (Fig. 6).

Eight or more bridging samples are recommended to conduct reference normalization. For additional guidance on how to select bridging samples, *see* **Note 9**.

NPX™ Explore software does not allow to perform reference sample normalization. This should be done by using the Olink Analyze R package. Refer to **Note 10** for further information.

3.5 Exploratory Data Analysis: Assessing the Overall Shape of the Data

Once the data has been cleaned up, the next step involves looking at the overall shape of the data and specific data points, with the aim of assessing whether the data should be used as is, or whether editing is needed. Some of the most common ways to do this include using

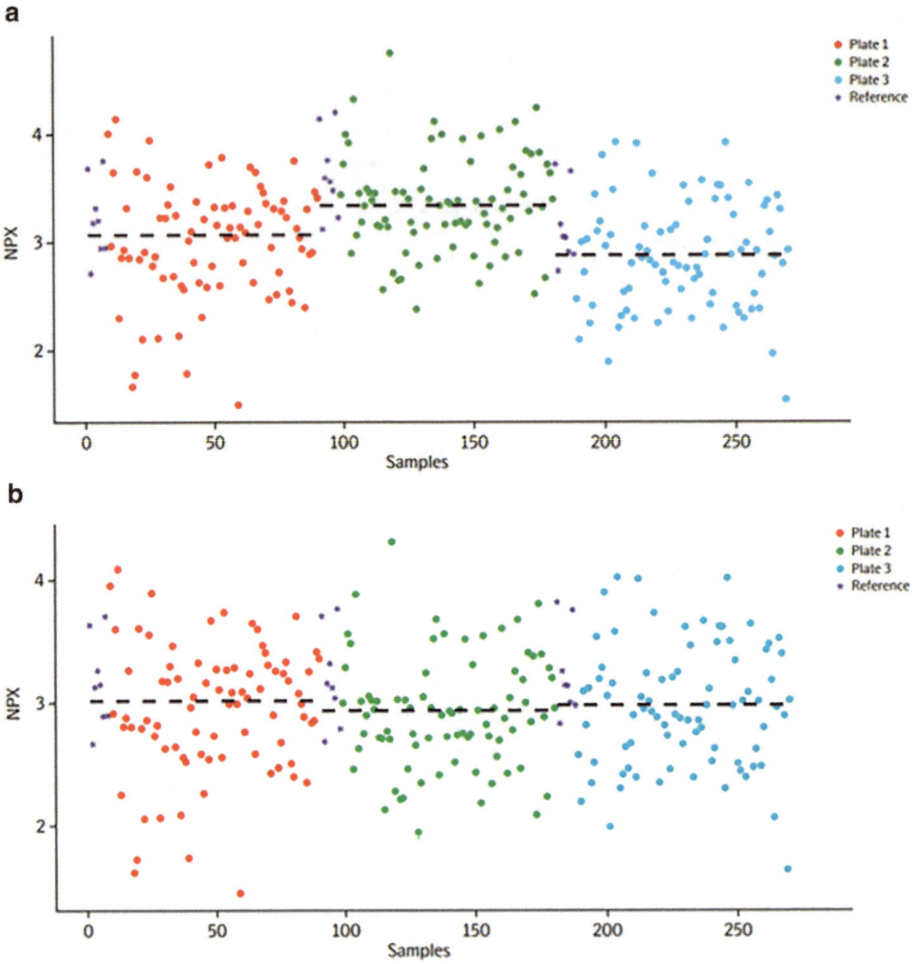

Fig. 6 NPX across three plates for each protein before (**a**) and after (**b**) reference sample normalization. After reference sample normalization, the median NPX improves but can still differ between plates, reflecting different compositions of sample groups

principal component analysis (PCA) plots and looking for outliers/odd, non-normally distributed data, other unusual configurations, or samples that are not within standard ranges.

Olink® Insight is an intuitive platform designed for exploratory data analysis. Figures 7, 8, 9, and 10 showcase visualizations generated from a sample dataset comparing two groups (treated vs. untreated), providing insights into the results obtained through Olink® Insight.

3.6 Statistical Analysis of NPX Data

Analyzing NPX data follows standard statistical approaches, as the log2 scale of NPX is directly comparable to linear data in most tests. Absolute protein concentrations for different proteins cannot be inferred from NPX. Comparisons should always be assay-specific and limited to samples within the same run or that has undergone

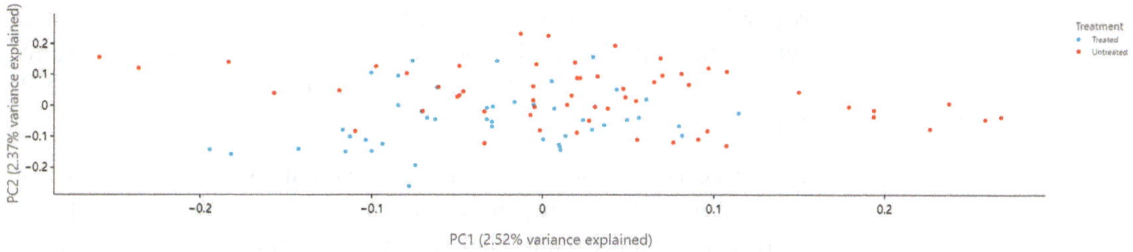

Fig. 7 PCA analysis. Scatterplot of 103 samples along two principal components

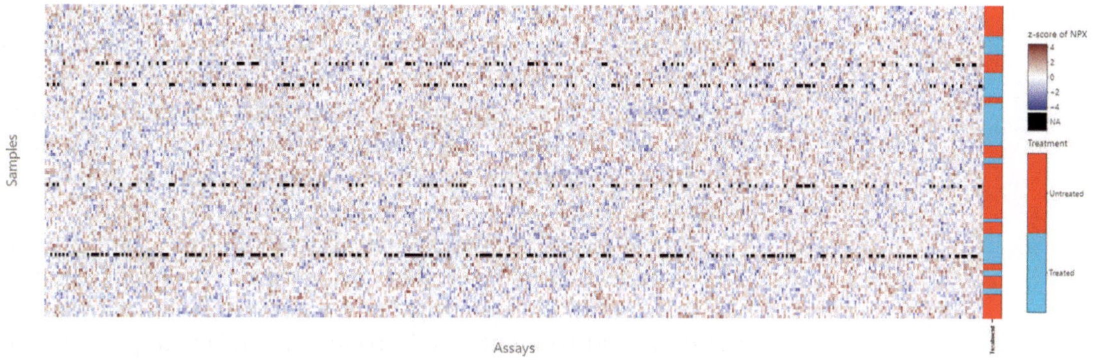

Fig. 8 This heatmap provides an overview of the protein level measurements for all 103 samples (rows) and 460 assays (columns) from all panels

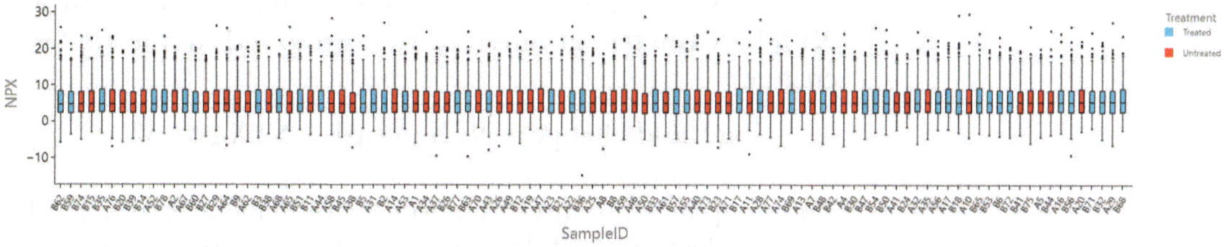

Fig. 9 The boxplots summarize the distributions of protein level in terms of median, first and third quartiles, upper and lower fences, and data points falling outside the fences

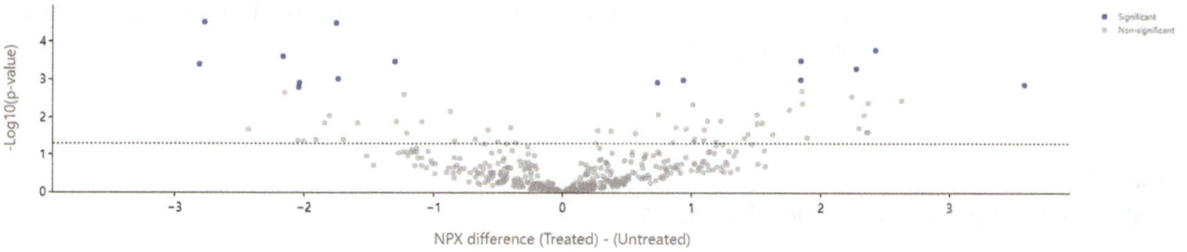

Fig. 10 Volcano plot comparing "Treated" to "Untreated." There are 15 out of 460 assays that have a statistically significant difference in means between the two groups

proper inter-plate normalization to avoid the risk of falsely interpreting shifts in the median between runs as a biological difference.

Below are key points for performing statistical analyses on NPX values:

1. *Statistical assumptions*: the same statistical methods and tests used for linear data, such as t-tests or ANOVA, can be applied to NPX data. Ensure that the assumptions of the statistical test (e.g., normality, variance homogeneity) are met.
2. *Comparison of groups*: To compare protein levels between two groups (e.g., treated vs. untreated, diseased vs. healthy), the mean NPX values can be compared using statistical tests like a t-test. For example, a t-test can assess whether differences in mean NPX between groups are statistically significant. Adjustments for multiple comparisons, such as the Benjamini-Hochberg correction, are critical when analyzing multiple proteins simultaneously to control the false discovery rate.

   Linear mixed models can be used to study patient protein level changes over time that are different between groups, while hierarchical clustering can be used to get an overview of the data, and to identify subgroups of similar samples or proteins based on protein profiles.
3. *Fold change*: Differences between groups are often expressed as fold changes, defined as the ratio of protein levels between two groups. Since NPX is already in log2 format, the difference in NPX between two groups directly corresponds to the log2 fold change

$$\text{log2 fold change} = \text{NPXgroup1} - \text{NPXgroup2}$$

This simplifies the interpretation of fold changes, as a 1 NPX difference reflects a two-fold change in protein concentration.

As an example for interpretation, Table 1 shows the results of a statistical test (e.g., t-test) on NPX values, showing the difference in mean NPX between two groups. *P*-values are adjusted for multiple testing using the Benjamini-Hochberg method. The interpretation of those results is as follows:

Table 1
Statistical example of a t-test on NPX values

Protein	Group 1 mean NPX	Group 2 mean NPX	Difference	*p*-value	Adjusted *p*-value
A	8.4	7.2	1.2	0.01	0.05
B	6.9	7.1	−0.2	0.20	0.50

- *Group 1 Mean NPX and Group 2 Mean NPX*: These columns show the average NPX values for each protein in the two groups (e.g., treated vs. untreated).
- *Difference:* This is the difference between the mean NPX values of the two groups. For Protein A, the difference is 1.2, meaning that Group 1 has a higher mean NPX value by 1.2 units compared to Group 2. Since NPX is in log2 scale, this corresponds to a 2^1.2 (roughly 2.3-fold) higher concentration of Protein A in Group 1. For Protein B, the difference is −0.2, meaning Group 2 has a slightly higher NPX value, though the difference is small.
- *p-value:* This indicates whether the difference in NPX between the two groups is statistically significant. For Protein A, the *p*-value is 0.01, meaning there is a 1% chance that the observed difference occurred by random chance, indicating the difference is statistically significant (typically, *p*-values <0.05 are considered significant). For Protein B, the *p*-value is 0.20, indicating no significant difference between the groups.
- *Adjusted p-value:* This corrects for multiple comparisons, reducing the risk of false positives. The Benjamini-Hochberg correction is used here. For Protein A, the adjusted *p*-value is 0.05, which still suggests the result is statistically significant. For Protein B, the adjusted *p*-value is 0.50, confirming the lack of significant difference after adjustment.

In this example, Protein A shows a statistically significant difference in concentration between Group 1 and Group 2, with Group 1 having a higher concentration (1.2 NPX difference, roughly a 2.3-fold increase). In contrast, Protein B shows no significant difference between the two groups.

3.7 Biological Interpretation

After completing data analysis, deeper insights can be gained by examining the biological relevance of the identified proteins. One approach is to explore how differentially expressed proteins are connected to specific diseases or pathways of interest. This process can be enhanced by leveraging databases that offer data-driven and literature-based annotations, providing information on pathway involvement and disease-related biomarkers, along with rankings or scores that highlight key connections.

Alternatively, researchers can explore pathway enrichment in a more quantitative manner, especially when looking to discover new pathways or generate novel hypotheses. Common methods for this include Over-Representation Analysis (ORA) and Gene Set Enrichment Analysis (GSEA). However, pathways should always be interpreted within the context of the specific data source; for example, if an inflammation-focused panel was used, the pathway analysis would be limited to proteins associated with inflammation.

Olink® Insight adds further depth to this interpretation by offering multiple layers of biological context, including tissue specificity, drug target data, biological processes, pathway information, and relevant publications. These annotations enhance the understanding of the biological significance of the findings.

4 Notes

1. We recommend MicrAmp® Fast Optical 96-well Reaction Plate Barcode (0.1 mL) from Applied biosystems.
2. We recommend using AXYGEN® Cyclear sealing film (ref: PCR-TS). Based on our experience, other films have proven less effective during shipping on dry ice, leading to the plate arriving slightly detached, which resulted in cross-contamination.
3. Long-term storage is an important pre-analytical factor to consider, as freeze-thaw cycles can directly affect the expression levels of certain proteins. Therefore, if samples have been exposed to varying numbers of freeze-thaw cycles, this should be accounted for as a variable in downstream analysis. However, relative differences between groups remain consistent for samples that have undergone the same number of freeze-thaw cycles. Even samples with multiple freeze-thaw events can still be processed with high-quality results.
4. For studies with limited sample availability, Explore 384 panels offer a more focused approach, reducing the risk of inflated type I errors due to multiple testing.

The rationale behind this comes from the fact that Olink is a multiplex test, meaning that more than one comparison takes place. All 92 proteins in Olink's panels are compared; this means that with a larger number of proteins being compared, the likelihood of random chance (type I error) leading to statistically significant results increases. To mitigate this risk of false positives in multiple hypothesis testing, the Bonferroni correction is employed. The Bonferroni correction adjusts the standard *p*-value based on the total number of tests performed:

$$\text{adjusted } p - \text{value} = p - \text{value}/n \text{ of tests}$$

Hence, the more assays tested, the more stringent the adjusted *p*-value becomes. This means statistically significant results become harder to find, potentially masking true differences. Increasing the number of tests (comparing more proteins) will require running more samples to maintain the same level of power. Considering the statistical implications, the most suitable Explore platform should be used based on the total sample size.

5. During Olink®'s protocol, there is an overnight incubation step that must be performed at 4 °C. It is important to monitor any temperature fluctuations in the fridge. To minimize disruptions, choose a fridge that is not frequently accessed during the incubation period. If possible, place the plate in the center of the fridge, as positioning it near the walls may expose it to temperature variations.
6. We recommend using the same thermocycler for PCRs to avoid technical variations. If the plate is difficult to fit, align it starting from one side. Proper positioning is critical to prevent uneven temperature distribution across wells, which could lead to technical errors and affect results.
7. CSF is a challenging sample matrix due to the direct influence of patient conditions on protein concentrations. For matrices other than serum and plasma, it is advisable to evaluate protein assays for susceptibility to the hook effect.

   To ensure that the proteins are measured in the optimal range for each assay, it is recommended to run pilot studies with few samples in three different starting concentrations (i.e., dilutions, commonly 1:2, 1:4 and 1:16). Power analysis can then be conducted, and such retrospective calculations may be useful when designing a scaled-up study.

   However, running a small pilot study might not be an option due to resource constraints. In this case, study samples can be run in the same plate with 6 samples at different starting concentrations. The buffer used for sample elution should also be included in the experiment in triplicate. If the results are unexpected, analyzing the buffers can help identify potential interfering reagents and provide valuable insights for further data analysis.

   Including technical replicates can further improve the estimation of CVs when using alternative matrices.
8. Olink® Insight is a platform for protein biomarker discovery. Calculation of statistical power for study design, overall shape evaluation of the data and biological interpretation can be performed in this platform. Available at: https://insight.olink.com/.
9. Bridging samples, which are used for reference sample normalization, are run across plates from different batches in continuous studies. It is recommended to select samples from the initial experiment as the reference and use them as bridging samples.

   The Olink Analyze R package helps in selecting suitable bridging samples by identifying those that passed quality control, have high detectability, and cover the dynamic range of the assays.

Olink advises using at least 8–16 bridging samples, ensuring a sufficient supply throughout the study. The selected samples should represent both cases and controls within the dataset, and where possible, ensure an adequate number of kits are available.

10. Olink® Analyze (R package) is a platform that provides functions ranging from randomization R package, handling Olink data including QC, various statistical tests and data visualization available at: https://github.com/Olink-Proteomics/OlinkRPackage. All information regarding the capabilities of this platform can be found at https://olink.com/software/olink-analyze#analyze.
11. SCs help troubleshoot by ruling out technical issues during protocol preparation. If the SCs pass all quality control checks but the rest of the samples fail, it indicates that the reagents were functioning properly, and the issue likely lies with the samples themselves. Although SC are plasma-based, they serve the same purpose when troubleshooting problems in samples from other matrices, such as CSF.
12. Using the function "olink_lod" Olink® Analyze (R package), users can calculate LOD for Explore 384, 3072 or HT based on:

- Negative Controls within the dataset: The Olink data must contain at least 10 Negative Controls to use this functionality.
- Or fixed LOD values: These values are provided by the user in an external CSV file, which can be downloaded from the olink.com website.

To assist users, there is a tutorial within the package that explains how to use the olink_lod function and provides additional context for LOD in Olink data. The tutorial, titled "Calculating LOD from Olink® Explore data," can be found on CRAN under the Vignettes section.

13. Randomization is crucial for obtaining accurate results when using Olink® technology. For studies involving a single plate that has been properly randomized, we recommend using intensity normalization rather than PC normalization for two key reasons. First, normalizing with the mean of a larger number of samples increases statistical power. Second, PC samples are synthetic pooled plasma, while the other samples may come from a different matrix, such as CSF. Intensity normalization ensures the use of real samples from the matrix of interest for normalization.

References

1. Assarsson E et al (2014) Homogenous 96-plex PEA immunoassay exhibiting high sensitivity, specificity, and excellent scalability. PLoS One 9(4). https://doi.org/10.1371/JOURNAL.PONE.0095192
2. Wik L et al (2021) Proximity extension assay in combination with next-generation sequencing for high-throughput proteome-wide analysis. Mol Cell Proteomics 20. https://doi.org/10.1016/J.MCPRO.2021.100168
3. Hansson O et al (2023) The genetic regulation of protein expression in cerebrospinal fluid. EMBO Mol Med 15(1). https://doi.org/10.15252/EMMM.202216359
4. Remsik J et al (2021) Inflammatory leptomeningeal cytokines mediate COVID-19 neurologic symptoms in cancer patients. Cancer Cell 39(2):276–283.e3. https://doi.org/10.1016/J.CCELL.2021.01.007
5. Göteson A et al (2021) Cerebrospinal fluid proteomics targeted for central nervous system processes in bipolar disorder. Mol Psychiatry 26(12):7446–7453. https://doi.org/10.1038/S41380-021-01236-5
6. Wang H et al (2022) Evaluation of neurofilament light chain as a biomarker of neurodegeneration in X-linked childhood cerebral adrenoleukodystrophy. Cells 11(5). https://doi.org/10.3390/CELLS11050913
7. Campbell K et al (2023) Identification of cerebral spinal fluid protein biomarkers in Niemann-Pick disease, type C1. Biomark Res 11(1). https://doi.org/10.1186/S40364-023-00448-X
8. Teunissen CE et al (2009) A consensus protocol for the standardization of cerebrospinal fluid collection and biobanking Supplemental data at www.neurology.org VIEWS & REVIEWS, Accessed: June 16, 2024. [Online]. Available: www.neurology.org
9. Wu Y, Li L (2016) Sample normalization methods in quantitative metabolomics. Elsevier B.V. https://doi.org/10.1016/j.chroma.2015.12.007

Chapter 13

High Throughput Proteomics Using the SomaLogic Platform

Hanjun Zhao and Laura Winchester

Abstract

There are two potential approaches for high-throughput protein measurement: affinity-based approaches with targeted assays and the unbiased approach used in mass spectrometry. Here, the focus is SomaLogic's SOMAscan affinity platform (https://somalogic.com/somascan-platform), a cutting-edge proteomics technology that allows for the precise measurement of proteins in biological samples such as CSF, plasma, serum, and other body fluids. It is a highly multiplexed, sensitive, quantitative and reproducible proteomic technology capable of measuring rapidly (high throughput), broadly (thousands of proteins simultaneously) and deeply (high-and low-abundance proteins) by employing SOMAmer (Slow Off-rate Modified Aptamer) reagents.

Key words Somalogic, Aptamer, Differential expression, Pathway

1 Introduction

There are two potential methodologies for high-throughput protein measurement approaches with targeted assays that complement the unbiased approach used in mass spectrometry. Highly multiplexed commercial platforms Olink and SomaLogic use affinity binding to profile proteome expression. Here, the focus is SomaLogic's SOMAscan affinity platform (https://somalogic.com/somascan-platform), a cutting-edge proteomics technology that allows for the precise measurement of proteins in biological samples such as CSF, plasma, serum, brain tissue, and other body fluids. It is a highly multiplexed, sensitive, quantitative and reproducible proteomic technology capable of measuring rapidly (high throughput), broadly (thousands of proteins simultaneously) and deeply (high-and low-abundance proteins) by employing SOMAmer (Slow Off-rate Modified Aptamer) reagents.

This chapter first describes the technology used and requirements for the platform itself. We discuss sample collection and storage in addition to the SOMAscan technology and assay process. Important to this process is the data use and analysis. The final

Enrique Santamaría and Joaquín Fernández-Irigoyen (eds.), *Cerebrospinal Fluid (CSF) Proteomics: Methods and Protocols*, Methods in Molecular Biology, vol. 2914, https://doi.org/10.1007/978-1-0716-4462-1_13,

sections of the chapter describe suggested quality controls methods and potential analysis approaches for large scale proteomics datasets.

2 Overview of SomaLogic SOMAscan Technology

2.1 The SomaScan Platform

SomaLogic's SomaScan assay leverages SOMAmer reagents, which are chemically modified single-stranded DNA aptamers, to selectively bind proteins. These aptamers have been engineered to exhibit high affinity and specificity for their protein targets. The SOMAscan platform is unique in that it combines the precision of nucleic acid-based binding technologies with the versatility of aptamers to detect proteins across a wide dynamic range.

The SomaScan 11 K Assay v5.0 profiles approximately 11,000 total protein measurements, covering over 10,000 unique human proteins, measured in small volumes of biological samples (typically ~50 μL of plasma or serum) with the potential to expand the multiplex further for more assays in updated formats. This broad coverage makes SOMAscan a valuable tool for understanding the proteomic landscape associated with various neurodegenerative diseases, including Alzheimer's disease (AD) and Parkinson's disease (PD).

The SOMAscan assays are conducted in-house by a specialized team.

2.2 Mechanism of SOMAscan Technology

SOMAmer Reagents are selected against proteins in their native folded conformations and are therefore generally found to require an intact, tertiary protein structure for binding SOMAmer. As such, unfolded and denatured—and therefore presumably inactive—proteins are not detected by SOMAmer Reagents. Crystal structures of SOMAmer Reagents bound to their cognate protein targets indicate that the modified nucleotides contribute extensively to intramolecular contacts within the SOMAmer Reagent and to intermolecular contacts with the protein targets.

The SOMAscan assay consists of three key stages:

Binding Phase: In the binding phase, the SOMAmer reagents are incubated with a biological sample. Each SOMAmer binds to its specific protein target based on the molecular affinity between the aptamer and the protein. This binding is highly selective, reducing off-target effects that are common in other proteomic assays.

Partitioning Phase: After binding, unbound proteins and other molecules are removed, and the protein-SOMAmer complexes are isolated. This is accomplished using a series of washing and separation steps that selectively retain only the SOMAmer-bound proteins.

Quantification Phase: In the final phase, the protein-SOMAmer complexes are separated, and the remaining SOMAmers are quantified using hybridization techniques similar to those used in DNA microarrays. Since each SOMAmer corresponds to a specific protein, the abundance of a protein in the original sample is inferred from the amount of SOMAmer detected in the final readout. The final step involves measuring the relative concentration of each SOMAmer, which is proportional to the amount of the target protein present in the sample.

The SOMAscan Assay Steps

1. SOMAmer Reagents are synthesized with a fluorophore, photocleavable linker, and biotin.
2. SOMAmer Reagents bound to streptavidin beads are used to capture proteins from a complex mixture of proteins.
3. Unbound proteins are washed away, and bound proteins are tagged with biotin.
4. UV light breaks the photocleavable linker, releasing complexes back into the solution.
5. Non-specific complexes dissociate while specific complexes remain bound.
6. A polyanionic competitor prevents the rebinding of non-specific complexes.
7. Biotinylated proteins (and bound SOMAmer Reagents) are captured on new streptavidin beads.
8. SOMAmer Reagents are released from the complexes by denaturing the proteins. Fluorophores are measured after hybridization to complementary sequences on a microarray chip. The fluorescence intensity detected on the microarray is related to the amount of available epitope in the original sample.

3 Preparing Cerebrospinal Fluid (CSF) Samples for SomaLogic Scan

The successful application of SomaLogic's SOMAscan platform for proteomic analysis depends heavily on the quality and preparation of the biological samples. Cerebrospinal fluid (CSF) is a particularly important biological fluid in the study of neurodegenerative diseases due to its close proximity to the central nervous system (CNS) [1]. This session will outline the key steps involved in preparing CSF samples for SOMAscan analysis, including collection, handling, and storage procedures, as well as special considerations for ensuring data accuracy and reliability.

3.1 Overview of CSF as a Biological Sample

Proteins present in CSF can reflect pathological processes occurring in the brain, making it a valuable source for biomarker discovery. Proteomic analysis of CSF presents greater challenges compared to other biofluids, such as blood plasma or serum, due to several key factors.

3.1.1 Low Protein Concentration

CSF has a significantly lower total protein concentration than plasma or serum. This makes it more difficult to detect low-abundance proteins that may be critical biomarkers. Sensitive and precise detection methods, such as SOMAscan assays, are required to capture these subtle proteomic signatures.

3.1.2 High Dynamic Range

Similar to plasma, the protein concentration in CSF can span several orders of magnitude, with a few highly abundant proteins potentially masking the detection of lower-abundance proteins. Dealing with this high dynamic range often requires strategies like protein fractionation to enrich less abundant but biologically significant proteins.

3.1.3 Protease Activity

CSF proteins can be rapidly degraded by proteases if not properly handled, leading to the potential loss of key proteins or misrepresentation of protein levels. Immediate sample processing, such as the use of protease inhibitors and cold storage, is critical to prevent proteolysis and preserve the integrity of the proteome.

3.1.4 Contaminants

Contamination from red blood cells (RBCs) or blood vessel leakage can introduce plasma proteins into CSF samples, potentially skewing the results. For example, hemoglobin from lysed RBCs can overwhelm the analysis, while plasma proteins may be mistakenly attributed to CSF-originating processes. Stringent quality control steps, such as visual inspection for blood contamination and appropriate centrifugation, are crucial to minimize these confounding factors.

3.1.5 Limited Sample Volume

Unlike blood, where large volumes can be collected, CSF samples are often limited in volume, especially in clinical or research settings involving vulnerable patient populations (e.g., elderly subjects). This limitation necessitates the use of highly efficient and sensitive proteomic platforms that can work with small sample volumes without compromising accuracy.

3.1.6 Post-translational Modifications

CSF proteins often undergo post-translational modifications (PTMs), such as glycosylation or phosphorylation, which can impact their detection and quantification. PTMs can also play critical roles in disease processes such as p-tau in AD, so specialized approaches may be needed to ensure their proper identification and characterization.

Due to these challenges, careful sample preparation is crucial to minimize technical artifacts and ensure the reliability of the proteomic data obtained from SOMAscan or other proteomic analysis methods. This includes the use of stringent protocols for sample collection, storage, and processing, along with advanced bioinformatics tools to filter out potential contaminants and noise. By addressing these complexities, proteomic analysis of CSF can yield valuable insights into neurological diseases, including AD, PD, multiple sclerosis, and brain tumors.

3.2 CSF Sample Collection

The process of CSF collection is known as lumbar puncture. During this procedure, a needle is inserted into the lumbar region of the spinal column to collect CSF from the subarachnoid space. Proper collection procedures are critical for obtaining high-quality samples suitable for proteomic analysis.

3.2.1 Standard CSF Collection Protocol

For proteomic studies using SOMAscan, smaller volumes (as little as 0.5–1 mL) are sufficient, but larger volumes ensure that there is enough sample for multiple analyses.

3.2.2 Immediate Handling of CSF Samples

1. *Aliquoting:* After collection, the CSF should be gently mixed by inverting the collection tube several times and immediately aliquoted into smaller volumes (typically 0.5 mL per aliquot) to avoid multiple freeze-thaw cycles, which can degrade proteins.
2. *Centrifugation:* To remove any cells or debris, the CSF samples should be centrifuged at 1000–1500 g for 10–15 min at 4 °C. This step ensures that the resulting sample contains only the soluble protein fraction and is free of particulate matter that could interfere with the SOMAscan assay.

3.3 CSF Sample Storage

Proper storage conditions are crucial for maintaining the integrity of CSF samples. Proteins in CSF can degrade rapidly at room temperature, so samples must be processed and stored promptly.

3.3.1 Immediate Freezing

1. *Snap-Freezing:* Immediately after centrifugation and aliquoting, CSF samples should be snap-frozen in liquid nitrogen or placed on dry ice. This rapid freezing helps to preserve the protein structure and prevents enzymatic degradation.
2. *Long-Term Storage:* For long-term storage, CSF samples should be kept at −80 °C. Storage at this temperature minimizes protein degradation and preserves the proteomic integrity of the samples for extended periods (years or longer).

3.3.2 Freeze-Thaw Cycles

CSF samples are highly sensitive to freeze-thaw cycles, which can cause protein denaturation and degradation. To avoid this, samples should be aliquoted into single-use volumes so that each aliquot can be thawed only once before analysis. Multiple freeze-thaw cycles should be avoided as much as possible.

4 Statistical Analysis of Proteomic Data

4.1 Statistical Analysis Overview

Statistical analysis plays a pivotal role in ensuring the rigor and validity of findings in large-scale omics studies, particularly when dealing with high-dimensional datasets. In this chapter, a series of well-established statistical methods were employed, including controlling batch effects and adjustments for multiple comparisons, to reduce the risk of false-positive results and ensure the comparability of results across experimental groups. Furthermore, the careful consideration of data normalization is paramount to ensure meaningful downstream analyses. Following the detailed quality control steps a four-stage analysis plan is discussed. Differential analysis to investigate comparative changes in protein expression is key in investigating the dataset. The exploration of these proteins using pathway enrichment, interaction analysis and clustering will allow the researcher to understand more about groups of proteins and potential disease mechanisms.

All analyses proposed in this chapter can be carried out using R Statistical Software (version 4.2.2). Example code is given in italic text. Given the complexity and high-dimensional nature of the data, specific strategies are suggested to reduce bias, correct for potential confounders, and maintain robustness in the face of missing or imbalanced data. Moreover, false discovery rate (FDR) correction is recommended to manage multiple comparisons, minimizing the likelihood of spurious associations that might arise due to the large number of variables under investigation. Requirements may vary according to the unique features of each dataset or cohort; however, the following steps can be adapted to carry out a guided analysis.

4.2 Quality Control

4.2.1 Importing Raw Data

SomaScan results are produced in a tab-delimited ASCII file with the filename extension ".adat"—an ADAT file. The ADAT file contains measurements for a series of analytes (columns) across a series of samples (rows) and includes analyte description and sample description information. The ADAT file is divided into 4 sections: Header Information, Column Data Description, Row Data Description, and Data Table Section. The Data Table Section is composed of three subsections: Analyte Description Information/Column Metadata, Sample Description Information/Row Metadata, and the SomaScan Assay Results section. The SomaDataIO R package provides a structured way to import, explore, and analyze SomaScan data stored in ADAT format in R.

4.2.2 Proteomic Data Handling

The quality control (QC) process is a critical step in omics data analysis because high-throughput technologies can introduce noise, variability, and errors into the data. Such issues could arise from technical variability, sample preparation, batch or machine errors, all of which necessitate comprehensive quality control measures.

The expression intensity values are stored in the expression matrix. Each row corresponds to an aptamer as a feature, while the columns represent separate samples from different groups. The following methods show how to examine and transform the data for further analysis.

Missing values are commonplace in omics datasets, often due to the low abundance of certain proteins that fall below the detection threshold. The threshold for excluding samples and features can be set to 70% missingness. This choice strikes a balance between removing unreliable data while retaining as much information as possible.

```
miss <- function(x){sum(is.na(x))/length(x)*100}

sam_mis_prop <- apply(input_protein_matrix, 1, miss)## calcu-
late missing data from samples
sample_rm <- sam_mis_prop >= 70 ## create logical vector with
samples passing filter
input_protein_matrix <- input_protein_matrix[!sample_rm,,drop
= TRUE] ## remove samples

var_mis_prop <- apply(input_protein_matrix, 2, miss) ## calcu-
late missing data from aptamers
var_rm <- var_mis_prop >= 70 ## create logical vector with
aptamers passing filter
input_protein_matrix <- input_protein_matrix[ ,var_rm, drop =
TRUE ]
dim(input_protein_matrix) ## print size of final data matrix
to obtain remaining aptamer and sample total
```

4.2.3 Assessing the Coefficient of Variation

To ensure reproducibility in proteomics, the low coefficient of variation (CV) is a critical criterion. CV (%) represents the relative dispersion of data around the mean. In the context of an analytical platform, this translates to technical variability, which is commonly referred to as "noise." The standard formula for calculating CV (%) is:

$$\text{CV}\ (\%) = (\text{Standard Deviation})/(\text{Mean}) \times 100\%$$

The SomaScan Assay has demonstrated a median CV of approximately 5%. This measure of performance stability enhances the reliability of the assay, making it well-suited for longitudinal studies and providing heightened confidence in assay reproducibility. As part of the quality control process, assays with a CV greater than 20% should be removed from further analysis.

4.2.4 Imputation of Missing Data

For the proteomic data, MICE [2], missForest [3], and KNN [4] are functions that can be used for imputing data. In this example, missing values were imputed using the K-nearest neighbor (KNN) method ($K = 10$). This method is a widely used algorithm in omics data processing due to its simplicity and effectiveness. The KNN algorithm imputes missing values by identifying the 10 most similar samples (i.e., neighbors) in terms of the available data and using their values to estimate the missing points. This imputation method assumes that samples with similar values for measured features will likely have similar values for missing ones. However, KNN imputation also comes with limitations, such as potential biases if the assumption of similarity does not hold.

```
library(VIM) **INSERT NEW LINE** Knn_imputed <- kNN(input_pro-
tein_matrix, k=10)
```

4.2.5 Log-Transformation and Normalization

To achieve normal distribution of the data, which is a prerequisite for many statistical tests, all biomarker levels were log-transformed. Log-transformation is crucial in high-dimensional datasets where the distribution of data is often skewed, as it reduces heteroscedasticity (unequal variance) and brings the data closer to a normal distribution. Normality is important because many statistical techniques, including linear regression and differential expression analysis, assume normally distributed residuals.

```
input_protein_matrix <- log10(apply(input_protein_matrix,
2, as.numeric))
```

In addition to log-transformation, standardization can be performed. Each protein was scaled to have a mean of 0 and a standard deviation of 1, a process known as z-score normalization. This step ensures that all features are on the same scale, preventing any one feature (e.g., proteins with large dynamic ranges) from disproportionately influencing the analysis.

```
input_protein_matrix <- scale(apply(input_protein_matrix,
2, as.numeric))
```

4.2.6 Batch Effect Correction with Combat Algorithm

Batch effects, or systematic variations in data due to experimental conditions rather than biological differences, are a well-known problem in omics studies. These effects can arise from different laboratory conditions, reagent batches, or machine calibration settings, potentially leading to false-positive results.

To detect a batch effect, we visualize the data by performing a principal component analysis (PCA). Here, we investigate the protein expression data for outliers and assess whether it clusters as

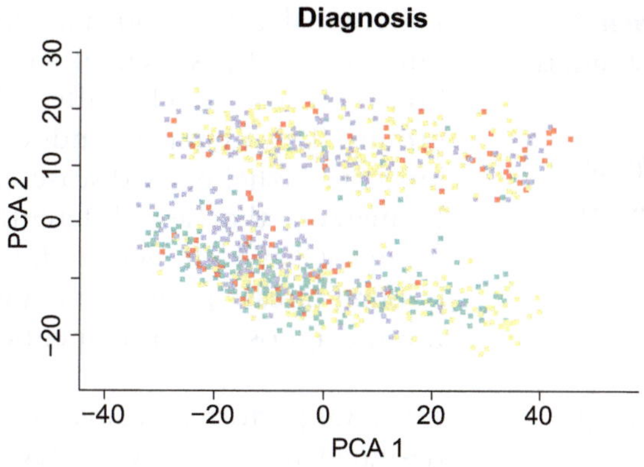

Fig. 1 PCA plot of all samples demonstrating a data substructure caused by experimental batches

anticipated, for instance, based on the group conditions (cases and controls, such as AD and CN) or whether clusters represent other variables (e.g., batch or sites of collection). In the example shown in Fig. 1, every point in the plot represents one sample, with the color indicating the diagnosis groups (AD vs CN).

```
pca <- prcomp(input_protein_matrix)
```

We see that the intensity distributions of the individual samples are quite different, indicating the need for a correction of batch effects. In this chapter, batch effects were mitigated using the Combat algorithm from the sva R package [5]. Combat is an empirical Bayes approach that adjusts for batch effects by estimating and removing the unwanted variability associated with batch factors, without sacrificing biological variability. Combat's strength lies in its ability to harmonize datasets from different batches while preserving the biological signal of interest. By doing so, it enhances the reliability and reproducibility of results across proteomic datasets.

```
modcombat = model.matrix(~1 + Diagnosis, data=input_protein_-
clinical_matrix)

temp_batch <- ComBat(dat = t(input_protein_matrix),
batch = input_protein_clinical_matrix$batch_information,
mod = modcombat)
full_df_combat <- as.data.frame(t(temp_batch))
```

4.3 Differential Expression Analysis (DEA)

4.3.1 Introduction to Differential Expression

Differential expression analysis (DEA) is a fundamental tool in omics research, as it enables the identification of proteins or metabolites whose levels differ significantly between experimental groups. This example conducted DEA to compare two groups of interest: Alzheimer's disease (AD) vs. control (CN). Identifying biomarkers that are differentially expressed between these two groups is key to understanding the molecular mechanisms underpinning disease progression and could lead to the development of novel diagnostic or therapeutic targets.

4.3.2 Methodology

DEA was performed using the limma R package [6]. The package is designed to analyze complex experiments involving comparisons between many experimental groups simultaneously while remaining reasonably easy to use for simple experiments. The main idea is to fit a linear model to the expression data for each feature. Empirical Bayes and other methods are used to borrow information across features for the residual variance estimation, leading to "moderated" t-statistics, and stabilizing the analysis for experiments with just a small number of samples.

The makeContrasts function created a matrix from a symbolic description of the contrast of interest. The lmFit function was used to fit linear models to the data, where the expression levels of proteins were modeled as a function of disease status (AD or CN) while controlling for potential confounders such as age and gender.

```
## Define sample group information
sample_groups <- factor(c("control", "control", "control",
"control",
" case", " case ", " case ", " case "))
## Create the design matrix
design_matrix <- model.matrix(~ 0 + sample_groups)
colnames(design_matrix) <- levels(sample_groups)

## Apply the model to the protein data
fit <- lmFit(t(full_df_combat), design)
fit_contrast <- contrasts.fit(fit, contrast.matrix)
fit_ebayes <- eBayes(fit_contrast)
## Generate Protein list
fullList <- topTable(fit_ebayes, number=nrow(fit_ebayes),
sort.by="logFC", coef = 1)
```

One of the main benefits of using the limma package is that it offers a solid framework for estimating coefficients in high-dimensional datasets with relatively few samples. Since a large number of proteins were being analyzed, traditional statistical tests often struggle with high variability in variance estimates, resulting in inflated type I and type II error rates. The empirical Bayes approach mitigates this issue by pooling information across all features to stabilize variance estimates, thus enhancing the identification of differentially expressed proteins.

4.3.3 Multiple Testing and FDR Control

Given the thousands of tests performed in omics datasets, the risk of false-positive results is high. False Discovery Rate (FDR) is a method used to correct for multiple testing in statistical analyses, commonly applied in genomics and proteomics, which involve large-scale data. When performing many hypothesis tests simultaneously (for example, testing thousands of proteins for differential expression), some results will appear significant by chance alone. FDR controls the expected proportion of false positives among the declared significant results.

To address this, the Benjamini-Hochberg [7] method is usually employed to control for the false discovery rate (FDR). This method ensures that the proportion of false discoveries (incorrectly identified biomarkers) is controlled, allowing researchers to confidently interpret the results. In this example, proteins with an adjusted p-value <0.05 were considered significantly differentially expressed.

4.3.4 Results Interpretation

Differential expression analysis resulted in the identification of several proteins that were differentially expressed between the AD and control groups. For a visualization of the differentially expressed genes, we created a volcano plot, which is commonly used to summarize the results of a differential expression analysis in a single figure. The example is shown in Fig. 2.

4.4 Pathway Enrichment Analysis with DE Proteins

4.4.1 Pathway Enrichment Overview

The goal of pathway enrichment analysis is to determine whether the differentially expressed proteins or metabolites are enriched for specific biological processes, molecular functions, or cellular components. This helps to link the observed changes in protein levels to biological mechanisms that might be driving disease. We can characterize the identified differentially expressed proteins more in detail by performing the GO enrichment analysis or KEGG enrichment analysis. Essentially the Gene Ontology (http://www.geneontology.org/) is a hierarchically organized collection of functional gene sets. The Kyoto Encyclopedia of Genes and Genomes (https://www.genome.jp/kegg/) is a database resource for understanding high-level functions which are generated by genome sequencing and other high-throughput experimental technologies.

4.4.2 Methodology: KEGG Pathways and the Hypergeometric Test

Pathway enrichment was conducted using the enricher function from the ClusterProfiler R package [8], with differentially expressed proteins serving as the input list. The GO and KEGG databases were used as the reference, providing a well-curated collection of pathways involved in various biological processes.

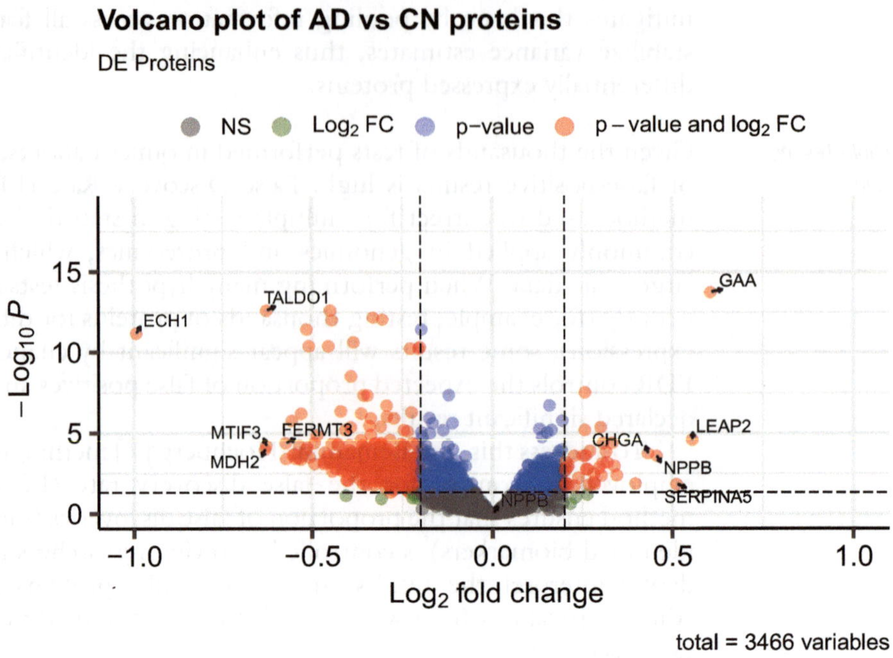

Fig. 2 Example Volcano plot of the Differentially Expressed proteins for AD cases compared to controls

In the ClusterProfiler R package, enrichment was evaluated using the hypergeometric test, which determines whether a specific pathway contains more differentially expressed proteins than would be expected by chance. The hypergeometric test is a statistical method used to assess whether the number of selected items (e.g., DE proteins) within a predefined group (e.g., a biological pathway) is significantly higher than expected by chance. This test is well-suited for this analysis because it calculates the probability of observing a given number of overlapping proteins between the differential expression list and the pathway list, considering the total number of features in the dataset.

```
ego <- clusterProfiler::enrichGO(
gene = DE_protein_entrez,
universe = annot_protein_entrez,
OrgDb = org.Hs.eg.db,
ont = "ALL",
pAdjustMethod = "fdr",
pvalueCutoff = 0.05,
readable = TRUE)

ekegg <- clusterProfiler::enrichKEGG(
gene = DE_protein_entrez,
universe = annot_protein_entrez,
organism = "hsa",
```

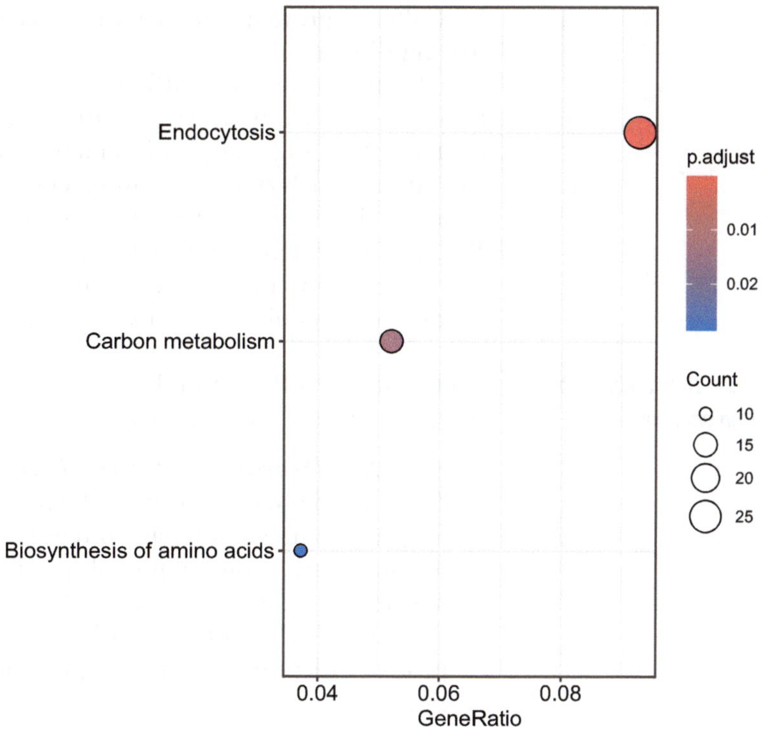

Fig. 3 Example dot plot of enriched KEGG pathways and their p–values

```
keyType = "kegg",
pAdjustMethod = "fdr",
pvalueCutoff = 0.05
qvalueCutoff = 0.2,
use_internal_data = FALSE)
```

4.4.3 Results and Biological Interpretation

The pathway enrichment analysis identified several key pathways that were overrepresented among the differentially expressed proteins. These pathways provide insight into the biological processes underlying AD. For instance, pathways involved in protein processing in the endoplasmic reticulum were found to be significantly enriched, as shown in Fig. 3, corroborating existing hypotheses about the molecular underpinnings of AD. The enrichment plot displays the results of the enrichment analysis as a graph, where the color represents the p-value of the pathway, and the size of the dot indicates the number of genes corresponding to each pathway.

4.5 Protein-Protein Interaction (PPI) Network Analysis

4.5.1 Introduction to Protein-Protein Interactions

Protein-protein interactions (PPIs) refer to the physical contact between two or more protein molecules as part of their biological functions. PPIs are fundamental to almost every process in a living organism, including signal transduction, metabolic pathways, gene regulation, and cellular architecture. In the context of neurodegenerative diseases like AD and PD, PPIs are crucial in understanding how altered proteins interact within cellular networks, potentially driving disease progression.

When proteins are differentially expressed in a disease state, they do not act in isolation. Instead, they interact with other proteins, forming complex interaction networks. Therefore, identifying and analyzing these interactions provides deeper insights into the biological mechanisms underlying diseases. Exploring PPIs among differentially expressed proteins will allow a researcher to map the functional relationships and identify key proteins that might serve as hub nodes or central players within the network.

4.5.2 Data Sources for PPI Information

To investigate PPIs, several well-established databases can be used to curate known interactions:

- *STRING (Search Tool for the Retrieval of Interacting Genes/Proteins)*: STRING is one of the most comprehensive resources for protein interaction data, integrating information from experimental data, computational prediction, co-expression, and knowledge from literature. In this example, STRING was used to extract high-confidence PPIs for the differentially expressed proteins identified from the proteomic analysis, as shown in Fig. 4.
- *BioGRID (Biological General Repository for Interaction Datasets)*: BioGRID is another database that stores protein interactions derived from high-throughput experiments. The database includes both physical and genetic interactions, which can provide additional insights into how proteins cooperate in biological pathways.

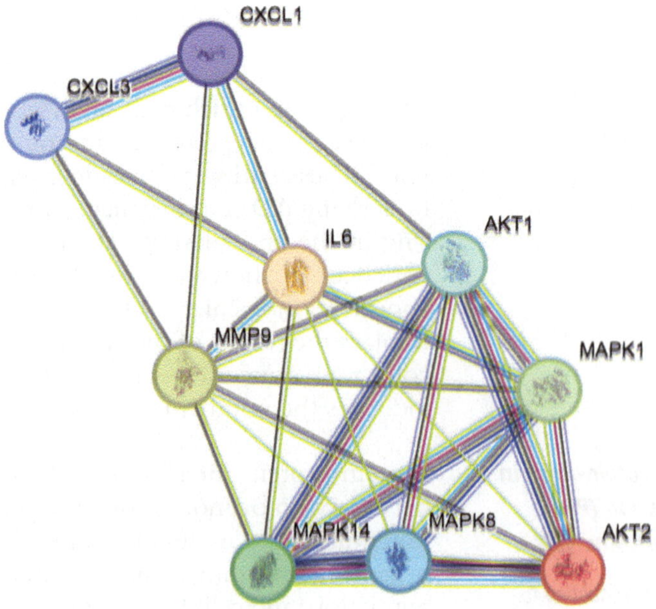

Fig. 4 Example plot of PPI network

- *IntAct and MINT*: These databases focus on experimentally verified interactions and provide datasets that support high-confidence interactions between proteins.

4.5.3 Methodology for PPI Network Construction

The PPI network for differentially expressed proteins was constructed using interaction data sourced from STRING. Here, proteins were represented as nodes, and interactions between them were represented as edges. A confidence score threshold of 0.7 (high confidence) for interactions retrieved from STRING is suggested. The STRING confidence score is based on various types of evidence, including experimental validation and computational prediction. This threshold ensured that only the most reliable interactions were included in the final network.

Once the PPI network was constructed, it was visualized using Cytoscape, an open-source platform that provides advanced network visualization and analysis tools. Cytoscape enables researchers to visualize large interaction networks, making it easier to identify key proteins, pathways, and modules.

4.6 Clustering with Weighted Gene Co-expression Network Analysis

Weighted Gene Co-expression Network Analysis (WGCNA) [9] identifies similar groups of proteins that show correlated patterns of expression across samples. WGCNA is a powerful tool for uncovering clusters (modules) of highly interconnected proteins and linking these clusters to clinical traits or phenotypes. The method has several other advantages; it identifies modules of co-expressed proteins without assuming direct physical interactions, and it helps prioritize hub proteins within modules, which may be key drivers of the biological processes underlying the disease.

4.6.1 WGCNA Workflow

The WGCNA method follows several key steps for constructing a co-expression network and identifying meaningful protein modules. The WGCNA R package should be used for the analysis, and the following steps outline the procedure applied in this study:

Step 1: Data Preparation.

WGCNA works with a format as the proteins are organized in the columns and the treatments in the lines. Proteins with high missing data or low variability were removed, and the remaining data were normalized to ensure comparability across samples.

Step 2: Network Construction and Soft Thresholding.

WGCNA constructs a co-expression network by calculating pairwise correlations between all proteins across the samples. To achieve scale-free topology (a characteristic of most biological networks), the pairwise correlations are transformed into adjacency scores using a soft-thresholding power. The soft thresholding approach ensures that the network emphasizes strong correlations while retaining weaker correlations for additional context.

Step 3: Topological Overlap Matrix (TOM).

Once the adjacency matrix is generated, it is transformed into a Topological Overlap Matrix (TOM), which measures the relative interconnectedness between pairs of proteins. The TOM considers not only the direct correlation between two proteins but also their shared neighbors in the network. This results in a more biologically meaningful representation of the co-expression relationships.

Step 4: Module Identification.

Proteins are grouped into modules based on their topological overlap scores. Modules are clusters of proteins with highly correlated expression patterns, and WGCNA uses hierarchical clustering to identify these groups. Each module is assigned a unique color for visualization purposes, with the dynamic tree-cut algorithm being employed to refine the module boundaries.

Step 5: Module-Trait Relationships.

One of the key strengths of WGCNA is its ability to correlate co-expression modules with external traits. By correlating module eigengenes (a summary expression profile of each module) with these traits, it is possible to identify modules that are significantly associated with the disease states.

Step 6: Hub Protein Identification.

Within each module, hub proteins are identified based on their connectivity. Hub proteins are highly interconnected with other proteins in the module and often play central roles in the biological processes represented by the module.

Hub proteins were identified by calculating the Module Membership (ME) for each protein, which measures the correlation between a protein's expression and the module eigengene. Proteins with high ME values are considered hubs and are likely to be key drivers of the biological processes occurring in that module.

4.6.2 Biological Interpretation of Modules

Each module should be subjected to functional enrichment analysis to uncover the biological processes, molecular functions, and pathways associated with the co-expressed proteins. These can be done by using GO and KEGG pathway analysis (see Subheading 4.4).

WGCNA can reveal several important findings that complement the results from traditional differential expression analysis:

1. *Disease-Associated Modules*: Multiple co-expression modules can be identified that are significantly associated with disease.

These modules represent clusters of proteins that are likely involved in key biological processes driving neurodegeneration.

2. *Hub Proteins as Potential Biomarkers*: Within the disease-associated modules, hub proteins can be identified that could serve as biomarkers or therapeutic targets. These hub proteins are central to the co-expression network and may play critical roles in the disease mechanisms.
3. *Biological Insights*: WGCNA provides insights into the biological processes underlying the observed protein expression changes.

References

1. Seino Y, Nakamura T, Kawarabayashi T, Hirohata M, Narita S, Wakasaya Y, Kaito K, Ueda T, Harigaya Y, Shoji M (2019) Cerebrospinal fluid and plasma biomarkers in neurodegenerative diseases. J Alzheimers Dis 68(1). https://doi.org/10.3233/JAD-181152
2. Azur MJ, Stuart EA, Frangakis C, Leaf PJ (2011) Multiple imputation by chained equations: what is it and how does it work? Int J Methods Psychiatr Res 20(1). https://doi.org/10.1002/mpr.329
3. Stekhoven DJ, Bühlmann P (2012) MissForest—non-parametric missing value imputation for mixed-type data. Bioinformatics 28(1). https://doi.org/10.1093/bioinformatics/btr597
4. Kowarik A, Templ M (2016) Imputation with the R package VIM. J Stat Softw 74(7). https://doi.org/10.18637/jss.v074.i07
5. Leek JT, Johnson WE, Parker HS, Jaffe AE, Storey JD (2012) The sva package for removing batch effects and other unwanted variation in high-throughput experiments. Bioinformatics 28(6). https://doi.org/10.1093/bioinformatics/bts034
6. Ritchie ME, Phipson B, Wu D, Hu Y, Law CW, Shi W, Smyth GK (2015) Limma powers differential expression analyses for RNA-sequencing and microarray studies. Nucleic Acids Res 43(7). https://doi.org/10.1093/nar/gkv007
7. Benjamini Y, Hochberg Y (1995) Controlling the false discovery rate: a practical and powerful approach to multiple testing. J R Stat Soc Series B (Methodological) 57(1). https://doi.org/10.1111/j.2517-6161.1995.tb02031.x
8. Yu G, Wang L-G, Han Y, He Q-Y (2012) clusterProfiler: an R package for comparing biological themes among gene clusters. OMICS 16(5). https://doi.org/10.1089/omi.2011.0118
9. Langfelder P, Horvath S, Langfelder P, Horvath S (2008) WGCNA: an R package for weighted correlation network analysis. BMC Bioinformatics 9:1–13. https://doi.org/10.1186/1471-2105-9-559

Chapter 14

A Comprehensive Workflow for Cerebrospinal Fluid Proteomics in the Study of Brain Tumors

Ankit Halder, Suhisna Dutta, and Sanjeeva Srivastava

Abstract

The cerebrospinal fluid (CSF) plays a critical role in maintaining homeostasis within the central nervous system (CNS). Detecting protein alterations in CSF associated with brain tumors offers a unique opportunity to enhance diagnosis, prognosis, recurrence monitoring, and treatment response. As an extracellular component of the CNS, it serves as a valuable reservoir of biomarkers that reflect changes associated with brain tumors—biomarkers that are often difficult to detect in the systemic circulation due to the blood-brain barrier. Advances in liquid biopsy techniques for CSF, coupled with high-throughput technologies, have the potential to significantly benefit patients. Here, we provide a comprehensive overview of methodologies for studying the CSF proteome to identify brain tumor-associated alterations.

Key words Cerebrospinal Fluid (CSF), Biomarkers, Proteomics, Mass Spectrometry, Data Dependent Acquisition (DDA), Data Independent Acquisition (DIA)

1 Introduction

The main organs of the central nervous system (CNS) are protected by the CSF and Several studies have elucidated the effectiveness of studying CSF in understanding the aggressiveness of brain tumors [1–5]. The minimally invasive technique, compared to tissue biopsy, can be more effective in understanding the changes associated with the onset or progression of disease [3]. Bolstered by breakthrough innovations in high-throughput omics technologies, a wide range of proteins could be identified from CSF, revealing numerous novel biomarkers [6]. The presence of abundant proteins makes it relatively difficult to quantitatively measure the less abundant proteins in CSF [7]. Antibody-based depletion of highly abundant proteins, such as albumin and immunoglobulins, can also lead to the loss of certain interactors, making it a less feasible method for depleting abundant proteins in CSF [8]. Thus, it becomes essential to preserve the natural matrix of CSF to

Enrique Santamaría and Joaquín Fernández-Irigoyen (eds.), *Cerebrospinal Fluid (CSF) Proteomics: Methods and Protocols*, Methods in Molecular Biology, vol. 2914, https://doi.org/10.1007/978-1-0716-4462-1_14,

understand the changes associated with the onset of disease. Apart from using strategies such as the enrichment of low-abundant proteins or deploying techniques like reverse-phase fractionation, the emergence of high-throughput instruments and data acquisition methods is playing a crucial role in increasing overall protein coverage without disrupting the natural matrix of CSF. The Data Independent Acquisition (DIA) technique plays a significant role in increasing protein coverage with high accuracy and reproducibility [9]. The DIA method, in which a range of precursor ions is fragmented and analyzed according to predefined isolation windows, is helping to achieve improved performance in terms of greater protein coverage, coupled with newer data analysis strategies and software [10]. The DIA method has been found to outperform Data Dependent Acquisition (DDA), a method in which only the top precursor ions are fragmented and analyzed, in quantifying low-abundant proteins in many instances [11]. Here, we provide a comprehensive workflow featuring both DDA and DIA methods for untargeted proteomics of CSF, highlighting its various applications in brain tumors and other brain-related disorders.

2 Materials

It is recommended to prepare reagents and solutions using distilled water. Certain reagents should be prepared with Milli-Q water and stored at room temperature unless stated otherwise. All materials related to LC-MS/MS submissions must be prepared using MS-grade quality reagents (*see* **Note 1**).

2.1 Sample Preparation

1. 6 M Urea: Dissolve 18.018 g of urea in 50 mL of Milli-Q water and vortex until a clear solution is formed.
2. PIC (Protein inhibitor cocktail): cOmplete™, Mini, EDTA-free Protease Inhibitor Cocktail. Dissolve 1 tablet in 1 mL of 1X PBS. 1 tablet is sufficient for around 50 mL of lysis buffer.

2.2 Bradford Estimation

1. BSA Stock (1 mg/mL) Preparation: Add 1 mg of BSA powder (HIMEDIA) to 1 mL of Milli-Q water. (Alternatively, 2 mg can be weighed and dissolved in 2 mL of water if weighing small amounts is difficult).
2. Bradford Reagent: Quick Start™ Bradford 1X Dye Reagent (Bio-Rad).
3. 96-well plate general purpose.
4. For reading, use SkanIt™ Software and Multi Detection plate reader (BIO-RAD), Thermo Scientific (Multiskan GO).

2.3 SDS-PAGE Gel

1. Tris-HCL 1.5 M (pH 8.8): Add 36.34 g of Tris-base Molecular Biology Grade (Calbiochem) to 100 mL of water and dissolve. Then, make up the volume to 200 mL. Adjust the pH to 8.8 by adding HCl.
2. Tris-HCL 1 M (pH 6.8): Add 24.22 g of Tris-base Molecular Biology Grade (Calbiochem) to 100 mL of water and dissolve. Then, make up the volume to 200 mL. Adjust the pH to 6.8 by adding HCl.
3. Ready Sol: 30% Acrylamide/Bis Solution (Bio-Rad).
4. 10% SDS: Add 10 g of SDS (Sigma-Aldrich) to 100 mL of Milli-Q water. Dissolve in two phases, first in 50 mL, then add the remaining water to make up the volume to 100 mL.
5. 10% Ammonium Per Sulphate (APS): Carefully weigh 30 mg of APS and dissolve it in 300 μL of Milli-Q water in a dark amber microcentrifuge tube. It is recommended to prepare APS fresh each time to prevent degradation.
6. TEMED: N,N,N′,N′-Tetramethyl ethylenediamine (Temed) Sigma-Aldrich.
7. 12.5% SDS-PAGE Resolving Gel: For each gel, add 2.5 mL of Milli-Q water, 3 mL of 30% Acrylamide/Bis Solution 29:1 (Bio-Rad), 1.9 mL of 1.5 M Tris-HCl (pH 8.8), 75 μL of 10% SDS, 75 μL of 10% APS, and 4 μL of TEMED. Mix the contents; transfer the mixture to the gel casting unit.
8. SDS- PAGE Stacking Gel: For each gel, add 2 mL of Milli-Q water, 500 μL of 30% Acrylamide/Bis Solution 29:1 (Bio-Rad), 375 μL of 1 M Tris-HCl (pH 6.8), 30 μL of 10% SDS, 30 μL of 10% APS, and 3 μL of TEMED. Mix the contents; transfer the mixture to the gel casting unit.
9. Staining Solution: In a proportion of 30 mL of methanol to 0.25 g of Coomassie Blue, 60 mL of distilled water, and 10 mL of glacial acetic acid, mix the ingredients. Filter using Whatman paper. Store in the dark at room temperature.
10. Destaining Solution: Add 70 mL of acetic acid to 530 mL of Milli-Q water. Then, add 400 mL of methanol to make 1 L of Destaining solution.
11. Loading Dye: pH 7.0, 187.5 mM Tris-HCl, 6% SDS (w/v), 30% glycerol, and 0.03% (w/v) bromophenol blue.
12. 10X Running Buffer: Dissolve 30.3 g of Tris-Base Molecular Biology Grade (Calbiochem), 144.1 g of glycine, and 10 g of SDS in 500 mL of Milli-Q water. Slowly add the remaining Milli-Q water to make a final volume of 1 L.
13. 1X Running Buffer: Take 100 mL of 10X Running Buffer, add 900 mL of Milli-Q water, and mix well to prepare 1X Running Buffer.

2.4 Gel Scanning

1. Multipurpose scanner for Gel scanning—EPSON A3 Transparency Unit.

2.5 In-Solution Digestion (Double Digestion)

1. 100 mM ABC (Ammonium Bicarbonate): Weigh 0.395 g of ammonium bicarbonate and transfer it into a 50 mL glass bottle. Add 50 mL of Milli-Q water and mix thoroughly until a clear solution is obtained.
2. 50 mM ABC (Ammonium Bicarbonate): Take 20 mL of 100 mM ABC and 20 mL of Milli-Q water in a 1:1 ratio. Mix well to prepare a 50 mM ABC solution.
3. 500 mM Tris(2-carboxyethyl)phosphine (TCEP).
4. 500 mM IAA (Iodoacetamide): Add 9.2 mg of IAA to 100 μL water. Need to be prepared fresh always.
5. Trypsin (0.25 μg/μL): Reconstitute 100 μg of lyophilized Pierce™ Trypsin Protease, MS Grade, in 400 μL of the dissolution buffer provided by the manufacturer.
6. Acetonitrile (ACN): 100% MS-Grade ACN (Thermo Scientific™ Acetonitrile (ACN), LC-MS Grade or similar).
7. Formic Acid (FA): Formic Acid, LC-MS Grade (Formic Acid, OPTIMA LC/MS Grade or similar).

2.6 Desalting of Digested Peptides

1. SEP-PAK Columns: Waters Sep-Pak C18 3 cc Vac Cartridge.
2. Conditioning Solution (90% MeOH): Combine 45 mL of MS-grade methanol, 5 mL of MS-grade water (Water, Optima™ LC/MS Grade, Fisher Chemical™ or similar), and 50 μL of trifluoroacetic acid (TFA) from (ADVENT). Gently mix the components to achieve a final volume of 50 mL of conditioning solution.
3. Loading Solution (0.1% TFA): In a suitable container, mix 50 mL of MS-grade water with 50 μL of trifluoroacetic acid (TFA) to prepare the loading solution.
4. Desalting Solution: Add 47.5 mL of MS-grade water, 2.5 mL of MS-grade methanol (Methanol, Optima™ LC/MS Grade, Fisher Chemical or similar), and 50 μL of TFA, then mix well to prepare the desalting solution.
5. Elution Solution: Take 15 mL of MS-grade acetonitrile (ACN), add 15 mL of MS-grade water, and 30 μL of TFA to achieve a total volume of 30 mL of elution solution.

2.7 Quantification of Desalted Peptides

1. 0.1% FA Water: Measure 2 mL of MS-grade water and add 2 μL of formic acid (FA). Vortex the mixture thoroughly to achieve a uniform solution.
2. Thermo Scientific™ μDrop plate.
3. SkanIt™ Software (Multiskan GO).

2.8 Mass Spectrometric Analysis

1. Orbitrap Fusion Tribrid Mass Spectrometer (Thermo Scientific).
2. DIA mode in the data acquisition technique.
3. DDA mode in the data acquisition technique.

2.9 Data Analysis

1. Primary analysis of raw data using DDA mode—Fragpipe v22.0 LFQ-MBR.
2. Primary analysis of raw data using DIA mode—Fragpipe v22.0 DIA-NN (library-based approach).

3 Methods

3.1 Sample Preparation

1. Collect CSF samples via lumbar puncture, performed by a trained healthcare professional. Transfer the samples on dry ice, and store them at −80 °C for future use.
2. Take out the CSF samples from −80 °C freezer and place them on ice at 4 °C to thaw completely.
3. Transfer 150–200 μL of thawed CSF from different patient samples into 1.5 mL micro-centrifuge tubes.
4. The samples should be sonicated on ice at 20% amplitude, using 5-s ON and 5-s OFF pulses for a total duration of 1 min and 30 s (*see* **Note 2**).
5. Concentrate the samples using a vacuum concentrator at medium speed and low heat.
6. Reconstitute the concentrated samples with 6 M urea to a final volume of up to 50 μL. Add Protease Cocktail Inhibitor Complex in the ratio of 1:10.
7. Total protein concentration to be measured from each protein sample using Bradford Assay.

3.2 Bradford Estimation

1. Weigh 1 mg of BSA and dissolve it in 1 mL of water to prepare a standard stock solution of 1 mg/mL.
2. Take 2 μL, 4 μL, 6 μL, 8 μL, and 10 μL from the stock solution to prepare the standard working solution.
3. Now, obtain 2 μL of the CSF samples reconstituted in 6 M urea and transfer them into microcentrifuge tubes.
4. Use a fresh microcentrifuge tube and add 1 mL of Bradford reagent to it. This will serve as a blank for accurate protein quantification.
5. Add 1 mL of Bradford reagent to each of the working solutions and the CSF samples.

6. After incubation for 10 min, take approximately 200 μL of each of the working solutions and samples and place them in 96-well plates.
7. Measure the absorbance at 595 nm.
8. Perform blank subtraction and obtain the equation to determine the concentration of the unknown sample.
9. Determine the concentration of the unknown sample using the equation (*see* **Note 3**).

3.3 SDS-PAGE Gel Run for Quality Check

1. Prepare the gel casting unit.
2. Quickly check that there are no leaks using distilled water.
3. Pour the formulated 12.5% SDS-PAGE resolving gel into the casting unit. Add IPA on top of the gel surface to prevent air contact and drying.
4. Leave the gel for about 30 min to polymerize.
5. After the resolving gel has polymerized, remove the IPA and pour the stacking gel mixture. Insert the comb carefully and allow it to polymerize.
6. Set up the SDS-PAGE gel electrophoresis unit and place the polymerized gel into the buffer tank.
7. Pour 1X SDS running buffer into the tank up to the mark designated for the number of gels.
8. Carefully remove the comb from the gel.
9. Prepare the samples according to the concentration for loading 5 μg onto the wells (*see* **Note 4**).
10. Connect the electrodes to the running unit and activate the power supply.
11. Switch off the electrophoresis when the tracking dye travels two-thirds of the way through the gel.
12. The successful gel run will ensure the accuracy of quantification and the quality of the proteins.

3.4 Gel Scanning

1. Scan the gel using a multipurpose scanner for Gel (Wipro GE Healthcare) and analyze the image using scant software (*see* **Note 4**).

3.5 In-Solution Digestion (Double Digestion)

1. Aliquot 30 μg of protein (*see* **Note 5**) from each CSF sample and adjust the total volume to 20 μL using 50 mM ABC.
2. Add 1.2 μL of TCEP to make a final concentration of 20 mM per 30 μg of protein and incubate for 1 h at 37 °C in a dry bath.
3. Add double volume of IAA (Iodoacetamide) with respect to the volume of TCEP; for 30 μg of protein, add 2.4 μL of 500 mM IAA.

4. Incubate the reaction mixture at room temperature for 20 min in the dark (*see* **Note 6**).
5. After incubation, check the pH and adjust it to 8.0 using 50 mM ABC. (*see* **Note** 7).
6. Add trypsin at a 1:30 ratio to the protein, corresponding to 1 μg of trypsin for every 30 μg of protein.
7. Incubate the reaction mixture at 37 °C overnight for approximately 16 to 18 h.
8. After incubation, add 100% ACN to achieve a final concentration of 80% ACN.
9. Add trypsin at a 1:100 ratio relative to the protein concentration.
10. Incubate the reaction mixture for 3 h at 37 °C.
11. Terminate the digestion by introducing formic acid (F.A.) after incubation, ensuring that the final concentration of F.A. is 2%.
12. Check the pH; if it is near pH 6, then the reaction is terminated (*see* **Note 8**).
13. Vacuum-dry the peptides and store them at −80 °C for further use.

3.6 Desalting of Digested Peptides

1. The vacuum-dried peptides are to be reconstituted in 1 mL of 0.1% TFA.
2. Vortex the reconstituted samples for 10–15 min.
3. Add 1 mL of conditioning solution (*see* Subheading 2.6.2) to the SEP-PAK column bed and repeat this step three times.
4. Pass 1 mL of loading solution (*see* Subheading 2.6.3) through the bed and repeat this step twice.
5. Pass the reconstituted sample through the column and repeat the process two more times.
6. Pass the desalting solution (*see* Subheading 2.6.4) 1 mL through the column bed and repeat two times.
7. Add 1 mL of 50% ACN elution solution (*see* Subheading 2.6.5) to the column and collect it in a new, fresh-labeled tube (*see* **Note 9**).
8. Vacuum-dry the desalted peptides and store them at −20 °C for further use.

3.7 Quantification of Desalted Peptides

1. Reconstitute the dried desalted peptides in 20 μL of 0.1% FA and vortex them thoroughly for 10 min.
2. Take 2 μL of the sample, along with the blank (0.1% FA), and place them into the Thermo Scientific™ μDrop plate for peptide purity assessment and quantification.

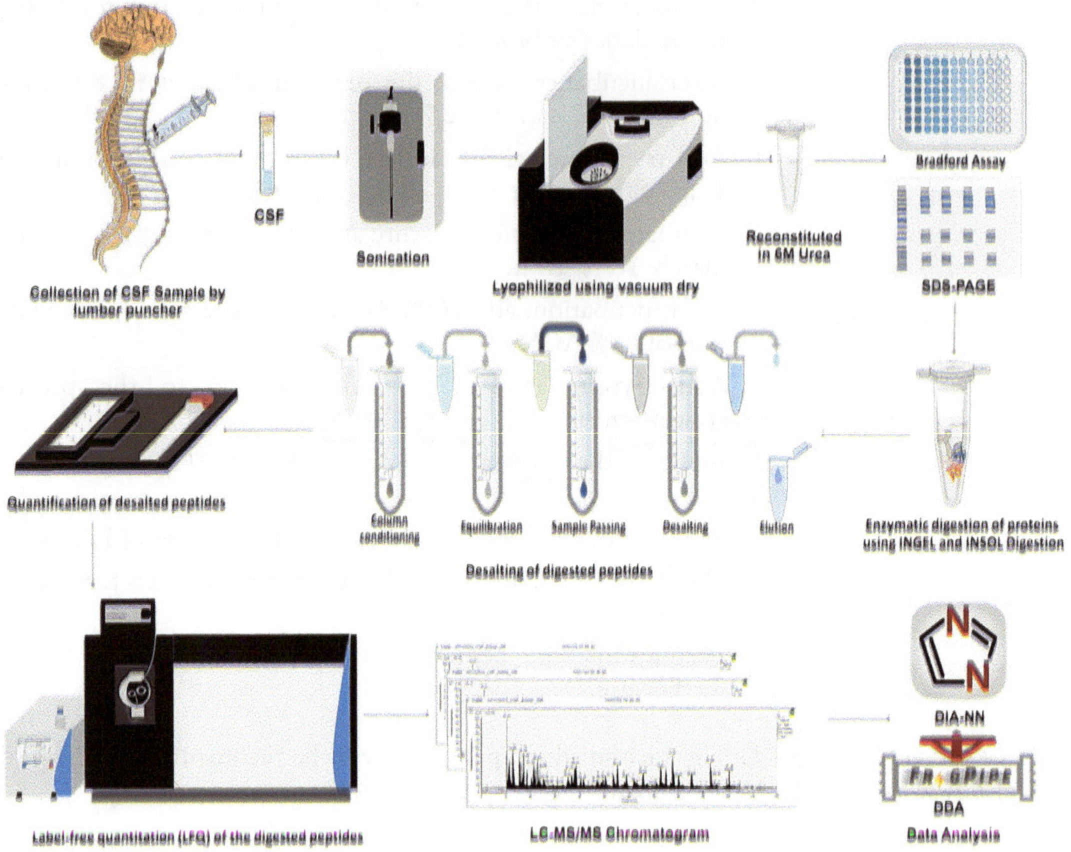

Fig. 1 The standard protocol for proteomics analysis of cerebrospinal fluid (CSF) is schematically represented. It begins with CSF sample collection, followed by protein isolation and fragmentation into peptides. The digested peptides are then desalted and analyzed using MS/MS through either Data-Dependent Acquisition (DDA) or Data-Independent Acquisition (DIA) methods. Protein identification and quantification are carried out through spectral processing

3. Measure the absorbance at 205 nm and 280 nm and quantification to be performed using the Scopes method [12].
4. Quantify the peptides and submit 1 μg of peptides for mass spectrometric analysis (*see* **Note 10**).

3.8 Mass Spectrometry Data Acquisition (DDA vs DIA)

1. Inject around 1 μg of the peptide in nLC (nano-LC) having precolumn and analytical column (Fig. 1).
2. Prepare an LC gradient over 120 min using solvent A (0.1% FA in water) and solvent B (80% ACN in 0.1% FA).
3. The flow-rate to be set at 300–350 nL/min.
4. The gradient might look as follows:

Flow-rate	~300–350 nL/min (Max. Pressure allowed 750 bar at equilibration stage, depending on column specification)
5% (v/v) B	0.0 min
7% (v/v) B	5.0 min
30% (v/v) B	91.0 min / 90.0 min
45% (v/v) B / 48% (v/v) B	106.0 min / 105.0 min
95% (v/v) B / 90% (v/v) B	110.0 min
95% (v/v) B / 90% (v/v) B	120.0 min

5. The eluted peptides to be analyzed in High-Resolution Mass Spectrometer (HRMS).
6. The settings for DDA mode in the data acquisition method for the Orbitrap Fusion Tribrid Mass Spectrometer can be set as follows:

Ion source type	NSI
Spray voltage	Static
Positive ion (V)	1800
Negative ion (V)	600
Sweep gas (Arb)	0
Ion transfer tube temp (°C)	320
LC peak Width (s)	15
Default charge state	2
Detector type (MS1)	Orbitrap
Orbitrap resolution	60,000
Scan range(m/z)	375–1800
AGC target	Custom
Normalized AGC target (%)	100
Maximum injection time (ms)	50
Microscans	1
Data type	Profile
Polarity	Positive

(continued)

Ion source type	NSI
Dynamic exclusion (exclude after n times)	1
Exclusion duration (s)	60
Mass tolerance (ppm)—Low-high	10
Exclude isotopes	True
Intensity threshold	5.0e4
Data-dependent mode	Cycle time
Time between master scans (sec)	3
$ddMS^2$ OT HCD Isolation mode	Quadrupole
Isolation window (m/z)	1.6
Collision energy mode	Fixed
HCD collision energy (%)	30
Detector type (MS2)	Orbitrap
Orbitrap resolution	15,000
Scan range mode	Define First Mass
First mass (m/z)	110
AGC target	Custom
Normalized AGC target (%)	100
Maximum injection time (ms)	30
Microscans	1
Data type	Centroid

7. The settings for DIA mode in the data acquisition method for the Orbitrap Fusion Tribrid Mass Spectrometer can be set as follows:

Ion source type	NSI
Spray voltage	Static
Positive ion (V)	1800
Negative ion (V)	1000
Sweep gas (Arb)	0
Ion transfer tube temp (°C)	320
LC peak width (s)	15
Default charge state	3

(continued)

Ion source type	NSI
Detector type (MS1)	Orbitrap
Use quadrupole isolation	False
Orbitrap resolution	60,000
Mass range	Normal
Scan range(m/z)	385–1015
AGC target	Custom
Normalized AGC target (%)	100
Maximum injection time (ms)	60
DIA precursor mass range (m/z)	385–1015
Multiplex ions	False
Isolation mode	Quadrupole
Isolation window (m/z)	26
Window overlap (m/z)	2
Number of scan events	25
Window placement optimization	On
Activation type	HCD
Collision energy mode	Fixed
HCD collision energy type	Normalized
HCD collision energy (%)	33
Detector type	Orbitrap
Orbitrap resolution	30,000
Mass range	Normal
Scan range mode	Define m/z range
Isolation window (m/z)	300–1500
RF lens (%)	60
AGC target	Custom
Normalized AGC target (%)	1000
Maximum injection time (ms)	60
Microscans	1
Data type	Centroid
Polarity	Positive
Source fragmentation	Disabled

(continued)

Ion source type	NSI
Loop control	N
N (number of spectra)	22

3.9 Primary Data Analysis: DDA Mode

1. The raw files can be directly analyzed or converted into standard file formats such as .mzML or .mzXML.
2. We are providing a standard workflow for primary data analysis using Fragpipe v22.0 [13], an open-source software.
3. Download Fragpipe v22.0 (*see* **Note 11**) on a PC that meets the minimum specified requirements.
4. Extract the compressed folders and install the software using the provided executable file.
5. Configure the software and download MSFragger, IonQuant, and diaTracer.
6. After the configuration is complete, select the LFQ-MBR workflow from the dropdown menu in the Workflow Settings tab (*see* **Note 12**).
7. After saving the changes, add the raw files for analysis from the designated path and assign them to Experiment/Bioreplicates.
8. Download the recently reviewed Human Database from UniProt [14] and decoy and contaminants database as well (*see* **Note 13**).
9. Navigate to the MSFragger option and provide the parameters as follows:

   Cleavage: Enzymatic; Enzyme Name: Trypsin
   Fixed modification: Carbamidomethylation.
   Set all other parameters as "default."
10. Run the validation tool and set it as "default."
11. Start the run, providing a designated folder for storing analyzed files.
12. After a successful run is complete, take the experimental annotation file and the combined_protein.tsv file for further analysis (Fig. 2) (*see* **Note 14**).

3.10 Primary Data Analysis: DIA Mode

1. The primary data analysis of DIA relies on the generation of spectral library search [15].
2. Download and configure the MSFragger, as stated previously.
3. Additionally, download DIA-NN [2], which is also included with Fragpipe, and install/upgrade EasyPOP as per the provided instructions.
4. Select the DIA_Speclib_Quant workflow from the Workflow Settings (*see* **Note 15**).

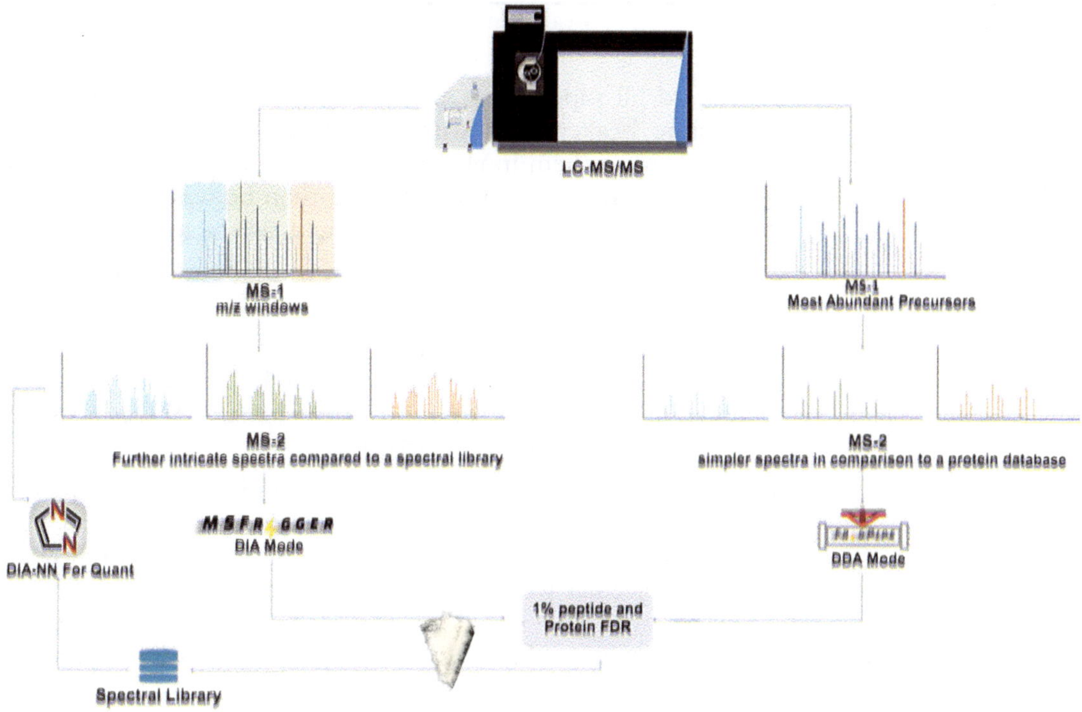

Fig. 2 Schematic representation of DDA mode of mass spectrometry versus data-independent acquisition (DIA) mode of MASS spectrometry data acquisition: DDA employs a survey scan to select abundant precursor ions for fragmentation, while DIA fragments all precursors simultaneously by dividing the mass range into wide precursor windows. It effectively offers a more comprehensive and reproducible data acquisition with higher computational intensity of analysis

5. Add the raw files generated from both DIA and DDA mode (if available) to generate a comprehensive spectral library, and specifically mention DIA/DDA in the space provided (*see* **Note 16**).
6. Assign the files to Experiment/Bioreplicates using “Custom,” then set the experiment and save.
7. Add the database as stated previously.
8. Add all other parameters as stated previously during the analysis of DDA mode.
9. In the Validation tab, run the validation tool in MSBooster.
10. In the “Quant (DIA)” tab, click the checkbox and add “Generate MSstats input.”
11. Start the run and provide a designated folder to store the analyzed files.
12. Take the analyzed file for further analysis (*see* **Note 17**).

4 Notes

1. Discard all pipette tips used to collect CSF samples into a separate beaker filled with disinfectant solution. Once collected, wrap them in aluminum foil and dispose of them in a biohazard bag. The beaker should then be washed with soap and water. The biohazard bags should also be autoclaved.
2. The samples should be kept on ice during sonication to prevent heat-induced degradation of proteins.
3. Determine the linear equation from the standard solution by adding a linear regression line with the intercept set to 0. The R^2 value should be as close to 1 (greater than 0.99) as possible. An example graph is provided (Fig. 3).
4. The protein should be run on SDS-PAGE to assess its quality. Sharp, distinct bands will confirm that no protein degradation has occurred. Additionally, the range of proteins, from low to high molecular weight, should be examined. An example of a gel run with CSF samples is provided. The presence of abundant proteins makes it challenging to identify low-abundant proteins (Fig. 4).
5. Dithiothreitol (DTT) also acts as a reducing agent and can be used if TCEP is unavailable.
6. IAA is photosensitive; therefore, it should be incubated in the dark.
7. It is an important step because trypsin exhibits maximal activity at a pH of approximately 8–8.5.
8. After the addition of formic acid (FA), the pH of the reaction medium will shift to 6–7, causing trypsin to lose its activity and stopping the catalytic reaction.

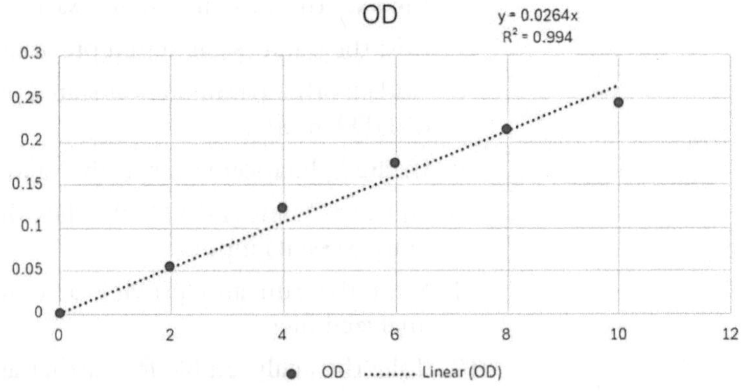

Fig. 3 The standard graph obtained after performing the Bradford Assay using standard stock solutions prepared with BSA

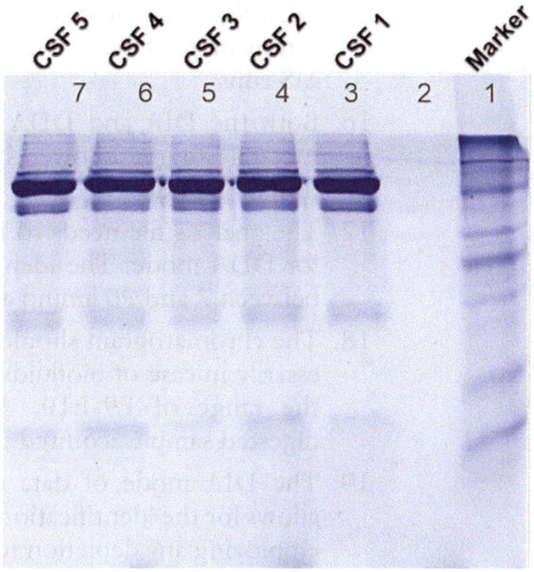

Fig. 4 SDS-PAGE Gel Analysis of CSF samples for quality assessment

9. The primary purpose of the desalting phase is to prevent interference from the high salt concentration in the CSF sample during the MS run.
10. The quantification is to be performed using the Scopes method [12].

    You can quantify the peptides using the given formula:

$$\text{Epsilon} = \text{E205}(27/[1 - (3.85^{*}\text{avg}\ (280/205)]$$

$$\text{Conc.of peptides}\ (\mu g/\mu L) = \text{Avg}\ (205)/(0.05^{*}\text{Epsilon})$$

11. FragPipe is freely available at https://fragpipe.nesvilab.org/.
12. LFQ-MBR: It improves the quality of label-free quantification by detecting missing peptides within a single sample, thereby increasing data consistency and completeness. It is widely used in proteomics-based studies to compare protein abundance.
13. The reviewed database should be used to avoid misleading results.
14. The analyzed file should be checked for contaminants, which should then be removed. Additionally, unique peptides need to be examined. A protein should be considered identified when at least one unique peptide is detected, provided it meets the stringent criterion of 1% FDR. The file can be further taken forward for statistical analysis.
15. DIA_SpecLib_Quant: It utilizes DIA data (along with optional DDA data) as input, constructs a spectral library using MSFragger-DIA, and subsequently quantifies the data with

DIA-NN. This process facilitates the comprehensive quantification of peptides using the spectral library generated from MS runs.

16. Both the DIA and DDA files will be utilized to generate a spectral library, followed by the identification and quantification of peptides.
17. The analysis file needs to be examined specifically as outlined for DDA mode. The identified peptides should preferably be between 7 and 30 amino acids in length.
18. The chromatogram should show a Gaussian pattern (not necessarily in case of biofluids) with maximal intensity ranging in the range of E9-E10. A typical chromatogram of in-sol digested sample acquired in DDA mode is provided (Fig. 5).
19. The DIA mode of data acquisition enhances coverage and allows for the identification of low-abundance proteins without employing any depletion techniques, which can lead to the loss of information about interacting proteins. A typical chromatogram of in-sol digested sample acquired in DIA mode is provided. The comprehensive library generated through gas-phase fractionation will lead to further increased coverage (Fig. 6).

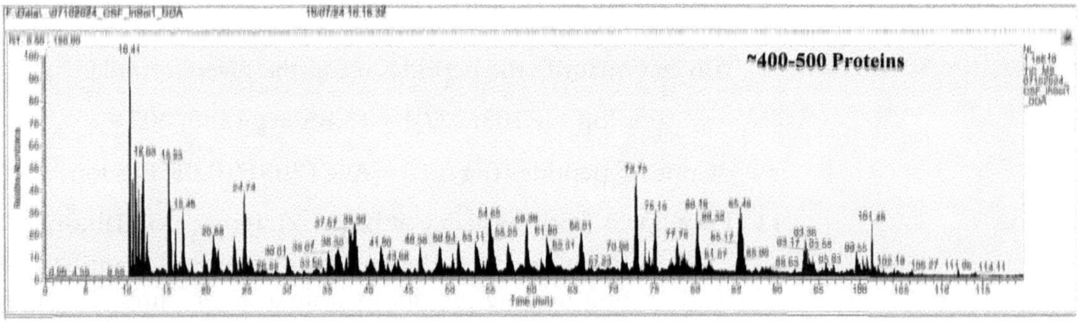

Fig. 5 LC-MS/MS Chromatogram of CSF sample in DDA mode

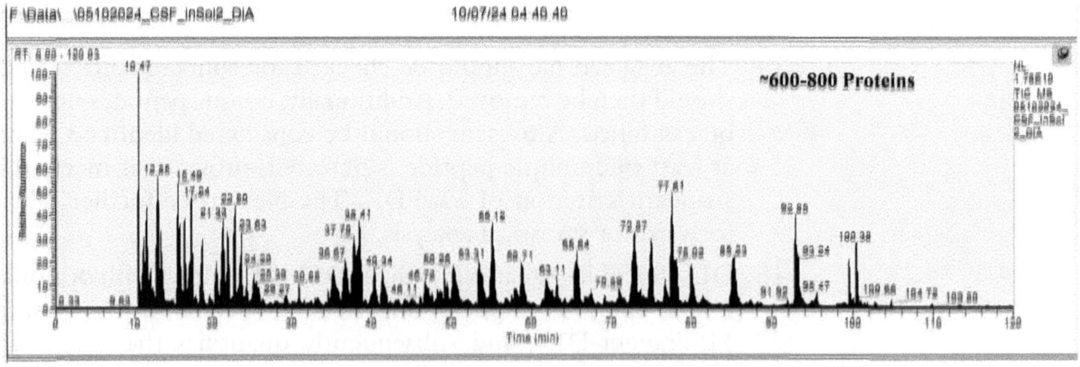

Fig. 6 LC-MS/MS Chromatogram of CSF sample in DIA mode of Data Acquisition

Acknowledgments

The study was supported through the MHRD-UAY Project (UCHHATAR AVISHKAR YOJANA) (UAY-(MHRD)), project #IITB_016 (2017) to S.S. MASSFIITB (Mass Spectrometry Facility at IIT Bombay) from the Department of Biotechnology (BT/PR13114/INF/22/206/2015). We also thank MERCK-COE (DO/2021-MLSP) for their extended support. AH was funded by the Ministry of Education, India, through the Prime Minister's Research Fellowship (PMRF) program. We are thankful to Mass Spectrometry Facility IIT Bombay (MASSFIITB).

References

1. Gahoi N, Malhotra D, Moiyadi A et al (2018) Multi-pronged proteomic analysis to study the glioma pathobiology using cerebrospinal fluid samples. Proteomics Clin Appl 12:e1700056. https://doi.org/10.1002/prca.201700056
2. Kim JW, Choi SA, Dan K et al (2024) Proteomic profiling of cerebrospinal fluid reveals TKT as a potential biomarker for medulloblastoma. Sci Rep 14:21053. https://doi.org/10.1038/s41598-024-71738-z
3. Mikolajewicz N, Khan S, Trifoi M et al (2022) Leveraging the CSF proteome toward minimally-invasive diagnostics surveillance of brain malignancies. Neuro-Oncol Adv 4: vdac161. https://doi.org/10.1093/noajnl/vdac161
4. Mirian C, Thastrup M, Mathiasen R et al (2024) Mass spectrometry-based proteomics of cerebrospinal fluid in pediatric central nervous system malignancies: a systematic review with meta-analysis of individual patient data. Fluids Barriers CNS 21:14. https://doi.org/10.1186/s12987-024-00515-x
5. Schmid D, Warnken U, Latzer P et al (2021) Diagnostic biomarkers from proteomic characterization of cerebrospinal fluid in patients with brain malignancies. J Neurochem 158:522–538. https://doi.org/10.1111/jnc.15350
6. Huang L, Shao D, Wang Y et al (2021) Human body-fluid proteome: quantitative profiling and computational prediction. Brief Bioinform 22:315–333. https://doi.org/10.1093/bib/bbz160
7. Jankovska E, Svitek M, Holada K, Petrak J (2019) Affinity depletion versus relative protein enrichment: a side-by-side comparison of two major strategies for increasing human cerebrospinal fluid proteome coverage. Clin Proteomics 16:9. https://doi.org/10.1186/s12014-019-9229-1
8. Günther R, Krause E, Schümann M et al (2015) Depletion of highly abundant proteins from human cerebrospinal fluid: a cautionary note. Mol Neurodegener 10:53. https://doi.org/10.1186/s13024-015-0050-7
9. Li J, Smith LS, Zhu H-J (2021) Data-independent acquisition (DIA): an emerging proteomics technology for analysis of drug-metabolizing enzymes and transporters. Drug Discov Today Technol 39:49–56. https://doi.org/10.1016/j.ddtec.2021.06.006
10. Halder A, Verma A, Biswas D, Srivastava S (2021) Recent advances in mass-spectrometry based proteomics software, tools and databases. Drug Discov Today Technol 39:69–79. https://doi.org/10.1016/j.ddtec.2021.06.007
11. Willems P, Fels U, Staes A et al (2021) Use of hybrid data-dependent and -independent acquisition spectral libraries empowers dual-proteome profiling. J Proteome Res 20:1165–1177. https://doi.org/10.1021/acs.jproteome.0c00350
12. Scopes RK (1974) Measurement of protein by spectrophotometry at 205 nm. Anal Biochem 59:277–282. https://doi.org/10.1016/0003-2697(74)90034-7
13. Yu F, Teo GC, Kong AT et al (2023) Analysis of DIA proteomics data using MSFragger-DIA and FragPipe computational platform. Nat Commun 14:4154. https://doi.org/10.1038/s41467-023-39869-5
14. The UniProt Consortium, Bateman A, Martin M-J et al (2023) UniProt: the universal protein knowledgebase in 2023. Nucleic Acids Res 51: D523–D531. https://doi.org/10.1093/nar/gkac1052
15. Yang Y, Liu X, Shen C et al (2020) In silico spectral libraries by deep learning facilitate data-independent acquisition proteomics. Nat Commun 11:146. https://doi.org/10.1038/s41467-019-13866-z

Part V

Other -Omics Applied to Cerebrospinal Fluid

Chapter 15

Mass Spectrometry Applied to Human Cerebrospinal Fluid Lipidome (II)

Laura Millán, Joaquín Fernández-Irigoyen, Enrique Santamaría, and Rebeca Mayo

Abstract

Lipidomics aims at characterizing lipid profiles and their biological role with respect to protein expression involved in lipid metabolism. Specifically, cerebrospinal fluid (CSF) lipidomics is offering a new perspective in the search for surrogate biomarkers to facilitate early diagnosis of psychiatric and neurodegenerative diseases. In this chapter, we describe a nontargeted approach to profile lipid molecular species present in human CSF using ultrahigh-performance liquid chromatography-electrospray ionization-time-of-flight mass spectrometry (UPLC-ESI-ToF-MS). This workflow complements the toolbox useful for the exploration and monitoring of neurodegenerative mechanisms associated with dysregulation in lipid metabolism.

Key words CSF, Mass spectrometry, Lipidomics

1 Introduction

It is estimated that 100,000 different lipid species may compose the human lipidome [1], varying in terms of molecular weight, the number/nature of carbon-carbon bonds, the polarity of the head group composition and the hydrophobic tails [2]. At the cerebral level, lipids play several roles in neural function, influencing the elastic properties and supramolecular organization of synaptic membranes and regulating the activity of proteins involved in intra- and transcellular signaling [3–5].

Potential cerebrospinal fluid (CSF) biomarkers may represent brain pathology relatively straightforward because major proteins and metabolites from the brain could be released and diffused into the CSF [6]. As the breakdown of cell membranes is a characteristic feature of neurodegeneration in brain disorders, the investigation of CSF metabolomic profiles may unveil specific lipid species released by damaged neurons or glial cells. Consequently, the application of lipidomics is emerging as a powerful approach that

Enrique Santamaría and Joaquín Fernández-Irigoyen (eds.), *Cerebrospinal Fluid (CSF) Proteomics: Methods and Protocols*, Methods in Molecular Biology, vol. 2914, https://doi.org/10.1007/978-1-0716-4462-1_15,

complements protein-centric workflows, detecting and quantifying specific lipid species that may serve as biomarkers to evaluate the risk of developing a neurodegenerative disease or monitor the disease progression [7, 8]. Although we are in a learning curve to obtain a comprehensive atlas of the lipidome of the human brain [9], the deployment of mass spectrometry coupled to different variants of chromatographic separations is considered the method of choice to analyze lipid profiles in a high-throughput manner. During the last years, several lipidomic platforms have been used to analyze the lipid composition of serum and CSF in the context of Alzheimer's disease [10–12], multiple sclerosis [13, 14], Parkinson's [15, 16], and amyotrophic lateral sclerosis [17].

In this chapter, we describe a workflow to determine the global lipidomic profile of human CSF samples using ultra-performance liquid chromatography-electrospray ionization time-of-flight mass spectrometry (UPLC-ESI-ToF-MS). Due to the wide concentration range of metabolites in CSF and in other biological matrixes and their extensive chemical diversity, it is not possible to analyze the full lipidomic profile with a unique platform. Therefore, metabolite extraction is carried out by fractionating the CSF samples into groups of species with similar physicochemical properties, using appropriate solutions of organic solvents and then analyzing the different extracts in specific analytical platforms. In this study, two analytical platforms are used to analyze the complete lipidomic profile of CSF samples, hereinafter referred to as lipidomic platform 1 and lipidomic platform 2.

2 Materials

2.1 Solvents, Reagents, and Standards

All chemicals and solvents should be of analytical, HPLC or HPLC-MS grade.

1. Acetonitrile.
2. Methanol.
3. Water.
4. Isopropanol.
5. Formic acid.
6. Ammonium formate.
7. Butylated hydroxytoluene.
8. Internal standard (IS) compounds: nonadecanoid acid; dehydrocholic acid; tryptophan-(indole-d5); 1-tridecanoyl-2-hydroxy-sn-glycero-3-phosphocholine, (13:0 Lyso PC); N-hexanoyl-D-erythro-sphingosylphosphorylcholine, SM (d18:1/6:0); 1,2-diheptadecanoyl-sn-glycero-3-phosphoethanolamine, (17:0 PE); 1,2-dinonadecanoyl-sn-glycero-3-

phosphocholine, (19:0 PC); N-heptadecanoyl-D-erythro-sphingosine, Ceramide (d18:1/17:0); tritridecanoylglycerol, (13:0 TG); triheptadecanoylglycerol, (17:0 TG); and cholesteryl laurate, ChoE (12:0).

2.2 Preparation of Internal Standard Solutions

1. IS Stock solutions: Prepare individually for each standard of lipidomic platforms 1 and 2, in the solvents indicated in Tables 1 and 2, respectively. Using a volumetric pipette, add the required amount of IS stock solution solvent to one vial containing pre-weighted quantities of the IS to reach IS Stock Solution concentration indicated in Tables 1 and 2. Store at −80 °C.
2. IS Intermediate solution for lipidomic platform 1. Using a volumetric pipette, place the required amount of the individual IS stock solutions in a volumetric flask to achieve a pool intermediate solution with all IS for platform 1 with the concentration indicated in Table 1. Make up to volume with chloroform/methanol (2:1). Store at −80 °C.
3. IS Working solution for lipidomic platform 1. Pipette 1 mL of IS Intermediate solution into a 100 mL flask. Pipette 1 mL of a 5 mg/mL solution of butylated hydroxytoluene (BHT) in ethanol to obtain a final concentration of BHT of 50 (μg/mL). Make up to 100 mL with methanol. Store at −20 °C.
4. IS Working solution for lipidomic platform 2. Using a volumetric pipette, place the required amount of the individual IS stock solutions in a volumetric flask to achieve a pool solution with all IS for platform 2 with the concentration indicated in Table 2. Make up to volume with chloroform/methanol (2:1). Store at −20 °C.

Table 1
Internal Standard Solutions Platform 1

IS	IS Stock Solution (μg/mL)	IS Stock Solution Solvent	IS Intermediate Solution in $CHCl_3$:MeOH (2:1) (μg/mL)	IS Working Solution in MeOH (μg/mL)
13:0 Lyso PC	10,000	$CHCl_3$	10	0.1
Dehydrocholic acid	5000	$CHCl_3$: MeOH (1: 1)	30	0.3
Nonadecanoic acid	10,000	$CHCl_3$	500	5.0
Tryptophan-(indole-d5)	5000	0.05% formic acid in water	200	2.0

Table 2
Internal Standard Solution Platform 2

IS	IS Stock Solution (μg/mL)	IS Stock Solution Solvent	Working IS Solution in $CHCl_3$:MeOH (2:1) (μg/mL)
SM(d18:1/6:0)	5000	$CHCl_3$	5
PE(17:0/17:0)	10,000	$CHCl_3$:MeOH:H_2O	50
PC(19:0/19:0)	10,000	$CHCl_3$	10
TG(13:0/13:0/13:0)	10,000	$CHCl_3$	5
TG(17:0/17:0/17:0)	10,000	$CHCl_3$	5
Cer(d18:1/17:0)	10,000	$CHCl_3$	10
ChoE(12:0)	10,000	$CHCl_3$	250

2.3 Mobile Phases for Chromatography

1. Lipidomic Platform 1. Mobile Phase A: 0.05% Formic acid in water. Add about 500 mL water in a 1 L volumetric flask. Using a volumetric pipette, transfer 500 μL formic acid to the flask. Mix and make up to 1 L with water.
2. Lipidomic Platform 1. Mobile Phase B: 0.05% Formic acid in acetonitrile. Add about 500 mL acetonitrile in a 1 L volumetric flask. Using a volumetric pipette, transfer 500 μL of formic acid to the flask. Mix and make up to 1 L with acetonitrile.
3. Lipidomic Platform 2. Mobile Phase A: Water:Acetronitrile (2:3) + 10 mM Ammonium Formate. Add 800 mL water in a 1 L graduated cylinder. Transfer about 750 mL to a 2 L volumetric flask. Weigh 1.26 g ammonium formate and dissolve it in a glass baker with the rest of water. Transfer the aqueous solution to the flask. Make up to 2 L with acetonitrile.
4. Lipidomic Platform 2. Mobile Phase B: Acetronitrile:Isopropanol (1:9) + 10 mM Ammonium Formate. Add about 1 L isopropanol to a 2 L volumetric flask. Weigh 1.26 g ammonium formate and dissolve it in 10 mL of water in a glass baker. Transfer the aqueous solution to the flask. Add 200 mL acetonitrile to a 250 mL graduated cylinder. Transfer the acetonitrile to the flask. Mix and make up to 2 L with isopropanol (*see* **Note 1**).
5. Before placing the mobile phases in the liquid chromatography system, eliminate bubbles by sonicating the solutions for 15 min (*see* **Note 2**).

2.4 Equipment

1. Centrifuge 5427 R (Eppendorf).
2. miVac sample concentrator (Genevac).
3. VXR basic vibrax (IKA).
4. Acquity UPLC – Xevo G2 QTof (Waters Corp.).
5. Acquity UPLC – LCT Premier XE (Waters Corp.).

2.5 Software

1. TargetLynx application manager for MassLynx 4.1 software (Waters Corp.) is used for chromatograms integration.
2. Package R v.3.1.1 (R Development Core Team, 2011; http://cran.r-project.org) is used for statistical analysis.

3 Methods

3.1 CSF Extraction

1. Samples are obtained by lumbar puncture (LP) following ethically approved protocols, using a dry clean tube without preservatives, immediately frozen, and stored at −80 °C until use (*see* **Note 3**).
2. Centrifuge at 2500 × g, for 10 min to remove cellular components and blood traces (*see* **Note 4**).

3.2 Metabolite Extraction

1. 150 μL of CSF are vortexed with 600 μL of ice-cold IS working solution for platform 1, containing BHT as antioxidant, 30 μL of ice-cold IS working solution for platform 2 and 570 μL of ice-cold chloroform (*see* **Note 5**).
2. All solvents are ice-cold (*see* **Note 6**).
3. Cap the tubes and shake vigorously at room temperature for 20 min using a vortex mixer.
4. Incubate for 1 h at 4 °C.
5. Centrifuge at 18,000 × g for 15 min at 4 °C.
6. For lipidomic platform 1: transfer 650 μL to a screw-top 1.5 mL micro-centrifuge tube and dry the extracts at 40 °C in a vacuum concentrator.
7. Dissolve the dried extracts in 50 μL of methanol containing BHT at 50 μg/mL, shaking vigorously at room temperature for 20 min using a vortex mixer (*see* **Note 7**).
8. Centrifuge at 18,000 × g for 5 min at 4 °C to precipitate any particulates and transfer the supernatant to a plate for analysis of lipidomic platform 1.
9. For lipidomic platform 2: transfer 650 μL (from **step 5**) to a screw-top 1.5 mL micro-centrifuge tube, add 50 μL of H_2O and vortex for a few seconds.
10. Centrifuge at 18,000 × g for 5 min at 4 °C (*see* **Note 8**).

11. Transfer 400 μL of the lower organic phase to a screw-top 1.5 mL micro-centrifuge tube and dry the extracts at 40 °C in a vacuum concentrator.
12. Dissolve the dried extracts in 50 μL of acetonitrile:isopropanol 1:1, shaking vigorously at room temperature for 10 min using a vortex mixer (*see* **Note** 7).
13. Centrifuge the tubes at 18,000 × g for 15 min at 4 °C to precipitate any particulates and transfer the supernatant to a plate for analysis of lipidomic platform 2.

3.3 Chromatography & Mass Spectrometry

1. Chromatographic separation and mass spectrometric detection conditions employed for each platform are summarized in Table 3 (*see* **Notes 9** and **10**).
2. Analysis of fatty acyls, bile acids, steroids and lysoglycerophospholipids is carried out with lipidomic platform 1 (*see* **Note 11**).
3. Analysis of glycerolipids, cholesterol esters, sphingolipids and glycerophospholipids is carried out with lipidomic platform 2 (*see* **Note 12**).

3.4 Data Processing

4. All data are processed using the TargetLynx application manager for MassLynx 4.1 software (Waters Corp.). A set of retention time and mass-to-charge ratio pairs are defined to each metabolite included in the analysis and are fed into the program. A list of chromatographic peak areas is then generated for each sample injection.
5. Data normalization is performed following the procedure described by Barr et al. [18]. The ion intensities detected for each peak are normalized, within each sample, to the sum of the peak intensities in that sample (*see* **Notes 13** and **14**).
6. Using this dual workflow, a total of 257 metabolic features are detected in human cerebrospinal fluid samples, including bile acids (6), fatty amides (10), acylcarnitines (3), glycerolipids (65), glycerophospholipids (111), non-esterified fatty acids (22), shpingolipids (33), and sterols (7).

4 Notes

1. Using 10 mL water helps to dissolve the ammonium formate and avoid precipitates when mixing the isopropanol and acetonitrile. The addition of solvents in the indicated order also prevents salt precipitates.
2. To remove air bubbles completely from mobile phases and thus avoid problems with retention time drifts, sonication of the solutions is not enough, and it is necessary to have an in-line degasser in the UHPLC system.

Table 3
UHPLC-MS analysis methods

	Platform 1	Platform 2
Column type	UPLC BEH C18, 1.0 × 100 mm, 1.7 μm	UPLC BEH C18, 2.1 × 100 mm, 1.7 μm
Flow rate	0.140 mL/min	0.400 mL/min
Solvent A	0.05% formic acid in water	Water:Acetronitrile (2:3) + 10 mM ammonium Formate
Solvent B	0.05% formic acid in acetonitrile	Acetonitrile:Isopropanol (1:9) + 10 mM ammonium Formate
(%B), time	0%, 0 min	40%, 0 min
(%B), time	50%, 2 min	100%, 10 min
(%B), time	100%, 13 min	40%, 15 min
(%B), time	0%, 18 min	40%, 17 min
Column temperature	40 °C	60 °C
Injection volume	2 μL	3 μL
Autosampler temperature	10 °C	10 °C
Source temperature	120 °C	120 °C
Nebulisation N_2 flow	600 L/h	1000 L/h
Nebulisation N_2 temperature	350 °C	500 °C
Cone N_2 flow	30 L/h	30 L/h
Ionisation	ESI −ve	ESI +ve
Capillary voltage	2.8 kV	3.2 kV
Cone voltage	50 V	30 V
Type of data	Centroied	Centroid
Scan time	0.2 s	0.2 s
Acquisition range	50–1000 Da	50–1200 Da

3. CSF is a body fluid that occupies the subarachnoid space and the ventricular system. The total CSF volume in humans is around 135 mL, and it is composed by a few numbers of cells (0–4 cells/μL), low protein concentration (400 μg/mL), and salts. The most common procedure for the collection of CSF is the lumbar puncture (LP), which consists of a needle insertion into the spinal canal. CSF can be contaminated by blood cells as a result of the needle prick (reddish samples). If this is the case, it might affect the analysis and special attention should be paid to the data interpretation.

4. It is also possible to use specific filters to remove all membrane and cellular debris (for instance, Millex- HV filters).
5. Considering CSF as an aqueous solution, the lipid content of human CSF is expected to be low. Compared to serum, it represents about 0.2% of serum total lipid content. Consequently, lipid extraction is considered the most critical step. Highly specific and sensitive methods are needed for monitoring different lipid families. Liquid-liquid extraction with a chloroform/methanol mixture is considered a robust and efficient lipid extraction method for large-scale global lipid profiling [19]. The addition of BHT in the extraction solvent limits lipid oxidation.
6. The most popular method for the removal of protein from biological fluids is the addition of organic solvent. The pre-cooled solvent also facilitates protein precipitation because it irreversibly denatures the three-dimensional structure of proteins.
7. Using solvents at room temperature to dissolve the dried extracts improves the metabolite recovery.
8. After the addition of water, two layers can be observed. The interphase is formed by the protein disk. The lower phase contains the most apolar lipids.
9. To guarantee exact mass, all spectra are mass corrected in real time by reference to leucine encephalin through an independent reference electrospray, sampled every 10 s.
10. In order to assess the data quality, it is essential to use quality control (QC) samples. In this study, two different reference serum samples are used, although other kinds of samples could also be considered, such as a pool of CSF samples, for instance. These QC samples are extracted and analyzed at the same time as individual samples and then are injected uniformly interspersed throughout the entire batch run. Both are used to assess how well the data pre-processing procedure corrected the different response factors between and within batches.
11. Although some steroids can be followed up with this method, this platform is not optimal for the analysis of steroids.
12. This platform is not optimal for the analysis of cholesterol ester metabolites.
13. Before applying this type of data normalization, it should be verified that there are no significant differences between the total intensities used for normalization of the sample groups compared in the study.
14. The other type of data normalization is also possible, applying different transformations of the metabolites, such as logarithmic or center-scale transformations, in order to convert non-normal data into data with a normal distribution

[20]. Selecting a proper pre-treatment method for the data is a crucial point in the analysis of metabolomic data because this extensively affects the results of the succeeding analyses [21].

Acknowledgments

The Proteomics Unit of Navarrabiomed is a member of Proteored, PRB3-ISCIII, and is supported by grant PT17/0019 of the PE I + D + i 2013-2016, funded by ISCIII and ERDF. Part of the work described here was funded by MCIU/AEI/10.13039/501100011033/FEDER, UE, with a grant ref. PID2023-152593OB-I00 to ES and JFI.

References

1. Shevchenko A, Simons K (2010) Lipidomics: coming to grips with lipid diversity. Nat Rev Mol Cell Biol 11:593–598. doi:10.1038/nrm2934
2. Brügger B (2014) Lipidomics: analysis of the lipid composition of cells and subcellular organelles by electrospray ionization mass spectrometry. Annu Rev Biochem 83:79–98. doi:10.1146/annurev-biochem-060713-035324
3. Piomelli D, Astarita G, Rapaka R (2007) A neuroscientist's guide to lipidomics. Nat Rev Neurosci 8:743–754. doi:10.1038/nrn2233
4. Gross RW, Han X (2011) Lipidomics at the interface of structure and function in systems biology. Chem Biol 18:284–291. doi:10.1016/j.chembiol.2011.01.014
5. Lauwers E, Goodchild R, Verstreken P (2016) Membrane lipids in presynaptic function and disease. Neuron 90:11–25. doi:10.1016/j.neuron.2016.02.033
6. Marques TM, van Rumund A, Bruinsma IB et al (2019) Cerebrospinal fluid Galectin-1 levels discriminate patients with parkinsonism from controls. Mol Neurobiol 56:5067–5074. doi:10.1007/s12035-018-1426-9
7. Wei J, Wong LC, Boland S (2024) Lipids as emerging biomarkers in neurodegenerative diseases. Int J Mol Sci 25. doi:10.3390/ijms25010131
8. van Kruining D, Luo Q, van Echten-Deckert G et al (2020) Sphingolipids as prognostic biomarkers of neurodegeneration, neuroinflammation, and psychiatric diseases and their emerging role in lipidomic investigation methods. Adv Drug Deliv Rev 159:232-244. doi:10.1016/j.addr.2020.04.009
9. Naudí A, Cabré R, Jové M et al (2015) Lipidomics of human brain aging and Alzheimer's disease pathology. Int Rev Neurobiol 122:133–189. doi:10.1016/bs.irn.2015.05.008
10. Proitsi P, Kim M, Whiley L et al (2017) Association of blood lipids with Alzheimer's disease: a comprehensive lipidomics analysis. Alzheimers Dement 13:140–151. doi:10.1016/j.jalz.2016.08.003
11. Wong MW, Braidy N, Poljak A et al (2017) Dysregulation of lipids in Alzheimer's disease and their role as potential biomarkers. Alzheimers Dement 13:810–827. doi:10.1016/j.jalz.2017.01.008
12. Dakterzada F, Jové M, Huerto R et al (2024) Cerebrospinal fluid neutral lipids predict progression from mild cognitive impairment to Alzheimer's disease. Geroscience 46:683–696. doi:10.1007/s11357-023-00989-x
13. Gonzalo H, Brieva L, Tatzber F et al (2012) Lipidome analysis in multiple sclerosis reveals protein lipoxidative damage as a potential pathogenic mechanism. J Neurochem 123:622–634. doi:10.1111/j.1471-4159.2012.07934.x
14. Pieragostino D, Cicalini I, Lanuti P et al (2018) Enhanced release of acid sphingomyelinase-enriched exosomes generates a lipidomics signature in CSF of multiple sclerosis patients. Sci Rep 8:3071. doi:10.1038/s41598-018-21497-5
15. Roede JR, Uppal K, Park Y et al (2013) Serum metabolomics of slow vs. rapid motor progression Parkinson's disease: a pilot study. PLoS One 8:e77629. doi:10.1371/journal.pone.0077629
16. Kwon DH, Hwang JS, Kim SG et al (2022) Cerebrospinal fluid metabolome in Parkinson's

disease and multiple system atrophy. Int J Mol Sci 23:1879. doi:10.3390/ijms23031879

17. Blasco H, Veyrat-Durebex C, Bocca C et al (2017) Lipidomics reveals cerebrospinal-fluid signatures of ALS. Sci Rep 7(1): 17652. doi:10.1038/s41598-017-17389-9
18. Barr J, Vázquez-Chantada M, Alonso C et al (2010) Liquid chromatography-mass spectrometry-based parallel metabolic profiling of human and mouse model serum reveals putative biomarkers associated with the progression of nonalcoholic fatty liver disease. J Proteome Res 9:4501–4512. doi:10.1021/pr1002593
19. Colsch B, Seyer A, Boudah S et al (2014) Lipidomic analysis of cerebrospinal fluid by mass spectrometry–based methods. J Inherit Metab Dis 38:53–64. doi:10.1007/s10545-014-9798-0
20. van den Berg RA, Hoefsloot HCJ, Westerhuis JA et al (2006) Centering, scaling, and transformations: improving the biological information content of metabolomics data. BMC Genomics 7:142. doi:10.1186/1471-2164-7-142
21. Martínez-Arranz I, Mayo R, Pérez-Cormenzana M et al (2015) Enhancing metabolomics research through data mining. J Proteomics 127:275–288. doi:10.1016/j.jprot.2015.01.019

Chapter 16

Analysis of Conformational Molecular Changes in Cerebrospinal Fluid Using Fourier Transform Infrared Spectroscopy in Neurodegenerative Diseases

Tanja Dučić and Jan Christoph Koch

Abstract

Studying changes in molecular conformations of cerebrospinal fluid (CSF) in neurological diseases and identifying disease-related secondary structure changes are promising potential biomarkers for early diagnosis and contribute to a better understanding of the pathophysiology. This protocol demonstrates that Fourier transform infrared spectroscopy (FTIR) characterization of CSF is a powerful analytical tool. When combined with chemometric analysis, infrared spectroscopy becomes an essential method for revealing central features of molecular structural changes, especially of protein secondary structure conformations associated with neurodegenerative diseases. This chapter describes a protocol for collecting FTIR spectra from CSF samples, and it covers instrumental options, sample preparation, and recent advancements in spectral data acquisition and interpretation.

Key words FTIR spectroscopy, Protein secondary structure of CSF Conformational changes, Biomacromolecules, Neurodegenerative diseases

1 Introduction

Evaluating cerebrospinal fluid (CSF) is a pivotal component of the diagnostic work-up in neurodegenerative diseases. The CSF, being in direct contact with affected neurons, contains specific protein markers of the disease process. This is well established for evaluation of diseases like Alzheimer's (AD) (elevated tau and decreased Amyloid-beta-42) [1] and Parkinson's disease (PD) (decreased alpha-synuclein) [2]. In Amyotrophic lateral sclerosis (ALS), specific findings in the CSF have also been reported, and other studies have shown cytotoxic effects of CSF from patients with ALS [3], indicating a role for the CSF in the disease process. Using an immune-infrared sensor it was possible to detect misfolded TDP-43 protein in human CSF samples and distinguish ALS patients from healthy controls and PD patients with a sensitivity

Enrique Santamaría and Joaquín Fernández-Irigoyen (eds.), *Cerebrospinal Fluid (CSF) Proteomics: Methods and Protocols*, Methods in Molecular Biology, vol. 2914, https://doi.org/10.1007/978-1-0716-4462-1_16,

and specificity between 80% and 90% [4]. Our recent study presents a comprehensive Synchrotron-based Fourier-Transform Infrared Spectroscopy (SR-FTIR) analysis of CSF from ALS patients compared to controls, identifying specific biomolecular signatures of ALS in CSF. This study provides valuable insights into the disease process, potential biomarkers and explores the method as a potential diagnostic tool [5].

Most clinical laboratory analyses are based on a spectroscopic measurement in the visible or UV spectrum or indirect measurements based on a dye, enzyme-catalyzed reaction, or antigen-antibody interaction with the use of high amounts of specific reagents (disadvantages: costs, the time-consuming handling procedures, analytical errors caused by the reagents or errors occurring during the assay) [6]. Fourier transform infrared (FTIR) spectroscopy offers a promising alternative technique for the determination of clinical parameters, and clinical biochemistry of tissues, cells and body fluids—as an advantageous tool for the diagnosis of diseases and monitoring of medical treatments [6]. FTIR is a non-invasive analytical technique complementing traditional laboratory-based analytical devices (*e.g.*, chemiluminescence, gas chromatography, mass spectrometry) compactly with meager costs per test [6].

The application of FTIR spectroscopy for non-destructive analysis of biological specimens is an expanding field. Besides analysis of the biofluids, it has the potential for cytological and histological diagnosis, providing a label-free tool for studying molecular composition and dynamics without disturbing the sample milieu [7]. This vibrational spectroscopy technique offers a biochemical fingerprint of the structure and function of examined specimens for biological materials. The important spectral regions typically observed are the fingerprint region (800–1500 cm^{-1}) and the amide I and amide II (amide I/II) region (1500–1700 cm^{-1}). In addition, the higher-wavenumber region (2550–3500 cm^{-1}) is connected with stretching vibrations, such as S-H, C-H, N-H, and O-H [7]. Conversely, the lower-wavenumber regions correspond to bending and carbon skeleton fingerprint vibrations, including C-H vibrations in lipids and carbohydrates as well as P=O in nucleic acids [8]. Together, these regions offer a biochemical pattern of the structure and function of the examined biological specimens. Figure 1 demonstrates a typical FTIR spectrum of CSF with molecular assignments.

FTIR spectroscopy could use several light sources, such as a conventional thermal (globar) or synchrotron radiation source for FTIR interferometric measurements, as well as alternative sources such as quantum-cascade lasers (QCLs) and filters, which eliminate the need for interferometers [7]. A globar source is made up of a silicon carbide rod that generates IR radiation, while synchrotron-based (SR) FTIR uses this IR spectral part of the synchrotron radiation.

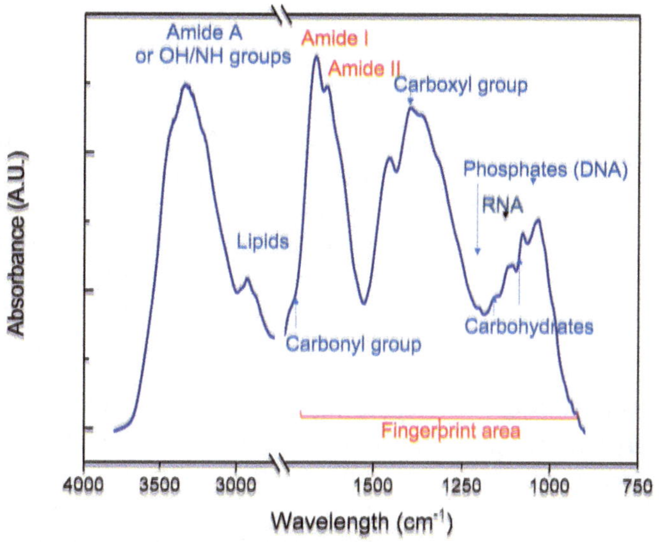

Fig. 1 The comprehensive FTIR averaged spectra of the CSF sample 4000–750 cm^{-1} including the regions of fingerprint, lipids, and OH/NH groups. The major bands in the fingerprint are assigned

SR-FTIR is a powerful, non-destructive technique that provides higher spatial resolution and spectral quality compared to classic FTIR [9]. A synchrotron radiation light source is approximately 100–1000 times brighter than current benchtop thermal sources, and it has a much smaller focus, *i.e.*, it illuminates a smaller area, typically on the cell size 10 × 10 μm with a high signal-to-noise ratio [10]. Thus, besides body fluids analysis, this setup makes it possible to achieve single-cell lateral spatial resolution with these up-to-date sources, allowing subcellular molecular distribution analysis. Synchrotron facilities for IR spectroscopy and microspectroscopy have expanded worldwide in response to increasing demand for beamtime. There are currently more than 20 operational or under-construction infrared beamlines worldwide, all easily accessible for routine use as they operate on a call-for-project basis [10]. On the other hand, other available sources that may be advantageous to individual studies include optic parametric oscillator lasers, QCLs, and free-electron lasers (FELs). However, these sources have been primarily used for other types of investigation, *e.g.*, gas sensing, while modern QCLs can cover broad wavelength regions [7].

SR-FTIR spectroscopy is increasingly being utilized in clinics due to its effectiveness, speed, and label-free macromolecular profiling of biological samples. As a non-invasive diagnostic tool, it holds potential for the diagnosis of neurodegenerative disorders. For example, the SR-FTIR approach provides insights into the neuropathology of ALS and could unveil potential biomarkers as, *e.g.*, antiparallel β-sheet content [11, 12]. Previous studies have successfully used SR-FTIR to analyze ALS-related biomolecules in various models and human samples, demonstrating its potential as a diagnostic tool [12, 13].

Until recently, biomarker discovery studies were performed through bottom-up proteomics utilizing mass spectroscopy, however, it would be important to concentrate on a more global picture of all biomolecule conformational changes in CSF. The FTIR of CSF is appealing to extensive interest due to its potential to resolve protein conformation and content, as this approach offers a non-invasive, label-free identification of specific biologically relevant molecules in a few microliters of CSF samples.

1.1 FTIR of Proteins' Secondary Structure in Neurodegenerative Diseases

Regarding protein's secondary structure, FTIR is a powerful method to study the molecular mechanisms of protein folding, *i.e.*, conformational changes, misfolding or aggregation. Misfolding can occur during biogenesis or post-translationally, leading to the loss of biological function or causing disease. These diseases are also called "protein conformational diseases" such as Alzheimer's disease (Amyloid β-peptide), Parkinson's (α-synuclein), Chorea Huntington's (PolyQ proteins), or ALS (TDP-43) [14]. The early diagnosis of neurodegenerative diseases is important since putative therapeutic interventions are most effective when applied early in the disease course. However, up to now diagnosis can only be achieved by the combination of clinical data and neurophysiological evidence together with symptom progression, often leading to a delay in diagnosis and limiting prompt intervention.

Although the pathogenesis of ALS is still not completely understood, intraneuronal aggregates of the protein TDP-43 are regarded as the pathological hallmark of the vast majority of ALS cases, suggesting that this protein might also serve as a potential biomarker [4] in this fatal neurodegenerative disease characterized by progressive motoneuron degeneration and muscle paralysis [15, 16]. On the other hand, PD is the second most frequent neurodegenerative disease with symptoms such as muscle rigidity, tremor and bradykinesia [17, 18]. Its pathological characteristics include the presence of intracytoplasmic inclusions, so-called Lewy bodies, which comprise fibrillar α-synuclein (αSyn) protein, and loss of dopaminergic neurons within the substantia nigra pars compacta.

In this protocol, we will use ALS and PD samples as examples to illustrate the workflow of the FTIR analysis of CSF. The data analysis protocol, coupled with statistical analysis, revealed differences in CSF from ALS and PD patients compared to the control group. FTIR analysis of protein conformation in CSF samples from ALS and PD patients showed significant differences in protein secondary structures, *i.e.,* protein conformations when compared to the control group. The correlation of spectroscopic data with molecular markers derived from multi-omics approaches and the ability to differentiate between disease-mimicking represents an important step in translating these findings into clinical applications.

2 Materials

2.1 CSF Samples

CSF samples could be obtained from the clinic tissue bank, *e.g.*, the University Medicine Göttingen (UMG), Germany. Permission of the local ethics committee has to be obtained before the initiation of the study (*e.g.*, Ethics Committee of the University Medicine Göttingen, No. 13/11/12), as well as written consent has to be provided by all patients. The study has to conform to the Code of Ethics of the World Medical Association (Declaration of Helsinki).

Usually, a group of patients underwent a lumbar puncture, where 10 mL of CSF is collected. Routine standard procedures include testing of WBC and RBC count, protein and lactate levels. Afterward, CSF should immediately be centrifuged at 2000 g; 20 min; 4 °C within 30 min of collection and frozen at −80 °C until further analysis.

2.2 CSF Parameters

Routine analysis of CSF parameters, including leukocyte and erythrocyte count, total protein, albumin, lactate, concentrations and CSF quotients of IgG, IgM, and IgA, as well as oligoclonal bands are usually evaluated in the local CSF lab. The contamination of blood in CSF presents a diagnostic challenge, as blood contamination can potentially alter red blood count and leucocyte counts, total protein concentration, specific protein fractions, and biochemical, microbiologic, and other tests. Samples with suspicion values, *e.g.*, leukocyte count >6/μL or a red blood count >20/μL in testing, should be excluded from the study.

There are some specific sample characterizations, for example, in the case of the Parkinson's disease cohort if all patients fulfill the UK Parkinson's Disease Society Brain Bank Diagnostic Criteria. There are additional paraclinical parameters that could be included in increasing diagnostic accuracy available (MRI, DatScan, or substantia nigra ultrasound). In the case of ALS patients, sample characterization includes diagnosis according to the revised El Escorial criteria and rated at least "clinically probable laboratory supported." Extensive examinations should be performed to rule out other diagnoses, including electrophysiological testing (electromyography, neurography, evoked potentials), MRI of the brain and spinal cord, and blood laboratory testing as well as determination of neurofilament phosphorylated heavy chain (pNfH) in the CSF.

In addition, control samples are also important to characterized.

2.3 FTIR Spectroscopy

For the FTIR measurements, the FTIR spectrometer and optionally microscope will be necessary. There are various standard equipment, and one of them is the 3000 Hyperion microscope coupled to a Vertex 70v spectrometer from Bruker (Germany). The detector for that system is a mercury cadmium telluride (MCT) cooled with liquid nitrogen. The spectroscope can be coupled with an internal source or synchrotron radiation. One of the synchrotron sources is, for instance, the MIRAS beamline at ALBA synchrotron, Barcelona, as the infrared light source. Typically, the FTIR setup uses a microscope coupled with an infrared spectrometer. FTIR microscope uses Schwarzschild's objectives, usually with a magnification of 36× or 15× and a matching condenser of the same type. The aperture could be set to a different size. High-resolution SR-FTIR microspectroscopy usually uses 10 × 10 μm or smaller aperture up to 3 × 3 μm, for example, to visualize putative oligomers and aggregates in CSF and smaller features in the bio-macromolecular profile. Visible light in the microscope allows the investigation of the sample prior to the measurement, as shown in Fig. 2.

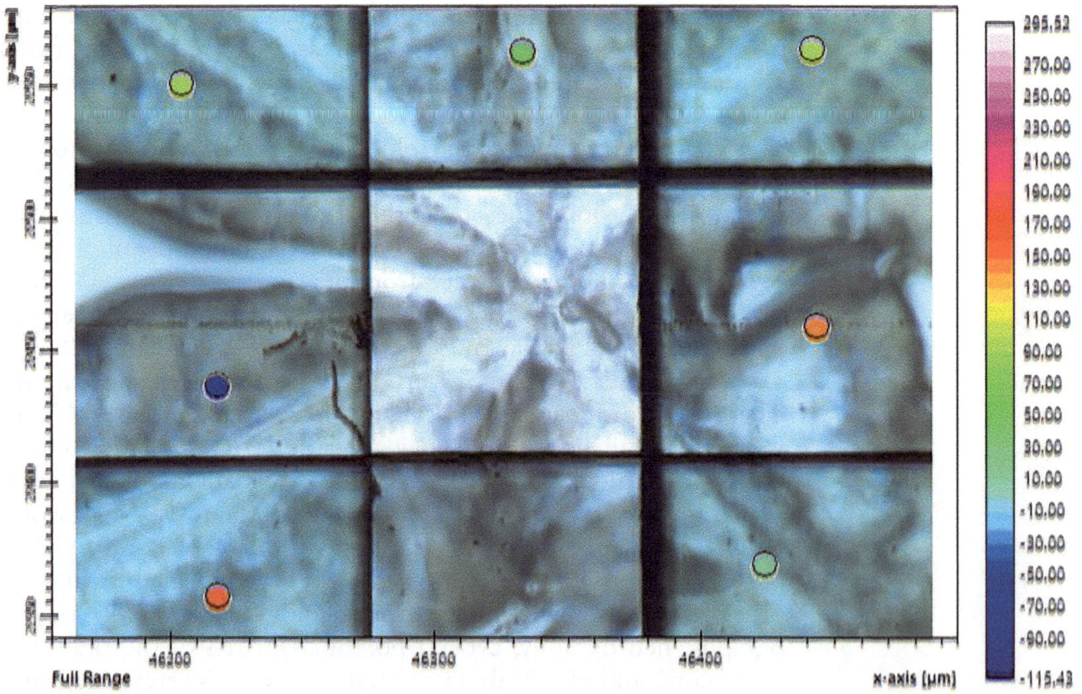

Fig. 2 Optical images of CSF sample after deposition on a CaF_2 IR window and water evaporation. The color points represent measured points with an aperture of 10 × 10 μm

3 Methods

3.1 Sample Preparation

The FTIR measurements are used to avoid invasive preparation protocols and prevent denaturation of the essential bio-macromolecules such as proteins, as well as other chemical changes by conventional fixation protocols. The CSF should be processed within 30 min after drawing, i.e., centrifuged (2000 g; 20 min; 4 °C) and frozen at −80 °C until further analysis (*see* **Note 1**). No other sample preparation procedures are undertaken (*see* **Note 2**).

A volume of 1–3 μL of the thawed CSF should be placed on the 0.5 mm thick CaF_2 or BaF_2 glass before the measurements. Allow to dry for 10–30 min at room temperature. CSF sample takes a typical fern shape when it is dry, as shown in image Fig. 2 (*see* **Note 3**).

Figure 2 shows optical images of CSF sample morphologies displaying fern-like morphology, i.e., samples after deposition on a CaF_2 IR window and water evaporation.

All measurements are performed at the ambient temperature with controlled humidity. In this example, we will show 33 samples from ALS patients, 19 for PDs and 32 samples from controls.

3.2 FTIR Measurements

The type of sample used greatly determines which type of IR spectroscopy is suitable and how it should be prepared for analysis. It is important to consider sample thickness and substrate choice for the best and most reproducible spectra.

An IR-transparent sample support such as BaF_2 or CaF_2 is needed for transmission measurements (*see* **Notes 4** and **5**).

Besides transmission mode, Attenuated Total Reflection (ATR) spectroscopy is also applicable for measuring biological materials. [7]. The main sample formats for clinical IR spectroscopy are biofluids, fixed cell and tissue samples, and live cells. Transmission and transflection sampling modes are usually used for analyzing biological specimens, while ATR-FTIR mode uses an internal reflection element with a high refractive index. The evanescent wave penetrates the sample, which must be in direct contact with the internal reflection elements such as diamond, zinc selenide, germanium or silicon (*see* **Note 6**).

A background single-beam spectrum provides the source intensity, as modified by the instrument. When a sample is placed in the beam path and the single beam is measured again, the additional effect of the sample absorbance can be theoretically determined. The absorbance, directly related to concentration by Beer's law, is obtained by taking the logarithm (to the base 10) of the ratio of these quantities. For point spectra, it is typical to retain a background spectrum for recording 5–10 sample spectra and with each different sample to reduce the effects of constantly changing atmospheric conditions.

3.3 The Spectroscopic Data Collection in Transmission Mode

Overall, as an example, 10 repetitions per measurement per sample are made and average spectra are used for further analysis. The data are collected in the mid-infrared range: 4000–800 cm^{-1}. Typical spectral resolution is 4 cm^{-1} with 128 or 256 co-added scans per spectrum (*see* **Notes 7** and **8**). The OPUS 8.2 (Bruker, Germany) software package can be used for data acquisition (*see* **Note 9**).

3.4 Data Processing

After data acquisition, processing involves several steps, including quality control, spectral pre-processing, feature extraction, and supervised or unsupervised classification, as shown in Fig. 3. Data processing aims to enhance the accuracy of subsequent multivariate analyses and improve data interpretability by addressing issues related to spectral data acquisition.

Data pre-treatments include de-noising, spectral correction, normalization, and other manipulations, with two or three methods often combined. The choice of pre-processing methods depends on the data collected, the analysis goal, the sample's physical state, and available resources. De-noising of IR spectra can be done using methods such as Savitzky-Golay smoothing, minimum noise fraction, or wavelet de-noising effectively eliminating high-frequency noise while preserving sharp peaks.

Another option is to decompose the spectra using principal component analysis (PCA) and reconstruct them from a few principal components, discarding those that mostly represent noise [7]. The complex data interpretation can lead to variabilities that affect the reliability of the quantitative results (*see* **Note 10**).

FTIR characterization of CSF serves as a powerful tool for diagnosing neurodegenerative disorders, with the need for evaluation on a large patient cohort.

Here we present methods to evaluate the CSF by using the SR-FTIR method, as well as the statistical methods for the analysis and evaluation of CSF proteins.

3.5 Spectral Pre-analysis

Before analyzing the spectra, several pre-treatment steps need to be carried out. First, the spectra should undergo quality control, including an assessment of absolute absorption, signal-to-noise ratio (SNR), and the level of water vapor. The atmospheric water and CO_2 could influence the data quality; thus, in the Opus software the application automatic "atmospheric subtraction" can be performed.

Additional pre-treatment steps involve baseline corrections to rectify spectra with sloped or varying baselines. There are possibilities to apply linear or Rubber band correction. Rubber band baseline correction subtracts a rubber band, which is stretched "bottom-up" at each spectrum, eliminating slopes. If necessary, smoothing to reduce background noise could be applied. Usually, synchrotron FTIR spectra do not require it.

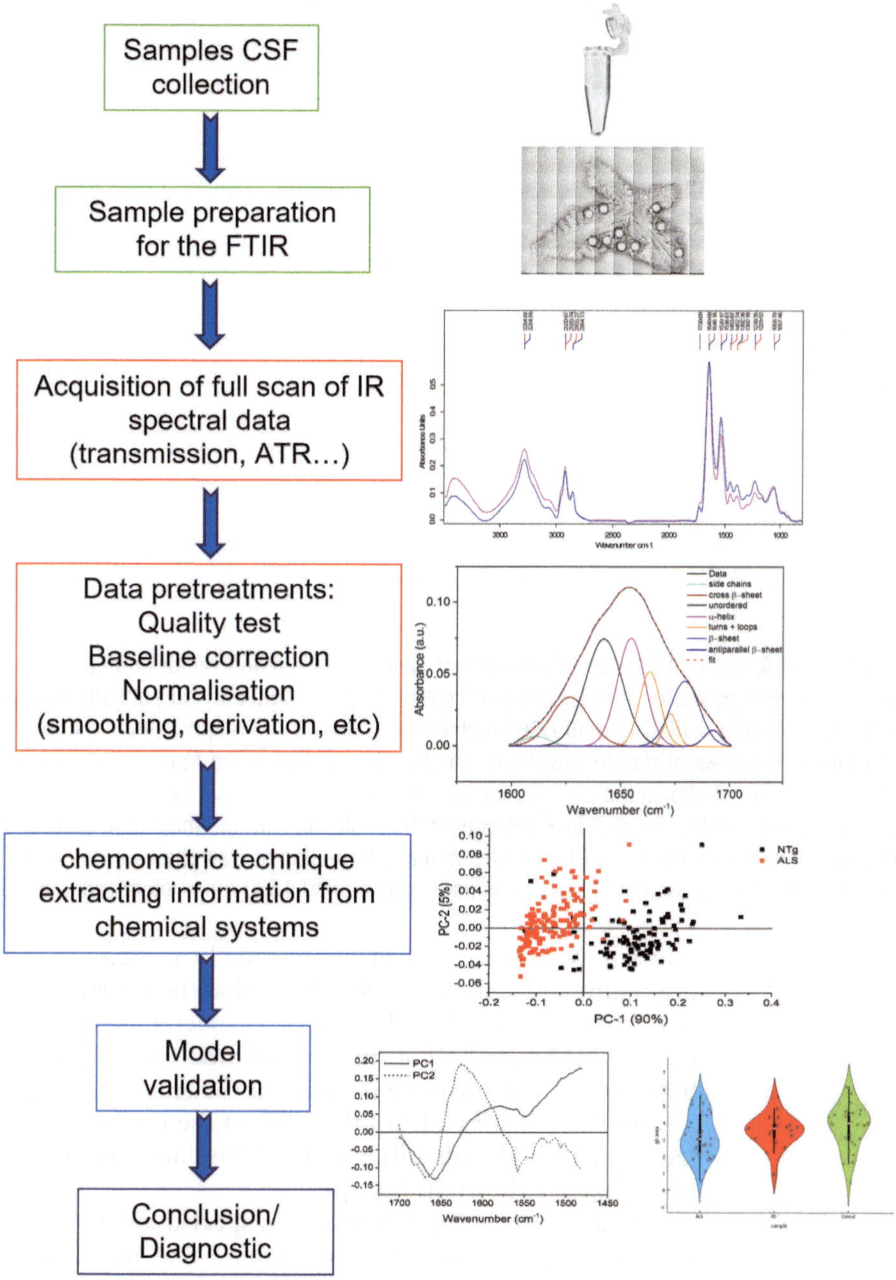

Fig. 3 FTIR spectroscopy workflow of CSF sample. The three major steps are sample preparation, FTIR spectral acquisition, and data analysis

However, the step of normalization is important as we have to take into account differences in absorption due to varying sample thickness or size. Therefore, when comparing the spectra of different samples, or surfaces, normalization is a crucial step. There are several options and vector normalization is one of the best options

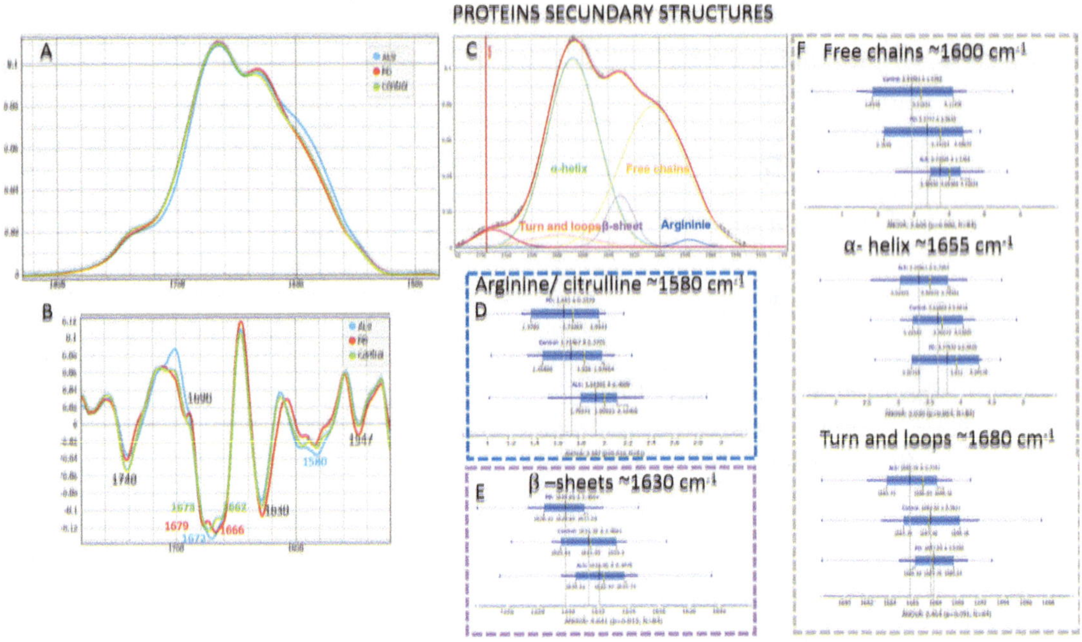

Fig. 4 The spectra of Amide I and Amide II after Rubber band correction and vector normalization (A) and the second derivate followed by vector normalization of the region 1480–1780 cm^{-1} (B) of FTIR averaged spectra of ALS (blue), PD (red) and control (green) CSF samples (N = 33, 19, 32 respectively). The main peaks are assigned. Graphs in **C** represent the deconvoluted spectra: the red line is the fit of the all-averaged spectra shown in black dash line; **D** shows the contribution of amino acids as Arginine, Glutamate or Citrulline in the protein spectra (blue line in **C**). The plot in **E** represents the ß-sheets with the maximum around 1630 cm^{-1} calculated from the fit (violet line in C). Chart F shows the contribution of free chains at 1600 cm^{-1} (orange line), α-helix at 1655 cm^{-1} (green line), and turns and loops around 1680 cm^{-1} (light red line in C)

for each spectrum. Normalization could be made to the amide I/II peak shifts and scales all the spectra so that their vertical minimum is at zero and the amide I/II peak of all spectra matches at the same height [7]. Amide I/II normalization entails adjusting all spectra to have the same absorbance intensity at the amide I/II peak.

An example of FTIR of the CSF, of the protein part of spectra (Amide I and II bands), Rubber band baseline corrected and vector normalized is shown in Fig. 4A.

To address overlapping bands in spectra, mathematical derivatives are effectively used to reduce their full width at half maximum (FWHM), thus potentially revealing subtle differences between them more clearly [7]. The Savitzki-Golay differentiation method successfully eliminates slopes and resolves overlapping bands. Although this method alters the shape of the spectra and increases noise levels, second-differentiated spectra could be very noisy. However, vector normalization is routinely applied following differentiation, and this approach does not require a reference peak as amide I/II normalization does. While each differentiation amplifies noise, the Savitzki-Golay differentiation algorithm (with implicit de-noising) is a widely favored solution. Following differentiation, vector normalization is applied to bring the Euclidean norm of each spectrum to unity [6].

Positions of the peaks in the spectra could be determined using the second derivative method by, *e.g.*, the Quasar software (Bioinformatics Laboratory of the University of Ljubljana [19, 20]), as shown in Fig. 4B. Here it is used the second derivative with 17 smoothing points, third polynomial order (Fig. 4B). After that, the vector normalization was implemented for proteins and esters region (1780–1500 cm^{-1}). Vector normalization was applied after differentiation, as this normalization technique does not require a reference peak [7].

3.6 Statistical Analysis

For diagnostic purposes, feature extraction (FE) plays a crucial role in data reduction to align the complexity of subsequent supervised classification with available data, mitigating the risks of overfitting or undertraining [6]. Principal component analysis (PCA) serves as a widely used form of unsupervised method. Determining the optimal number of PCA factors may involve ordering them based on p values from statistical tests or considering the explained variance percentage. Feature selection (FS) within FE is particularly noteworthy as it bestows biological interpretability on the classification system. Notable FS methods include forward selection and variance analyses. Another approach to FS involves utilizing spectral features derived from a biochemical understanding of the issue [6].

Clustering involves categorizing different objects (*e.g.*, spectra) using a distance measure. Common clustering methods, such as hierarchical cluster analysis (HCA) and k-means clustering (KMC), are frequently employed in FTIR studies for tissue morphology identification [6] (*see* **Note 11**).

Supervised classification entails creating a classification function from training data using machine-learning techniques. Well-regarded techniques include artificial neural networks, support vector machines, linear discriminant classifiers, and Bayesian inference-based methods. When selecting a classifier, important considerations encompass accuracy, ease of training, computational time, spatial resolution considerations, and software availability [6].

The PCA analysis can be accomplished, for example, by using the free-of-charge Quasar software (Bioinformatics Laboratory of the University of Ljubljana [19], with the spectroscopy package [20]. The same software is also used to perform PCA analysis on the different wavenumber regions and for the deconvolution of the peaks.

For instance, Fig. 4 shows the protein analysis of three groups of ALS, PD and control CSF samples. The main protein conformation changes in the Amide I and Amide II area of FTIR spectra were assigned after the second derivatization (Fig. 4B). By calculating the second derivative of the protein region in the spectral data, the changes in the slope of the absorption peaks are enhanced. This process helps to more clearly identify the minima positions, which correspond to the exact locations of peaks or troughs in the spectrum. The second derivative highlights subtle inflection points that are not always visible in the original data, making it easier to pinpoint specific molecular features and improve the accuracy of analysis, especially when distinguishing overlapping peaks or detecting small spectral shifts (as shown in Fig. 4B).

An example of the deconvolution of a single spectrum is illustrated in Fig. 4C. The deconvolution analysis showed that the difference between the two groups of patients is most pronounced for specific conformational changes, *i.e.*, the turns and loops secondary structure. The positions of Gaussian curves were utilized as center positions for the band deconvolution of the protein region in each spectrum. The most prominent bands in the CSF samples appear at ~1660, ~1671, and ~1630 cm^{-1}, corresponding to α-helix, turn and loops and β-sheet structure [21, 22], respectively. Besides, the spectral profile of Amide I and II differed from CSF from different origins, showing significant changes at ~1580 cm^{-1} as well as the one at ~1600 cm^{-1}, corresponding to free amino acids chains, specifically to arginine, glutamate or aspartate [23].

The areas under all mentioned peaks were integrated for each spectrum and presented as box plots (Fig. 4D-F).

3.7 Spectral Interpretation

For the accurate spectral interpretation comprehensive database is required. Each peak in the spectral range is associated with a specific vibration of chemical bonds in molecules. However, it could be a time-consuming task to find the meanings of different unknown peaks. There are detailed studies reported to date in the literature that indicate that most of the scientists have mainly used the previously published articles to define the data acquired from the spectra collected as in [8, 24]. In biological sample interpretation, where a wide range of chemical bonds and functional groups can be attributed to every single peak, finding appropriate meanings can prove the clinical importance of the technique and achieved results can be one of the most important steps in completing spectroscopic labor [24].

Regarding the protein analysis, for illustration, Fig. 4 shows Amide I and II area, where PD samples showed shift of β structure towards lower wave numbers (~1629 cm^{-1}). This change probably indicates the presence of protein cross β-structure and aggregates in the form of fibrils [25] (Fig. 4E). On the other hand, ALS samples showed a peak at 1580 cm^{-1}, potentially associated with glutamate, arginine, citrulline, or aspartate [23]. These findings suggest a possible enrichment in these amino acid side chains, raising questions about the possible imbalance in amino acid metabolism. Furthermore, ALS samples contained mostly proteins in turn and loops secondary structure (Fig. 4F).

4 Notes

1. The number of samples to be analyzed per condition must be considered before the experiment. The number of measured samples directly affects statistical significance and accuracy in the final results. For CSF analysis, it is also recommended to have multiple biological replicates. At least 3 biological replicates (biofluid aliquots on the crystal) and 3 technical replicates (spectra collected from each aliquot) are recommended.
2. Sample preparation and pre-treatment are crucial, as the residues or contamination could directly influence the results. Samples should be processed quickly after drawing and other pre-analytical factors like fasting status and time of the day at sample collection should be taken into account.
3. Sample preparation for FTIR measurements involves placing 1-3 μL of the sample onto a CF_2 IR slide and drying it in an inert atmosphere, such as a nitrogen-purged chamber. Make sure that the sample is dry before the measurements, as water makes artifacts in the Amide I region.
4. Be sure that the sample is optically in the focus prior to the FTIR measurements.
5. When measuring in transmission mode, check condenser alignment in x, y and z directions by using the lower apertures of the microscope and focus on it by using the VIS mode of the microscope. After that, check the IR signal.
6. In addition to the transmission FTIR described here, other IR techniques could be applied as the ATR method. Please keep in mind the following procedure for acquiring spectra using ATR-FTIR spectroscopy:
   (i) Ensure that the ATR-FTIR crystal is clean and dry before use; if required, it can be washed with ethanol or distilled water.
   (ii) Collect background spectrum with a resolution of 4 cm^{-1} using 128 co-added scans. It is very important that the background spectrum is collected first. It should be re-collected upon changes in the environmental conditions. Background collection between each sample is recommended.
   (iii) Place 1 μL of the biofluid on the crystal and allow it to dry. Direct contact of the sample with the crystal is a very important parameter for ATR-FTIR analysis. One μL of the sample is dry within 8–10 min.

(iv) Acquire sample spectrum with a resolution of 4 cm^{-1} using 128 co-added scans. It is common to collect multiple technical replicate samples.

(v) After sample spectral acquisition wash the crystal with ethanol and ensure the crystal is dry.

7. If the MCT detector is used for spectra acquisition, filling it with liquid nitrogen is required; thus, allowing the detector to stabilize for around 20 min is essential. At that time peak-to-peak values should be optimal. After that, each 12–18 h, depending on the detector, should be refilled with liquid nitrogen.
8. Optimizing the aperture of the instrument depends on the sample size and resolution needed for the experiment. In the case of FTIR imaging, the step size should be equal to or smaller than the aperture size, ideally divided by 2, to achieve overlapping each pixel twice.
9. Other software can be used to perform the data acquisition, as well as the data treatment, depending on the instrument used.
10. PCA analysis is an unsupervised statistical method that allows dimensionality reduction without a significant loss of information.
11. Hierarchical clustering is also an unsupervised method; however, the dissimilar distances calculated between the samples could be used.

Acknowledgments

The authors thank the ALBA Synchrotron Facility for beamtime allocation and ALBA In-house research project "Synergetic multimodal FTIR and X-ray spectro-microscopical approach for 3D cell culture evaluation" and the Spanish Ministry of Science and Innovation (MCIN/AEI/10.13039/501100011033). We thank Jennifer Heinemann and Gudrun Leyerer for their support in organizing the UMG CSF collection (LIX) and Elisabeth Barski for technical support. We are grateful to all patients who donated their biomaterials for this research.

References

1. Mattsson-Carlgren N, Grinberg LT, Boxer A et al (2022) Cerebrospinal fluid biomarkers in autopsy-confirmed Alzheimer disease and frontotemporal lobar degeneration. Neurology 98: e1137. https://doi.org/10.1212/WNL.0000000000200040
2. Katayama T, Sawada J, Takahashi K, Yahara O (2020) Cerebrospinal fluid biomarkers in parkinson's disease: a critical overview of the literature and meta-analyses. Brain Sci 10. https://doi.org/10.3390/brainsci10070466
3. Oeckl P, Weydt P, Thal DR et al (2020) Proteomics in cerebrospinal fluid and spinal cord

suggests UCHL1, MAP2 and GPNMB as biomarkers and underpins importance of transcriptional pathways in amyotrophic lateral sclerosis. Acta Neuropathol 139:119. https://doi.org/10.1007/s00401-019-02093-x
4. Beyer L, Günther R, Koch JC et al (2021) TDP-43 as structure-based biomarker in amyotrophic lateral sclerosis. Ann Clin Transl Neurol 8:271. https://doi.org/10.1002/acn3.51256
5. Dučić T, Koch JC (2023) Synchrotron-based Fourier-transform infrared micro-spectroscopy of cerebrospinal fluid from amyotrophic lateral sclerosis patients reveals a unique biomolecular profile. Cells 12:1451. https://doi.org/10.3390/cells12111451
6. De Bruyne S, Speeckaert MM, Delanghe JR (2018) Applications of mid-infrared spectroscopy in the clinical laboratory setting. Crit Rev Clin Lab Sci 55:1
7. Baker MJ, Trevisan J, Bassan P et al (2014) Using Fourier transform IR spectroscopy to analyze biological materials. Nat Protoc 9: 1771–1791. https://doi.org/10.1038/nprot.2014.110
8. Malek K, Wood BR, Bambery KR (2014) FTIR imaging of tissues: techniques and methods of analysis. In: Optical spectroscopy and computational methods in biology and medicine. Springer, Netherlands, Dordrecht, pp 419–473
9. Dumas P, Martin MC, Carr GL (2020) IR spectroscopy and spectromicroscopy with synchrotron radiation. In: Synchrotron light sources and free-electron. Accelerator Physics, Instrumentation and Science Applications, Lasers
10. Miller LM, Dumas P (2018) Infrared spectroscopy using synchrotron radiation. In: Encyclopedia of biophysics
11. Bourassa MW, Brown HH, Borchelt DR et al (2014) Metal-deficient aggregates and diminished copper found in cells expressing SOD1 mutations that cause ALS. Front Aging Neurosci 6:110. https://doi.org/10.3389/fnagi.2014.00110
12. Dučić T, Stamenković S, Lai B et al (2019) Multimodal synchrotron radiation microscopy of intact astrocytes from the hSOD1 G93A rat model of amyotrophic lateral sclerosis. Anal Chem 91:1460–1471. https://doi.org/10.1021/acs.analchem.8b04273
13. Kreuzer M, Stamenković S, Chen S et al (2020) Lipids status and copper in a single astrocyte of the rat model for amyotrophic lateral sclerosis: correlative synchrotron-based X-ray and infrared imaging. J Biophotonics 13:e202000069. https://doi.org/10.1002/jbio.202000069
14. Hauser K (2013) Infrared spectroscopy of protein folding, misfolding and aggregation. In: Encyclopedia of biophysics
15. Boillée S, Vande Velde C, Cleveland DWW (2006) ALS: a disease of motor neurons and their nonneuronal neighbors. Neuron 52:39
16. Robberecht W, Philips T (2013) The changing scene of amyotrophic lateral sclerosis. Nat Rev Neurosci 14:248
17. Dexter DT, Carayon A, Javoy-Agid F et al (1991) Alterations in the levels of iron, ferritin and other trace metals in Parkinson's disease and other neurodegenerative diseases affecting the basal ganglia. Brain 114(Pt 4):1953–1975
18. Gasser T (2007) Update on the genetics of Parkinson's disease. Mov Disord 22(Suppl 1): S343–S350. https://doi.org/10.1002/mds.21676
19. Demšar J, Curk T, Erjavec A et al (2013) Orange: data mining toolbox in python. J Mach Learn Res 14:2349–2353
20. Toplak M, Birarda G, Read S et al (2017) Infrared Orange: connecting hyperspectral data with machine learning. Synchrotron Radiat News 30:40–45. https://doi.org/10.1080/08940886.2017.1338424
21. Bhatia NK, Srivastava A, Katyal N et al (2015) Curcumin binds to the pre-fibrillar aggregates of Cu/Zn superoxide dismutase (SOD1) and alters its amyloidogenic pathway resulting in reduced cytotoxicity. Biochim Biophys Acta Proteins Proteomics 1854:426–436. https://doi.org/10.1016/j.bbapap.2015.01.014
22. Sarroukh R, Cerf E, Derclaye S et al (2011) Transformation of amyloid β(1–40) oligomers into fibrils is characterized by a major change in secondary structure. Cell Mol Life Sci 68: 1429–1438. https://doi.org/10.1007/s00018-010-0529-x
23. De Meutter J, Goormaghtigh E (2021) Amino acid side chain contribution to protein FTIR spectra: impact on secondary structure evaluation. Eur Biophys J 50:641. https://doi.org/10.1007/s00249-021-01507-7
24. Movasaghi Z, Rehman S, ur Rehman I (2008) Fourier transform infrared (FTIR) spectroscopy of biological tissues. Appl Spectrosc Rev 43:134–179. https://doi.org/10.1080/05704920701829043
25. Araki K, Yagi N, Aoyama K et al (2019) Parkinson's disease is a type of amyloidosis featuring accumulation of amyloid fibrils of α-synuclein. Proc Natl Acad Sci USA 116: 17963. https://doi.org/10.1073/pnas.1906124116

Chapter 17

Elucidation of the Human Cerebrospinal Fluid Metabolome by Ion Mobility Mass Spectrometry and NMR Spectroscopy

Jannik Sprengel, Georgy Berezhnoy, and Christoph Trautwein

Abstract

In the last decades, the prevalence of neurodegenerative disorders such as Parkinson's disease, dementia and stroke has significantly increased. The lack of early-stage biomarkers hereby is a major reason for poor outcomes and limited treatment options. Metabolism alterations are a major driver and readout of neurological implications; however, it can be assessed only to a very minor extent by non-invasive imaging technologies. By contrast, typical diagnostic biofluids like serum or urine reflect metabolic changes on the systemic scale and are not specific to the brain. That's why the analysis of cerebrospinal fluid (CSF) using metabolomic techniques offers a great perspective to better stratify patients with neurological implications. However, as metabolites can readily degrade and CSF collection in a clinical environment is challenging, several confounders have to be considered. Furthermore, no single technique can measure the full CSF metabolome. Hence in this work we used directly collected and quenched CSF for a multi-modal metabolomics profiling approach to define the ex vivo baseline of the human CSF metabolome by robust Nuclear Magnetic Resonance (NMR) spectroscopy and latest state of the art liquid chromatography (LC) ion mobility spectrometry mass spectrometry (IMS-MS).

Key words Liquor, CSF, Metabolomics, NMR spectroscopy, Mass spectrometry, timsTOF

1 Introduction

The human brain is not only embedded in a highly complex vascular network, providing major nutrients through the bloodstream, but also connected to the ventricles and spine. Inside these cavities, cerebrospinal fluid (CSF) portrays a valuable diagnostic matrix to investigate pathological processes and is already used for clinical lab analysis in urgent neurological complications. Current routine diagnostics of CSF cover basic laboratory parameters such as pH, inflammation markers, immune cell counts, and protein content. Further, major brain metabolites like glucose and lactate are absolutely quantified to have a readout of altered brain energy metabo-

Enrique Santamaría and Joaquín Fernández-Irigoyen (eds.), *Cerebrospinal Fluid (CSF) Proteomics: Methods and Protocols*, Methods in Molecular Biology, vol. 2914, https://doi.org/10.1007/978-1-0716-4462-1_17,

lism. So indeed, even existing CSF tests include some kind of "targeted metabolomics" but the obtained information is very sparse.

To overcome this limitation, additional mass spectrometry (MS) or nuclear magnetic resonance (NMR) spectroscopy approaches can be used to generate a more holistic readout of the different metabolic processes occurring inside the brain. One major challenge herein is the availability of different technologies and the analytical challenges for very small or very large metabolites. Furthermore, the absolute quantitation of CSF metabolites is a prerequisite for giving CSF samples a reliable diagnostic meaning. Due to the different types of MS and NMR systems available worldwide, very different CSF metabolomes have been reported in the past. Furthermore, there is evidence that CSF is not a very stable sample matrix.

Naturally, CSF is a vital analyte to monitor neurological disease and aging processes; however, compared to other biofluids, it is relatively difficult to obtain, and the collection comes with a certain risk for strong headaches. This factor usually means that CSF samples are often strictly limited for research and analysis. Note that when analyzing CSF samples using distinct analytical techniques, one must be aware that several preanalytical factors may influence the overall obtained profiles of the CSF metabolome [1]. Major factors that contribute to metabolite alterations are time delays in preparation and quenching, storage and ambient temperature, and also repeated freeze and thaw cycles.

At +4 °C, the CSF metabolome was shown to be relatively stable, but at room temperature it can change significantly [2, 3]. Noga et al. (2018) found if samples are refrigerated, centrifuged and frozen within 2 h of sampling, most amines can be considered stable [4]. Another important aspect is cell removal by centrifugation as in certain pathologies there can be a high background of, e.g., immune cells whose metabolites contribute to the overall CSF metabolome if not removed beforehand. It was found that if the cells are spun down by centrifugation as soon as the sample is collected, the CSF metabolome is stable for 2 h at ambient room temperature [3, 5]. Regarding the protein background removal, some sources suggest that CSF centrifugation is not necessary [2, 6], while the study by Albrecht at al. (2020) suggested a procedure that allows even sensitive and low-concentration metabolites to remain stable for up to 48 h by maintaining a biological pH and keeping samples refrigerated during collection. In addition [2], they provided an experimental description of how the time-related degradation of CSF samples can be studied by NMR time series experiments. Finally, there is always a chance of CSF contamination with blood cells, so a visual inspection of CSF for clear appearance is one of the first quality checks prior to starting any metabolomics analysis.

These different confounders are only partially discussed in the literature, and in many published studies using CSF samples, it remains unclear to what extent they have been considered. As a consequence, for some years now, researchers have been stressing the requirement for a stable procedure for the collection and optimal storage of CSF if metabolomics investigations shall be performed [1, 7], also considering that certain metabolites with inherent volatility or chemical instability might get lost [6].

To date, different studies have tried to characterize the human CSF metabolome by combining different metabolomics techniques. Herein, the first comprehensive work was published by Wishart et al. (2008), who identified a total of 308 metabolites by a combination of NMR, GC-MS and LC-MS approaches [6]. Here, besides a core of common metabolites, NMR provided a large set of metabolites that are unique (21 compounds), whereas LC-MS demonstrated 4 unique metabolites, and GC-MS rendered a targeted list of 12 unique metabolites [6]. A more recent study from 2020 used LC and flow injection analysis (FIA) high-resolution MS (HRMS) methods, and using a set of 408 targeted metabolites, 196 metabolites were unambiguously detected in CSF samples [8]. Herein, a relatively low median coeffcient of variation was obtained while assessing the effect of prolonged storage of CSF samples by the LC-HRMS technique, which was 2.5%. Finally, a single triple quadrupole MS analysis kit focused on a targeted panel of 180 compounds can characterize the CSF metabolome, including sphingomyelins [8–10].

These studies show that there is ongoing technological progress towards the full annotation of the human CSF metabolome and also toward a higher throughput as shown in a novel developed method by matrix-assisted laser desorption ionization mass spectrometry imaging (MALDI-MSI) techniques that allows the analysis of pure human CSF on a MALDI plate [11]. So, the purpose of this work was to apply robust NMR at 600 MHz together with UHPLC ion mobility mass spectrometry to freshly collected human CSF and come up with a baseline list of metabolites that can be unambiguously identified in a multi-modal metabolomics approach.

2 Materials

CSF samples were freshly collected at the clinical ward upon clinical implication (normalization of brain pressure) and directly transported on dry ice to a −80 °C freezer where they were stored for 3 months until analysis of this study (*see* **Note 10**).

2.1 Materials for NMR Analysis

2.1.1 Chemicals

1. Potassium phosphate dibasic (anhydrous, 99.99%).
2. Deuterium oxide (99% by the ^{2}H isotope).
3. Sodium azide (99.5%).
4. TSP—Trimethylsilylpropanoate sodium salt (98% by the ^{2}H isotope).
5. Demineralized water, e.g., millipore or LC-MS grade.

2.1.2 Consumables

1. 1.5 mL or 2.0 mL safe-lock tubes (e.g. Eppendorf).
2. Pipette tips.
3. NMR tubes (with autosampler fitting cap for automated loading).
4. Quality Control tubes (e.g., Bruker IVDr QC solutions).

2.1.3 Equipment

1. Mechanical pipettes (10–100 μL and 100–1000 μL).
2. 600 MHz NMR IVDr (In-Vitro Diagnostics research, e.g., Bruker).
3. Auto-sampling system (e.g., Bruker SampleJET).
4. NMR 5 mm Triple Resonance (TXI) room temperature probe.

2.2 Materials for LC-MS Analysis

2.2.1 Chemicals

1. Acetonitrile LC-MS or Ultra-LC-MS grade (≥99.5%) (*see* **Note 1**).
2. Ultrapure water (*see* **Note 2**).
3. Formic acid, LC-MS grade (≥99%) (*see* **Note 3**).
4. Ammonium formate, LC-MS grade (≥98%).
5. Mobile phase A for RP-LC: 0.1% formic acid in ultrapure water (v:v).
6. Mobile phase B for RP-LC: 0.1% formic acid in acetonitrile (v: v).
7. Weak wash for RP-LC: ultrapure water.
8. Strong wash for RP-LC: acetonitrile.
9. Mobile phase A for HILIC: 10 mM ammonium formate and 0.1% formic acid in ultrapure water (v:v).
10. Mobile phase B for HILIC: 10 mM ammonium formate and 0.1% formic acid in 90% acetonitrile (v:v).
11. Weak wash for RP-LC: 90% acetonitrile (v:v).
12. Strong wash for RP-LC: ultrapure water.

2.2.2 Consumables

1. 1.5 mL-Eppendorf tubes and pipette tips (*see* **Note 4**).
2. HPLC vials with caps and 200 μL inserts for HPLC vials.
3. Glass pipettes (graduated and/or volumetric).

4. RP-LC column: Intensity Solo HPLC Column C18, 100 × 2 mm, 2 μm (Bruker) or equivalent.
5. HILIC column: ACQUITY UPLC BEH Amide Column (150 × 2.1 mm, 1.7 μm, Waters, Milford, MA, USA) or equivalent.

2.2.3 Equipment

1. Evaporator, e.g., Concentrator plus (Eppendorf).
2. Mechanical pipettes (10–100 μL and 100–1000 μL).
3. Centrifuge with cooling function and at least 15,000 rpm.
4. Vortex-mixer.
5. UHPLC system, e.g., Elute plus (Bruker).
6. ESI source, e.g., VIP-HESI (Bruker).
7. Mass spectrometry analyzer with ion mobility unit, e.g., tim-sTOF Pro 2 (Bruker).

3 Methods

3.1 NMR Parameters

1. Sampling volume (inside NMR tube—CSF and with buffer): 600 μL.
2. Auto-sampling system holder temperature (+ 6 °C).
3. NMR probe recording temperature (300 K, + 27 °C).
4. One-dimensional NOESY sequence (32 scans, 7 min 39 s long).
5. Two dimensional JRES pulse program (2 scans, 4 min 14 s long).

3.2 Sample Preparation for NMR

1. Thaw up samples from −80 °C storage to room temperature, ca. 30 min (*see* **Note 11**).
2. Add 150 μL Buffer solution (deuterated phosphate buffer with sodium azide, *see* **Note 12**) to a safe-seal tube (of either 1.5 or 2.0 mL volume).
3. Add CSF aliquot to safe-seal tube (750 μL). (*see* **Note 13**)
4. Gently mix the resulting mixture (CSF NMR sample solution) by rotating by hand for approximately 1 min.
5. Transfer 600 μL from the resulting mixture to an NMR tube (5 mm diameter size) and placed within the NMR autosampler holder.

3.3 LC Parameters

1. Injection volume: 5 μL.
2. Autosampler temperature: 4 °C.
3. Oven temperature: 50 °C.

4. Flow rate (RP-LC): 0.4 mL/min.
5. Gradient (RP-LC): 0–2 min, 5% B; 2–10 min, 5–30% B; 10–13 min, 30–60% B; 13–14 min 60–98% B; 14–15.6 min, 98% B; 15.6–16 min, 98–5% B; 16–18 min 5% B.
6. Flow rate (HILIC): 0.5 mL/min.
7. Gradient (HILIC): 0–3 min, 100% B; 3–10 min, 100–85% B; 10–14 min, 85–50% B; 14–15 min 50% B; 15–15.1 min, 50–100% B; 15.1–23 min, 100% B.

3.4 Ion Source Parameters

1. End plate offset: 500 V.
2. Capillary voltage: 4500 V.
3. Nebulizer pressure: 2 bar.
4. Dry Gas Temperature: 230 °C.
5. Dry gas Flow: 8.0 L/min.
6. Sheath Gas Temperature: 400 °C.
7. Sheath Gas Flow: 4.0 L/min.

3.5 IMS-MS Parameters

The settings are given here for the timsTOF Pro 2 instrument we used for our analyses. For other instruments, terminology and optimal settings might differ. Consult the manufacturer manuals or the support for optimal settings on your instrument.

1. Acquisition mode: PASEF (dda MS/MS).
2. Number of PASEF ramps: 2.
3. Polarity: positive and negative.
4. Mass range: 50–1300 Da.
5. Mobility range ($1/K_0$): 0.10–1.50 V*s/cm^2.
6. Tims Ramp time: 100 ms.
7. Collision RF: 450 Vpp.
8. TOF transfer time: 65 μs.
9. Pre Pulse Storage: 3 μs.
10. TIMS Stepping: Enabled (two steps).
11. Collision Energy: 10 eV (MS), 20/50 eV (MS/MS Stepping).

3.6 Sample Preparation for LC-MS

1. Thaw the samples at 4 °C (e.g., in the fridge).
2. Transfer 100 μL of CSF into a 1.5 mL Eppendorf tube.
3. Create at least one procedural blank per batch by transferring 100 μL of ultrapure water into an Eppendorf tube and treating it the same as the regular sample (i.e., performing all steps described below).
4. Add 300 μL of ice-cold (−20 °C) acetonitrile to the mixture and vortex for 1 min.
5. Incubate at −20 °C for 10 min.
6. Centrifuge at 4 °C for 15 min with 14,000 rpm.

7. Transfer 50 μL of the supernatant into a HPLC vial with insert. This sample is used for HILIC measurements.
8. Transfer 10 μL to the pooled QC for HILIC measurements (*see* **Note 8**).
9. Transfer 300 μL of the remaining supernatant from centrifugation into a 1.5 mL Eppendorf tube and evaporate to dryness at an evaporator (*see* **Note 5**).
10. Reconstitute the sample in 60 μL of ultrapure water:acetonitrile 9:1 (v:v) and vortex for 10 s.
11. Centrifuge at 14,000 rpm at 4 °C for 15 min.
12. Transfer 50 μL of the supernatant into an HPLC vial with insert. This sample is used for RP measurements.
13. Transfer 5 μL to the pooled QC for RP measurements.

3.7 LC-IM-MS Measurements

1. Calibrate the gas flow for the ion mobility analyzer. This should be done before every batch of analyses.
2. If necessary, perform the calibration procedure for the mass spectrometer in the polarity that is going to be measured.
3. If necessary, perform the calibration procedure for the ion mobility analyzer.
4. Prepare solvents for either RP-LC or HILIC measurements. The solvents can be used for several days, but should be checked for contamination both visually as well as in blank runs without injection.
5. Set up the sample table by starting with a few blank injections, followed by several injections of the pooled QC to condition the column. Samples should be then analyzed in randomized order, with regular injections of pooled QC after every 5–10 sample injections to account for the linearity of the signals. The sequence should end with a pooled QC measurement, followed by at least one solvent blank injection (i.e., injection of pure solvent). (*see* **Note 6**)
6. Make sure that standard solutions for both m/z and mobility calibration are injected during each sample measurement. Perform re-calibration on each sample and re-measure samples where the error is >1 ppm after performing a calibration procedure.
7. When changing polarity during a continuous batch, add a pre-run time of 25–30 min before the first injection in the new polarity mode.

3.8 Data Processing

We performed data processing of our raw data using Metaboscape 2024b (Bruker Corporation, Billerica, MA, USA) with the following parameters.

Table 1
Tolerances and scorings

	Narrow	Wide	Unit
m/z	2.0	5.0	ppm
Retention time	0.1 (not used)	1.0 (not used)	min
mSigma	20	1000	
MS/MS scores	900	600	
CCS	1.0	5.0	%

1. Processing algorithm: T-REX 4D.
2. Peak detection: Feature Signal: Intensity; Intensity threshold: 1000 counts; Minimum 4D peak size: 100 points (*see* **Note 7**).
3. Enable recursive feature extraction; Minimum 4D Peak Size (recursive): 75 points.
4. Annotation criteria.

The following criteria apply to annotations via Metaboscape 2024b (Bruker Corporation, Billerica, MA, USA), where each parameter will be evaluated by the difference from the library value in a "narrow" or a "wide" match. Annotations should only be considered valid if accuarte mass (m/z) and at least one additional parameter (RT, mSigma (= isotopic pattern), MS/MS, CCS) fit the criteria. When using other annotation tools, terminology and scoring values may differ (Table 1).

In our case, annotations were performed in Metaboscape by matching with MS/MS spectra from spectral libraries and CCS values reported in CCS databases (where applicable) or the "CCS predict" algorithm of Metaboscape. The resulting annotations are therefore to be considered "Level 2" annotations according to Sumner et al. [9].

Towards NMR-based metabolomics, spectral data (from 7 min NOESY experiments) were taken as one-dimensional proton spectra profiles and fit with a library of 35 metabolites using commercial software for NMR mixture analysis (ChenomX NMR Suite 10.0, Edmonton, Canada, 2023).

4 Notes

1. CSF samples used in this study were collected freshly at the clinical ward and directly frozen on dry ice. This is crucial, as reaching the temperature of −80 °C causes metabolic quenching [12, 13], which is beneficial to the goal of completely unbiased metabolomic research. Of course, this perfect setting

often cannot be encountered in a clinical setting where the threshold of 2 h at room temperature for metabolic and proteomic stability often is passed [1].

2. In our experience, acetonitrile may start to show contamination peaks in chromatography just a few days after initial opening. We therefore recommend ordering/opening bottle sizes according to the expected solvent usage. Also, if possible, the original solvent bottle should be used as the reservoir for the mobile phase to reduce additional sources of contamination.
3. Ultrapure water can be bought pre-bottled (e.g., ROTISOLV® Ultra LC-MS from Carl Roth, Karlsruhe, Germany) or freshly prepared using a water purification system (e.g., a Milli-Q® systems from Merck Milipore, Darmstadt, Germany). In our experience, the water purification system, if maintained properly, exhibits several advantages: no storage is needed for solvent bottles; the possibility to freshly produce water before use minimizes contamination (e.g., accumulation in stored open bottles); and the best monetary value per liter if used regularly.
4. When pipetting formic acid (or any other acid), the use of plastic pipette tips should be avoided to prevent both contamination of the sample and damage to the mechanical pipette. Use glass pipettes instead.
5. To reduce contamination from additives and production aid residues of plastic pipette tips, only high-quality pipette tips should be used. If possible, non-autoclaved tips should be used. Alternatively, pipette tips can be pre-flushed at least three times with solvent to extract potential leachable prior to getting in contact with the sample.
6. If multiple aliquots of the same CSF sample are not available and one sample needs to be thawed up several times, between one to three freeze-and-thaw cycles have been reported to be reliable [12, 14].
7. For NMR analysis, the use of an azide salt that is added to the deuterated buffer prevents bacterial growth and hereby prevents further biological degradation, e.g., if contamination occurs [13, 15].
8. It is recommended to include in every study a pooled CSF quality control. A good example of such approach is given in [16].
9. HILIC in positive mode is able to cover a broad range of phospholipids due to interactions of the polar head group with the stationary phase. Therefore, separation of phospholipids mainly happens among the lipid subclasses (e.g., phosphatidylcholines, phosphatidylethanolamines, sphingomyelins,

etc.) in shorter gradients, while separation of the distinct isomers requires a longer run time [10].

10. We are using the Eppendorf Concentrator plus at the following settings: Mode = V-AL, T = off, t = 90 min.
11. The blanks at the beginning should also include the procedural blanks and are used to assess for contaminants that are present in the LC-MS system or were obtained during sample preparation. The solvent blanks at the end are used to check for possible carryover effects from metabolites.
12. Depending on the peak picking algorithm, a manual check of the extracted peak shapes is advised. If you notice a high amount of unacceptable peaks for your extracted features, the processing parameters should be adjusted to reduce unnecessary feature generation.

5 Annotated Metabolites timsTOF LC-MS and NMR

Tables 2, 3, and 4 show the annotated CSF metabolites for RP-LC-IMS-MS in positive and negative mode, HILIC-IMS-MS in positive and negative mode and NMR spectroscopy.

6 Conclusions

In general, metabolomic profiling of human CSF by LC-MS is a promising research field for developing new diagnostic and prognostic biomarkers that highlight differences in phenotypic changes in neurological disorders or complications [17, 18]. Also, NMR spectroscopy-based CSF metabolomics proved to be a suitable tool to investigate neurodegenerative disorders and even compare CSF findings with blood serum metabolomics [19]. Whenever possible, a multi-modal metabolomics approach will provide the most comprehensive readout, as the combination of RP-LC, HILIC and NMR as complementary techniques resulted in a coverage of metabolites that could not be reached by relying on only one of these techniques alone. However, the inclusion of any additional analysis technique, such as dedicated Lipidomics methods or GC-MS, should lead to an even better coverage of the metabolome when analyzed in an untargeted manner. Such works can be quite impactful [10, 20], so we believe that novel instrumental and methodological approaches will further move the field forward, and the implementation of ion mobility mass spectrometry CSF metabolic profiling herein is one important cornerstone.

Table 2
Annotated features from RP-LC measurements with name, CAS identifier (if available) and Annotation Quality (AQ) score

Name	CAS	Positive (+)/ negative (-)	AQ
9-HpODE	5502-91-0	–	20110
(2E)-4-Hydroxynon-2-enoic acid	95087-42-6	–	20220
2-Hydroxypalmitic acid	764-67-0	–	20221+
2-Hydroxyphenylacetate	611-71-2	–	20202
2-Ketohexanoic acid	2492-75-3	–	20201+
2-Mercaptobenzothiazole	149-30-4	–	20220
2-Methylglutarate	18069-17-5	–	20202
(Z)-3-(4-iodophenyl)-2-sulfanylprop-2-enoic acid	179528-45-1	–	20100
3,6-Ditert-butyl-1,2-benzenediol	15512-06-8	–	20200
3-Chlorobenzoic acid	535-80-8	–	20100
3-Dehydroepiandrosterone sulfate	651-48-9	–	20120
3-Hydroxybenzaldehyde	100-83-4	–	20201
3-Hydroxycaprylic acid	88930-08-9	–	20221+
3-Hydroxyvaleric acid*	**10237-77-1**	–	20201+
3-Methyl-2-oxovalerate	1460-34-0	–	20201
4-Amino-5-(2-methylfuran-3-yl)-4H-1,2,4-triazole-3-thiol	560998-03-0	–	20010
4-Chlorobenzoic acid	74-11-3	–	20220
4-Hydroxyphenyllactic acid	306-23-0	–	10221+
4-Octylphenol	1806-26-4	–	20121+
9-(2,3-Dihydroxypropoxy)-9-oxononanoic acid	109421-77-4	–	20200
9,10,13-TriHOME	29907-57-1	–	20221+
9,10-DHOME	125356-86-7	–	20202+
9-Hydroxynonanoic acid	3788-56-5	–	20202+
9-OxoODE	54232-59-6	–	20220
Adrenic acid	28874-58-0	–	20102
α-Linolenic acid	463-40-1	–	20101
Androsterone sulfate	2479-86-9	–	20202+
Arachidonic acid	506-32-1	–	20220
Azelate	123-99-9	–	20202

(continued)

Table 2
(continued)

Name	CAS	Positive (+)/ negative (-)	AQ
Cefdinir	91832-40-5	–	20200
Cefotaxime	63527-52-6	–	20220
Curvulinic acid	19053-94-2	–	20110
Docosahexaenoic acid	6217-54-5	–	20201+
D-Tryptophan	153-94-6	–	20201
Glycochenodeoxycholate	16564-43-5	–	20202
Hexadecanedioic acid	505-54-4	–	20202+
Hippuric acid	495-69-2	–	20121
Indolelactic acid	1821-52-9	–	20201+
Lauric acid	143-07-7	–	20202
Linoleic acid	60-33-3	–	20202
LPA 18:3		–	*20200*
LPE 18:2		–	*20211*
LPE 20:4		–	*20212*
LPS 14:3		–	*10202*
LysoPC (13:0)	20559-17-5	–	20201+
N-Glycocholate	475-31-0	–	20102
Oleic acid	112-80-1	–	20202
Palmitoleic acid	373-49-9	–	10202
Paullinic acid	5561-99-9	–	20202
PE-Cer 30:6;O3		–	20100
Phenylacetyl L-Glutamine	28047-15-6	–	20221
Phthalic acid	88-99-3	–	20201+
Propofol β-D-glucuronide	114991-26-3	–	20221+
Sebacate	111-20-6	–	20202
Suberate	505-48-6	–	20201
Tetradecanedioic acid	821-38-5	–	20222+
Thromboxane B2	54397-85-2	–	20102
Uric acid*	69-93-2	–	20221
Ursodeoxycholic acid	128-13-2	–	20102
1,2-Dilauroyl-sn-glycero-3-phosphocholine	18194-25-7	+	20102

(continued)

Table 2
(continued)

Name	CAS	Positive (+)/ negative (-)	AQ
1,2-Dipalmitoyl-sn-glycero-3-phosphocholine*	63-89-8	+	20102
1-Methyladenosine*	15763-06-1	+	20011+
1-Methylguanine	938-85-2	+	20102+
1-Methylxanthine	6136-37-4	+	20221+
2-Hydroxybutyric acid	**600-15-7**	+	20201
3-Benzylhexahydropyrrolo[1,2-a]pyrazine-1,4-dione	14705-60-3	+	20222+
3-Hydroxybutyric acid	**625-08-01**	+	2,202+
3-Methylxanthine	1076-22-8	+	20122
5-Methyltetrahydrofolate	68792-52-9	+	20210+
7α-Hydroxy-3-oxo-4-cholestenoic acid	1246298-65-6	+	20210+
Allose	2595-97-3	+	20202
Benzoic acid	65-85-0	+	10102
Bilirubin	635-65-4	+	20201+
Butyrylcarnitine*	25576-40-3	+	20221+
Cortisol	50-23-7	+	20212
Caffeine*	**58-08-2**	+	20221
Choline*	62-49-7	+	20220+
Creatine*	**57-00-1**	+	20202
Creatinine*	**60-27-5**	+	20202+
D-Pantothenic acid*	79-83-4	+	20222
Dilauryl thiodipropionate	123-28-4	+	20111+
Dimethyl sulfone	**67-71-0**	+	20102+
Entecavir	142217-69-4	+	20201+
Hypoxanthine*	68-94-0	+	20201
Indoleacrylic acid	1204-06-4	+	20201+
Indoline	496-15-1	+	20211+
Isovalerylcarnitine*	31023-24-2	+	20201+
L-Leucine*	**61-90-5**	+	20222
L-Methionine	63-68-3	+	20022+
L-Propionylcarnitine*	20064-19-1	+	20221+

(continued)

Table 2
(continued)

Name	CAS	Positive (+)/ negative (-)	AQ
L-Tyrosine*	**60-18-4**	+	20202
m-Coumaric acid	588-30-7	+	20201^{+}
N,N-Dimethylguanosine	2140-67-2	+	20221^{+}
N-Acetylaspartylglutamic acid	3106-85-2	+	20201^{+}
Normetanephrine	97-31-4	+	20102^{+}
Norvaline	6600-40-4	+	20102^{+}
O-Acetylcarnitine*	5080-50-2	+	20102
Octanoylcarnitine	25243-95-2	+	20121^{+}
Paraxanthine*	611-59-6	+	20202
Phenol	108-95-2	+	20101^{+}
PS 36:01		+	*20102*
PS 38:02		+	*20102*
Pyridoxamine	85-87-0	+	20102^{+}
Ribitol	488-81-3	+	20002
SPB 14:3;O2		+	*20200*
Succinyladenosine	4542-23-8	+	20201^{+}
Theobromine	83-67-0	+	20221
Theophylline	58-55-9	+	20221
Topiramate*	97240-79-4	+	20222^{+}
Uridine*	58-96-8	+	20202
Vanylglycol	534-82-7	+	20200
δ-Valerolactam	675-20-7	+	20211^{+}
Cinnamic acid	140-10-3	Both	20201 20201^{+}
L-Phenylalanine	**63-91-2**	Both	20212^{+} 20221
L-Tryptophan	73-22-3	Both	20211 20221

Metabolites marked with an asterisk (*) were also annotated in HILIC measurements. The polarity setting with which the feature was detected is also given as + (positive), – (negative) or "both." The Annotation Quality (AQ) showcases which parameters matched with literature values (digit 1: m/z, digit 2: RT (not used), digit 3: mSigma, digit 4: MS/MS spectra, digit 5: CCS value) at which confidence (2 = "narrow," 1 = "wide," 0 = no match). Matching with predicted CCS values is marked with a superscript "+". Lipids in *italic* were annotated through the" Lipid species" prediction algorithm provided by Metaboscape. Metabolites that were also annotated using NMR are marked in **bold**

Table 3
Annotated features from HILIC measurements with name, CAS identifier (if available) and Annotation Quality (AQ) score

Name	CAS	Positive (+)/ negative (-)	AQ
2-Hydroxy-4-methylvaleric acid	26671-68-1	–	20221+
3-Hydroxymethylglutarate	503-49-1	–	20202
3-Hydroxyvaleric acid*	10237-77-1	–	20201+
Acetazolamide	59-66-5	–	20220+
Anserine	10030-52-1	–	20202
Aconitic acid	4271-99-2	–	20201
Citric acid	**77-92-9**	–	20,221
D-Galactonic acid	13382-27-9	–	20222+
D-Glucuronolactone	32449-92-6	–	20202+
D-Lysine	923-27-3	–	20002+
D-Mannitol	69-65-8	–	20221+
Dodecylbenzenesulfonic acid	121-65-3	–	20111+
D-Threo-isocitric acid	6061-97-8	–	20200
Erythrose	1758-51-6	–	20200+
Glyceric acid	473-81-4	–	20200+
Inositol	643-12-9	–	20201
Lactic acid	**79-33-4**	–	20201
Ascorbic acid	**50-81-7**	–	20221
L-Ribonic acid	710941-59-6	–	20201
L-Threonic acid	7306-96-9	–	20221
Mesaconic acid	498-24-8	–	20211
N-Acetylaspartic acid	997-55-7	–	20011+
N-Acetylneuraminate	131-48-6	–	20202
N-Formyl-L-methionine	4289-98-9	–	20221
Paracetamol sulfate	32113-41-0	–	20201+
Uric acid*	69-93-2	–	20221
Xanthine	69-89-6	–	20221
3-Hydroxyisovaleroylcarnitine	99159-87-2	+	20121+
16:0 Lyso PC	17364-16-8	+	20122
	63-89-8	+	20202

(continued)

Table 3
(continued)

Name	CAS	Positive (+)/ negative (-)	AQ
1,2-Dipalmitoyl-sn-glycero-3-phosphocholine*			
1-Methyladenosine*	15763-06-1	+	20221+
1-Stearoyl-2-oleoyl-sn-glycero-3-phosphoethanolamine	6418-95-7	+	20102
2-(β-Glucosyl)glycerol	10588-30-4	+	20102
2-Hydroxypyridine	142-08-5	+	20001
2-Nitrobenzylamine	1904-78-5	+	20222+
3-Hydroxyanthranilic acid	548-93-6	+	20220
3-Hydroxybutyrylcarnitine	875668-57-8	+	20201+
5,6-Dimethylbenzimidazole	582-60-5	+	20102+
7-Methylguanine	578-76-7	+	20102
Betaine	107-43-7	+	20202
Butyrylcarnitine*	25576-40-3	+	20221+
Caffeine*	**58-08-2**	+	20221
Choline*	62-49-7	+	20220+
Creatine*	**57-00-1**	+	20202
Creatinine*	**60-27-5**	+	20,200+
Deoxycarnitine	407-64-7	+	20002
D-Glucosamine	3416-24-8	+	10202
Glucose	**50-99-7**	+	20202
2-Aminooctanoic acid	644-90-6	+	20201+
D-Pantothenic acid*	79-83-4	+	20122
Eplerenone	107724-20-9	+	20101+
Glufosinate	77182-82-2	+	20202+
Glutarylcarnitine	102636-82-8	+	20102+
Hecogenin	467-55-0	+	20101+
Hypoxanthine*	68-94-0	+	20201
Isovalerylcarnitine*	3*1023-24-2	+	20021+
L-Alanine	**56-41-7**	+	20201+
L-Carnitine	6645-46-1	+	20202

(continued)

Table 3 (continued)

Name	CAS	Positive (+)/ negative (-)	AQ
L-Leucine*	**61-90-5**	+	20102
L-Lysine	56-87-1	+	20022
L-Pipecolic acid	3105-95-1	+	20201+
L-Propionylcarnitine*	20064-19-1	+	20121+
L-Tyrosine*	**60-18-4**	+	20202
N,N,N-Trimethyllysine	55528-53-5	+	20202
N-Acetyl-D-glucosamine	7512-17-6	+	20102
N-Acetylgalactosamine	1811-31-0	+	20200+
N-Acetyl-L-Histidine	2497-02-1	+	20222+
N-Acetylputrescine	18233-70-0	+	20002
N-alpha-Acetyl-L-asparagine	4033-40-3	+	20201+
N,N-Dimethyl-L-arginine	30315-93-6	+	20221+
Niacinamide	98-92-0	+	20200+
O-Acetylcarnitine*	5080-50-2	+	20202
Oleoyl ethylamide	85075-82-7	+	20110
Paraxanthine*	611-59-6	+	20202
PC (18:1/18:1) (Δ9-trans)	56782-46-8	+	20102
PC 30:0		+	*20221*
PC 32:00		+	*20200*
PC 32:01		+	*20202*
PC 34:02		+	*20110*
PC 35:01		+	*20202*
PC 36:01		+	*20102*
PC 36:4		+	*20122*
PC 38:2		+	*20102*
PC 38:4		+	*20101*
PC 38:5		+	*20101*
PC 38:7		+	*20101*
PC 40:4		+	*20201*
PC 40:5		+	*20112*
PC 40:7		+	*10112*

(continued)

Table 3 (continued)

Name	CAS	Positive (+)/ negative (-)	AQ
PC O-34:1		+	*20112*
PC O-34:2		+	*20112*
PC O-36:2		+	*10102*
PC O-36:4		+	*20112*
PC(18:1(9Z)/16:0)	59491-62-2	+	10112
PE (O-36:03)		+	*20102*
PE (O-38:05)		+	*20212*
PE 34:0		+	*20122*
PE 36:03		+	*10112*
PE 38:02		+	*20102*
PE 38:04		+	*20121*
PE 40:06		+	*20101*
PE O-38:7		+	*20100*
PE O-40:5		+	*20101*
PE O-40:8		+	*20100*
Proline betaine	32039-73-9	+	10121+
SM 32:1;O2		+	*20211*
SM 34:01		+	*20202*
SM 34:2;O2		+	*20102*
SM 36:01		+	*20202*
SM 36:02		+	*20102*
SM 38:01		+	*20202*
SM 38:2;O2		+	*20112*
SM 40:01		+	*20001*
SM 40:02		+	*20202*
SM 42:03		+	*20101*
Topiramate*	97240-79-4	+	20211+
Uridine*	58-96-8	+	20002+
L-Arginine	74-79-3	Both	20222 20221+
β-Pseudouridine	1445-07-4	Both	

(continued)

Table 3 (continued)

Name	CAS	Positive (+)/ negative (-)	AQ
			20201 20202+
Fructose	**57-48-7**	Both	20200+ 20002
Galactitol	608-66-2	Both	20201+ 20202
Glutamine	**56-85-9**	Both	20202+ 20202
***myo*-Inositol**	**87-89-8**	Both	20201 20201
Pyroglutamate	98-79-3	Both	20011+ 20202

Metabolites marked with an asterisk (*) were also annotated in RP-LC measurements. The polarity setting with which the feature was detected is also given as + (positive), – (negative) or "both." The Annotation Quality (AQ) showcases which parameters matched with literature values (digit 1: m/z, digit 2: RT (not used), digit 3: mSigma, digit 4: MS/MS spectra, digit 5: CCS value) at which confidence (2 = "narrow," 1 = "wide," 0 = no match). Matching with predicted CCS values is marked with a superscript "+". Metabolites that were also annotated using NMR are marked in **bold**

Table 4
Annotated metabolites from NMR profiling at 600 MHz and their associated CAS identification number

Name	CAS
2-Hydroxybutyric acid	**600-15-7**
2-Hydroxyisovaleric acid	4026-18-0
3-Hydroxybutyric acid	**625-72-9**
3-Hydroxyisobutyric acid	**2068-83-9**
3-Hydroxyisovaleric acid	**625-08-1**
Acetic acid	64-19-7
Acetoacetate	541-50-4
Acetone	67-64-1
Alanine	**56-41-7**
Alpha-ketoisovaleric acid	759-05-7
Ascorbic acid	**50-81-7**
Caffeine	**58-08-2**

(continued)

Table 4
(continued)

Name	CAS
Citric acid	**77-92-9**
Creatine	**57-00-1**
Creatinine	**60-27-5**
Dimethyl sulfone	**67-71-0**
Ethanol	64-17-5
Formic acid	64-18-6
Fructose	**57-48-7**
Glucose	**50-99-7**
Glutamine	**56-85-9**
Histidine	71-00-1
Isoleucine	73-32-5
Isopropanol (artefact disinfectant)	67-63-0
Lactic acid	**79-33-4**
Leucine	**61-90-5**
Mannose	3458-28-4
Phenylalanine	**63-91-2**
Propylene glycol (artefact disinfectant)	57-55-6
Propanol (artefact disinfectant)	71-23-8
Pyruvic acid	127-17-3
Tyrosine	**60-18-4**
Urea	57-13-6
Valine	72-18-4
***myo*-Inositol**	**87-89-8**

Metabolites that were also annotated using IM-MS are marked in **bold**

Acknowledgments

The authors want to greatly thank Daniele Bucci, Prof. Kathrin Brockmann and Prof. Thomas Gasser from the University Hospital Tübingen, 72076 Tübingen, Germany, for helping in the organization, collection and cryogenic transport of fresh CSF samples. The authors furthermore greatly acknowledge the support by Hartmut Schäfer and Claire Wegner from Bruker BioSpin GmbH & Co. KG, 76275 Ettlingen, Germany, as well as Sven Meyer and Stefanie

Wernisch from Bruker Daltonics GmbH & Co. KG, 28359 Bremen, Germany. CT reports an active research grant by Bruker Switzerland AG.

References

1. Rosenling T, Slim CL, Christin C, Coulier L, Shi S, Stoop MP et al (2009) The effect of preanalytical factors on stability of the proteome and selected metabolites in cerebrospinal fluid (CSF). J Proteome Res 8(12): 5511–5522. https://doi.org/10.1021/pr9005876
2. Albrecht B, Voronina E, Schipke C, Peters O, Parr MK, Díaz-Hernández MD et al (2020) Pursuing experimental reproducibility: an efficient protocol for the preparation of cerebrospinal fluid samples for NMR-based metabolomics and analysis of sample degradation. Meta 10(6):251. https://doi.org/10.3390/metabo10060251
3. Fomenko MV, Yanshole LV, Tsentalovich YP (2022) Stability of metabolomic content during sample preparation: blood and brain tissues. Meta 12(9):811
4. Noga MJ, Zielman R, van Dongen RM, Bos S, Harms A, Terwindt GM et al (2018) Strategies to assess and optimize stability of endogenous amines during cerebrospinal fluid sampling. Metabolomics 14(4):44. https://doi.org/10.1007/s11306-018-1333-0
5. Rosenling T, Stoop MP, Smolinska A, Muilwijk B, Coulier L, Shi S et al (2011) The impact of delayed storage on the measured proteome and metabolome of human cerebrospinal fluid. Clin Chem 57(12):1703–1711. https://doi.org/10.1373/clinchem.2011.167601
6. Wishart DS, Lewis MJ, Morrissey JA, Flegel MD, Jeroncic K, Xiong Y et al (2008) The human cerebrospinal fluid metabolome. J Chromatogr B 871(2):164–173. https://doi.org/10.1016/j.jchromb.2008.05.001
7. Wuolikainen A, Hedenström M, Moritz T, Marklund SL, Antti H, Andersen PM (2009) Optimization of procedures for collecting and storing of CSF for studying the metabolome in ALS. Amyotroph Lateral Scler 10(4): 229–2U8. https://doi.org/10.1080/17482960902871009
8. Carlsson H, Abujrais S, Herman S, Khoonsari PE, Åkerfeldt T, Svenningsson A et al (2020) Targeted metabolomics of CSF in healthy individuals and patients with secondary progressive multiple sclerosis using high-resolution mass spectrometry. Metabolomics 16(2):26. https://doi.org/10.1007/s11306-020-1648-5
9. Koal T, Klavins K, Seppi D, Kemmler G, Humpel C (2015) Sphingomyelin SM(d18:1/18:0) is significantly enhanced in cerebrospinal fluid samples dichotomized by pathological amyloid-β 42, tau, and phospho-tau-181 levels. J Alzheimers Dis 44:1193–1201. https://doi.org/10.3233/JAD-142319
10. Mandal R, Guo AC, Chaudhary KK, Liu P, Yallou FS, Dong E et al (2012) Multi-platform characterization of the human cerebrospinal fluid metabolome: a comprehensive and quantitative update. Genome Med 4(4):38. https://doi.org/10.1186/gm337
11. Vallianatou T, Nilsson A, Bjärterot P, Shariatgorji R, Slijkhuis N, Aerts JT et al (2023) Rapid metabolic profiling of 1 μL crude cerebrospinal fluid by matrix-assisted laser desorption/ionization mass spectrometry imaging can differentiate De Novo Parkinson's disease. Anal Chem 95(50):18352–18360. https://doi.org/10.1021/acs.analchem.3c02900
12. Xu D, Dai X, Zhang L, Cai Y, Chen K, Wu J et al (2024) Mass spectrometry for biomarkers, disease mechanisms, and drug development in cerebrospinal fluid metabolomics. TrAC Trends Anal Chem 173:117626. https://doi.org/10.1016/j.trac.2024.117626
13. Vuckovic D (2012) Current trends and challenges in sample preparation for global metabolomics using liquid chromatography–mass spectrometry. Anal Bioanal Chem 403(6): 1523–1548. https://doi.org/10.1007/s00216-012-6039-y
14. Haijes HA, Willemse EAJ, Gerrits J, van der Flier WM, Teunissen CE, Verhoeven-Duif NM et al (2019) Assessing the pre-analytical stability of small-molecule metabolites in cerebrospinal fluid using direct-infusion metabolomics. Meta 9(10):236
15. Want EJ, Wilson ID, Gika H, Theodoridis G, Plumb RS, Shockcor J et al (2010) Global metabolic profiling procedures for urine using UPLC–MS. Nat Protoc 5(6):1005–1018. https://doi.org/10.1038/nprot.2010.50
16. Ntranos A, Park H-J, Wentling M, Tolstikov V, Amatruda M, Inbar B et al (2021) Bacterial neurotoxic metabolites in multiple sclerosis

cerebrospinal fluid and plasma. Brain 145(2): 569–583. https://doi.org/10.1093/brain/awab320

17. Klinke G, Richter S, Monostori P, Schmidt-Mader B, García-Cazorla A, Artuch R et al (2020) Targeted cerebrospinal fluid analysis for inborn errors of metabolism on an LC-MS/MS analysis platform. J Inherit Metab Dis 43(4):712–725. https://doi.org/10.1002/jimd.12213
18. Carlsson H, Rollborn N, Herman S, Freyhult E, Svenningsson A, Burman J et al (2021) Metabolomics of cerebrospinal fluid from healthy subjects reveal metabolites associated with ageing. Meta 11(2):126
19. Berezhnoy G, Laske C, Trautwein C (2023) Metabolomic profiling of CSF and blood serum elucidates general and sex-specific patterns for mild cognitive impairment and Alzheimer's disease patients. Front Aging Neurosci 15. https://doi.org/10.3389/fnagi.2023.1219718
20. Nielsen JE, Maltesen RG, Havelund JF, Færgeman NJ, Gotfredsen CH, Vestergård K et al (2021) Characterising Alzheimer's disease through integrative NMR- and LC-MS-based metabolomics. Metabol Open 12:100125. https://doi.org/10.1016/j.metop.2021.100125

Chapter 18

An LC-MS/MS Multiplexed Method to Quantify 14 Sphingolipids in Human Cerebrospinal Fluid (CSF)

Yadira Perez-Paramo, Rathna Veeramachaneni, Emily Parkhurst, Dawn Dufield, W. Rodney Mathews, and Veronica G. Anania

Abstract

Sphingolipids are bioactive lipids comprised of a sphingoid base linked to various long-chain fatty acids and phosphocholine and, in some cases, may also contain one or more sugars. A few species of sphingolipids share similar structures and may even have identical chemical composition but differ in their biological function. For this reason, they can only be accurately quantified using liquid chromatography coupled with mass spectrometry (LC-MS). This technique can separate sphingolipids based on their hydrophobicity and quantify them based on their mass-to-charge ratio (m/z). This is especially important when measuring isomeric species that have identical m/z but differ in their retention time. Here, we describe the development of a new LC-MS method to measure 14 sphingolipid species, including 7 ceramides (C14, C16, C18, C20, C24, C18:1, C24:1), 4 glycosphingolipids (C16 Glucosyl(ß) Ceramide (d18:1/16:0, C16 Galactosyl (ß) Ceramide (d18:1/16:0), C16 Lac Cer, C18 Lac Cer) and 3 sphingosines (Sphingosine (Sph), Glucosyl (ß) Sphingosine (d18:1), Galactosyl(ß) Sphingosine (d18:1)) in human cerebrospinal fluid (CSF). Using this 7-min method, chromatographic resolution of isomeric species glucosylceramide (Glc Cer) and galactosylceramide (Gal Cer) was achieved. This resolution is important as these two sphingolipid classes have distinct biological functions with important implications in neurological diseases like multiple sclerosis.

Key words Sphingolipids, LC-MS, Isomers, Glc Cer, Gal Cer, CSF, Neurodegenerative diseases

1 Introduction

Sphingolipids are bioactive lipids comprised of a sphingoid base linked to various long-chain fatty acids ceramides (Cer), phosphocholine (sphingomyelins), and some species also contain one or more sugars (glycosphingolipids, cerebrosides). Sphingolipids are one of the most abundant lipid species in the brain, with sphingomyelin species representing around 20% of total brain lipids

The original version of the chapter has been revised. A correction to this chapter can be found at https://doi.org/10.1007/978-1-0716-4462-1_23

Enrique Santamaría and Joaquín Fernández-Irigoyen (eds.), *Cerebrospinal Fluid (CSF) Proteomics: Methods and Protocols*, Methods in Molecular Biology, vol. 2914, https://doi.org/10.1007/978-1-0716-4462-1_18,

[1]. The CNS myelin sheath, which provides nutrients and protection to axons, is rich in lipids (70–85% of dry weight); it is mainly comprised of cholesterol, glycerophospholipids, and glycosphingolipids at a ratio of 4:4:2 [2, 3]. The central nervous system (CNS) is rich in sphingolipids and changes in sphingolipids and their metabolism are involved in various neurological diseases such as Alzheimer's disease, Parkinson's disease, amyotrophic multiple and lateral sclerosis, and Gaucher disease [4–7].

Glycosphingolipids are major components of oligodendrocyte plasma membranes and myelin, and they are essential for the stability and maintenance of myelin during aging. The sugar attached to the sphingosine base determines the subclass of glycosphingolipid; for example, galactosylceramide (Gal Cer) contains galactose, glucosylceramide (Glc Cer) contains glucose, and lactosylceramide (Lac Cer) contains lactose. It is important to highlight that Gal Cer and Glc Cer are structural isomers with the same molecular weight but different structures (differing only in their hydrophobicity properties/retention time). Interestingly, Gal Cer species are enriched in myelin (24%), and their accumulation in the brain has been used to measure myelin biogenesis [8]. While Glc Cer species are ubiquitously present in cellular membranes, they are especially enriched in lipid rafts and likely play a key role in membrane organization [9]. Other important Glc Cer functions include regulating endocytic trafficking and controlling cytoskeletal dynamics [10, 11]. Myelin damage is the hallmark of several neurological disorders and therefore it is critical to have an analytical method that can discern between these biologically distinct species.

Cerebrospinal fluid (CSF) is an ideal matrix to use to better understand CNS disorders as it is an accessible disease-proximal biofluid at the interface between the blood and brain. Previously established methods to measure sphingolipids in CSF have been hindered by their inability to resolve glucosyl and galactosyl isomers [5, 12–15]. Therefore, new methods with higher sensitivity, specificity, and better resolution of isomeric lipid species are needed to inform the biological interpretations of changes in these lipid profiles in the context of health and disease.

Here, we report a 7-min LC-MS method to measure 14 sphingolipid species, including the simultaneous and specific quantification of the isomers C16 Glc Cer and C16 Gal Cer in CSF.

2 Materials

2.1 General Labware

1. Clean bench.
2. Centrifuge.
3. Vortex.
4. Nitrogen evaporator.

5. Pipettes.
6. −80 freezer.

2.2 Sample Preparation

1. Water (H_2O; MS grade; Midland Scientific).
2. Acetonitrile (ACN; MS grade; Midland Scientific).
3. Methanol (MeOH; MS grade; Midland Scientific).
4. Formic Acid (FA; MS grade; Sigma Aldrich).
5. Ammonium formate (MS grade; Sigma Aldrich).
6. Standards and internal standards for ceramides, hexosylceramides, and sphingosines (Avanti Polar Lipids). C16 lactosyl (b) ceramide d3 (Matreya LLC). C16 lactosyl (b)ceramide (Cayman Chemical) (Table 1).
7. Analytical column HALO HILIC 4.6× 150 mm, 2.7 μm (Mac-Mod Analytical Inc.).
8. State-of-the-art analytical LC system: UPLC Nexera LC-30 AD Nexera ×2 ultra-high-performance liquid chromatography system (Shimadzu).
9. State-of-the-art triple quadrupole mass spectrometer: Sciex 5500 operated in multiple reaction monitoring (MRM) mode (Sciex).
10. Mobile phase A (MPA): 100:0.5% ACN: FA (volume for volume), i.e., 99.5 mL of ACN plus 0.5 mL of FA.
11. Mobile phase B (MPB): 50 mM ammonium formate in 60: 40:0.5 H_2O: MeOH: FA (volume for volume), i.e., 59.75 mL of H_2O, plus 39.75 mL of MeOH, plus 0.5 mL of FA.

3 Methods

3.1 Stocks, Calibration Curve, and Quality Control Sample Preparation

1. Prepare a working buffer solution (MPAB) consisting of 97.5: 2.5 (volume for volume) MPA: MPB.
2. Standards and internal standard stock solutions were prepared at 1 and 100 μg/mL, respectively, in methanol and stored at − 80 °C for long-term storage.
3. The standard stocks were prepared as follows: a stock solution of 14-mix each at 1 μg/mL prepared 97.5 MPA: 2.5 MPB.
4. Internal standard (ISTD): A mix of 2 deuterated standards (C16 and C18 Lac Cer) were combined to create a mixture that was 100 μg/mL for each lipid in MPAB. The remaining 8 deuterated standards are combined at 10 μg/mL for each lipid in MPAB. A stock solution of all 10 deuterated standards was prepared by adding 2.5 μL of the Lac Cer 2 lipid mix with 2.5 μL of the 8 lipid mix with 49.5 μL of 97.5 MPA: 2.5 MPB to give an overall concentration of 11.25 μg/mL of Lac Cer species and 1.125 μg/mL of the remaining 8 deuterated standards.

Table 1
List of standards and internal standards (Modified from Perez-Paramo et al. [6])

Family	Standard product nomenclature	Abbreviated name	Internal standard product nomenclature	Abbreviated name
Ceramides	C14 Ceramide (d18:1/14:0)	C14 Cer	C16 Ceramide-d7 (d18:1-d7/16:0)	IS C16 d7
	C16 Ceramide (d18:1/16:0)	C16 Cer	C16 Ceramide-d7 (d18:1-d7/16:0)	IS C16 d7
	C18 Ceramide (d18:1/18:0)	C18 Cer	C18 Ceramide-d7 (d18:1-d7/18:0)	IS C18 d7
	C20 Ceramide (d18:1/20:0)	C20 Cer	C24 Ceramide-d7 (d18:1-d7/24:0)	IS C24 d7
	C24 Ceramide (d18:1/24:0)	C24 Cer	C24 Ceramide-d7 (d18:1-d7/24:0)	IS C24 d7
	C18:1 Ceramide (d18:1/18:1(9Z))	C18:1 Cer	C24:1 Ceramide-d7 (d18:1-d7/24:1(15Z))	IS C24:1 d7
	C24:1 Ceramide (d18:1/24:1(15Z))	C24:1 Cer	C24:1 Ceramide-d7 (d18:1-d7/24:1(15Z))	IS C24:1 d7
Glycosylceramides	C18 Lactosyl(ß) Ceramides	C18 Lac Cer	N-Omega-CD3-Hexadecanoyl-Lactosylceramide-d3	IS C16 LacCer d3
	C16 Lactosylceramides (d18:4/16:0)	C16 Lac Cer	N-Omega-CD3-Hexadecanoyl-Lactosylceramide-d3	IS C16 LacCer d3
	C16 Glucosyl(ß) Ceramide (d18:1/16:0)	C16 Glc Cer	C18 Glucosyl(ß) Ceramide-d5	C18 GluCer d5
	C16 Galactosyl (β) Ceramide (d18:1/16:0)	C16 Gal Cer	C18 Galactosyl (β) Ceramide-d35	C18 GalCer d35
Sphingosines	Sphingosine (d18:1) D-erythro-sphingosine	Sph	Sphingosine-d7	IS Sph d7
	Glucosyl(ß) Sphingosine (d18:1)	Glc Sph	Glucosyl(β) Sphingosine-d5	IS GluSph d5
	Galactosyl (β) Sphingosine (d18:1)	Gal Sph	Galactosyl (β) Sphingosine-d5	IS GalSph d5

5. Pipette 20.5 μL of the standard stock (1 μg/mL 16 mix) + 79.5 μL of MPAB to generate the first point in the calibration curve (204.8 ng/mL); then perform 11 consecutive serial dilutions down to 0.1 ng/mL in MPAB buffer to prepare the calibration curve.

6. Prepare the quality control (QC) samples at 0, 1.25, 5, 20, and 80 ng/mL in MPAB for surrogate matrix QCs and in CSF for matrix QCs. Prepare 80 ng/mL by pipetting 10.4 μL of stock (1 μg/mL 14 mix) + 119.6 μL of MPAB/CSF. Prepare 20 ng/mL by pipetting 32.8 μL of 80 ng/mL QC + 98.4 μL of

MPAB/CSF. Prepare 5 ng/mL by pipetting 31.2 μL of 20 ng/mL QC + 93.6 μL of MPAB/CSF. Prepare 1.25 ng/mL by pipetting 25 μL of 5 ng/mL QC + 75 μL of MPAB/CSF.

3.2 CSF Sample Preparation for Sphingolipid LC-MS Analysis

This protocol describes the sample preparation protocol to perform sphingolipid analysis for downstream LC-MS analysis (Fig. 1).

1. Pipette 20 μL of CSF, surrogate matrix (MPAB), or QC samples into a 96-well plate (*see* **Notes 1**, **2**, and **3**).
2. Add 120 μL of 10-mix working internal standard in MPAB. This will allow protein precipitation. Seal the plate with aluminum foil.
3. Vortex the plate vigorously for 1 min.
4. Centrifuge the plate at 3000 rpm for 5 min at room temperature.
5. Transfer 120 μL of supernatant into a new 96-well plate.
6. Dry down the supernatant at 45 °C under nitrogen.
7. Reconstitute the dried samples with 100 μL of MPAB.
8. Mix well the samples by vortexing.
9. Analyze the samples using the LC-MS method described in the next section.

3.3 LC-MS Analysis

This protocol describes the analysis of sphingolipids using a UPLC Nexera LC-30 AD Nexera ×2 ultra-high-performance liquid chromatography system (Shimadzu) coupled to a Sciex 5500 operated in multiple reaction monitoring (MRM) mode (Sciex). Equivalent instrumentation can be used.

1. To analyze the reconstituted samples by LC-MS, load 25 μL of the extracted samples onto the analytical column (HALO HILIC 4.6× 150 mm, 2.7 mm) at 50 °C.
2. Separate the sphingolipids using a gradient of MPA and MPB with 1% to 25% of MPB in 2.7 min, followed by a linear gradient to 40% MPB from 2.7 to 3.5 min, a linear gradient to 45% MPB from 3.5 to 4.0 min and a linear gradient to 1% of MPB from 4.0 to 4.1 min, and a re-equilibrium step of 1.9 min to the initial 1% MPB at a flow rate 1.0 mL/min.
3. The triple quadrupole was set in an MRM with a positive turbo spray ion drive mode with the following parameters: curtain gas

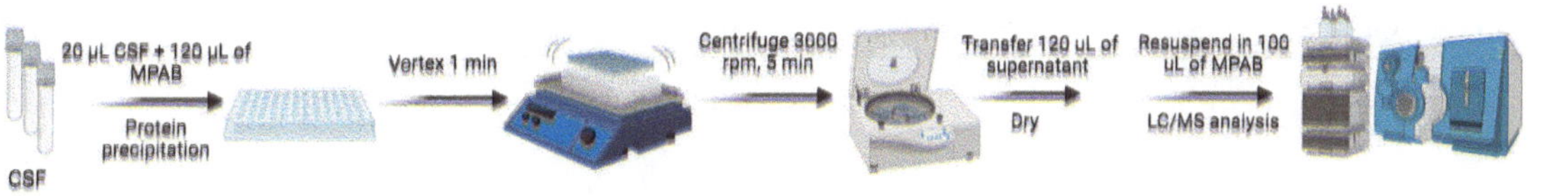

Fig. 1 CSF sample preparation protocol graphic overview. (Created in Biorender.com)

Table 2
Mass spectrometry parameters (Modified from Perez-Paramo et al. [6])

Abbreviated name	Q1 mass (Da)	Q3 mass (Da)	Dwell time (msec)	DP (volts)	EP	CE (volts)	C XP
C14 Cer	510.8	264.3	10	90	10	35	12
C16 Cer	538.8	264.3	10	90	10	35	12
C18 Cer	566.7	264.3	10	90	10	35	12
C20 Cer	594.7	264.3	10	90	10	35	12
C24 Cer	650.7	264.3	10	90	10	35	12
C18:1 Cer	564.7	264.3	10	90	10	35	12
C24:1 Cer	648.8	264.3	10	90	10	35	12
C16 Glc Cer	700.7	264.3	10	70	10	50	12
C18 Glc Cer	728.7	264.3	10	80	10	50	12
C24:1 Glc Cer	810.7	264.3	10	90	10	50	12
C16 Gal Cer	700.7	264.3	10	70	10	50	12
C16 Glc Cer	728.7	264.3	10	80	10	50	12
C24:1 Gal Cer	810.7	264.3	10	90	10	50	12
C16 Lac Cer	862.7	264.3	10	110	10	55	12
C18 Lac Cer	890.7	264.3	10	140	10	60	12
Sph	300.4	282.3	10	40	10	15	12
GluSph	462.4	282.3	10	110	10	30	12
GalSph	462.4	282.3	10	110	10	30	12
IS C16 d7	545.8	271.3	10	90	10	35	12
IS C18 d7	573.7	271.3	10	90	10	35	12
IS C24 d7	657.7	271.3	10	90	10	35	12
IS C24:1 d7	655.8	271.3	10	90	10	35	12
IS C18 Glc Cer d5	733.6	269.3	10	90	10	50	12
IS Sph d7	307.4	289.3	10	40	10	15	12
IS Glu Sph d5	467.3	287.3	10	110	10	30	12
IS Gal Sph d5	467.3	287.3	10	110	10	30	12
IS C16 Lac Cer d3	865.7	264.3	10	110	10	55	12
IS C18 Gal Cer d35	763.7	264.3	10	80	10	50	12

20, collision gas high, ion spray voltage 5500 V, temperature 500 °C, ion source gas 1 40, and ion source gas 2 60.

4. The Q1/Q3 transitions, dwell time, declustering potential, collision energy, entrance potential, and collision cell exit potential for each analyte are listed in Table 2.
5. A representative LC-MS/MS chromatogram of the chromatographic separation of the 14 analytes can be found in Perez-Paramo et al. [6].

3.4 Sample Data Processing/Analysis

1. Use Sciex OS v3.3 higher software for data analysis.
2. Integrate the peak area for both the internal standard and the analyte.
3. Calculate the analyte concentration using the area under the curve for the analyte and compare it to that of the calibration curve (*see* **Notes 4, 5, and 6**).

4 Notes

1. CSF samples should be stored in 1.5 mL BHT (Butylated Hydroxy Toluene) containing microtubes and at −80 °C until their analysis to avoid any artificial lipid oxidation.
2. Avoid freeze-thaw cycles to minimize sphingolipid degradation/biotransformation.
3. Manipulate all the sample preparation on ice.
4. Peak areas for C24 Cer and C18 Cer were corrected for natural stable isotope M12 contamination from the unsaturated species (C24:1 and C18:1 Cer). For C24 Cer the total peak area was corrected due to coelution, and for C18 Cer, 35% of the total peak area was corrected since only partial coelution is observed.
5. Chromatographic separation of the Glc Cer and Gal Cer isomers should be 0.07 min (4.2 s); *see* fig. 3 in Perez Paramo et al. [6] verified this separation throughout runs.
6. Once methods have been established, it is important to define the linear range and intra/inter-day assay coefficient of variance as these parameters will change on different instrumentation.

References

1. Colsch B, Seyer A, Boudah S et al (2015) Lipidomic analysis of cerebrospinal fluid by mass spectrometry—based methods. J Inherit Metab Dis 38:53–64
2. O'Brien JS, Sampson EL (1965) Lipid composition of the normal human brain: gray matter, white matter, and myelin. J Lipid Res 6:537–544
3. Cermenati G, Mitro N, Audano M et al (2015) Lipids in the nervous system: from biochemistry and molecular biology to patho-physiology.

Biochim Biophys Acta Mol Cell Biol Lipids 1851:51–60

4. Wang X, Li H, Sheng Y et al (2024) The function of sphingolipids in different pathogenesis of Alzheimer's disease: a comprehensive review. Biomed Pharmacother 171:116071
5. de Oliveira EML, Montani DA, Oliveira-Silva D et al (2019) Multiple sclerosis has a distinct lipid signature in plasma and cerebrospinal fluid. Arq Neuropsiquiatr 77:696–704
6. Perez-Paramo YX, Dufield D, Veeramachaneni R et al (2024) Development of an LC-MS/MS method to measure sphingolipids in CSF from patients with multiple SclerosisS. Mol Pharmacol 105:121–130
7. Vos M, Klein C, Hicks AA (2023) Role of ceramides and sphingolipids in Parkinson's disease. J Mol Biol 435:168000
8. Schmitt S, Cantuti Castelvetri L, Simons M (2015) Metabolism and functions of lipids in myelin. Biochim Biophys Acta Mol Cell Biol Lipids 1851:999–1005
9. Varela ARP, Couto AS, Fedorov A et al (2016) Glucosylceramide reorganizes cholesterol-containing domains in a fluid phospholipid membrane. Biophys J 110:612–622
10. Sillence DJ, Puri V, Marks DL et al (2002) Glucosylceramide modulates membrane traffic along the endocytic pathway. J Lipid Res 43: 1837–1845
11. Raju D, Schonauer S, Hamzeh H et al (2015) Accumulation of glucosylceramide in the absence of the beta-glucosidase GBA2 alters cytoskeletal. Dynamics 1–24
12. Checa A, Khademi M, Sar DG et al (2015) Hexosylceramides as intrathecal markers of worsening disability in multiple sclerosis. Mult Scler 21:1271–1279
13. Péter M, Török W, Balog AP et al (2020) Cerebrospinal fluid lipidomic biomarker signatures of demyelination for multiple sclerosis and Guillain—Barré syndrome. Sci Rep 10:1–13
14. Pieragostino D, Alessandro MD, Rossi C et al (2015) An integrated metabolomics approach for the research of new cerebrospinal fluid biomarkers of multiple sclerosis. Mol Biosyst 11: 1563–1572
15. Vidaurre OG, Haines JD, Katz I et al (2014) Cerebrospinal fluid ceramides from patients with multiple sclerosis impair neuronal bioenergetics. Brain J Neurol 137:2271–2286

Part VI

Bioinformatics Applied to Cerebrospinal Fluid

Chapter 19

Unraveling Potential Tauopathy Biomarkers in Cerebrospinal Fluid: Insights from Olfactory Tract Data

Mercedes Lachén-Montes, Paz Cartas-Cejudo, Elena Anaya-Cubero, Adriana Cortés, Joaquín Fernández-Irigoyen, and Enrique Santamaría

Abstract

Due to the increase in life expectancy, the prevalence of neurodegenerative diseases is estimated to grow considerably in the next decades. Unfortunately, there are no current curative treatments for the majority of these disorders and this might be due to the lack of early diagnosis. Nowadays, diagnosis of neurodegenerative disorders is mainly based on neuroimaging and clinical symptoms, although postmortem neuropathologic confirmation remains the gold-standard diagnostic technique. In this sense, cerebrospinal fluid (CSF) proteome is considered a valuable molecular repository for diagnosing and targeting the neurodegenerative process. On the other hand, olfactory dysfunction is an early symptom that affects Alzheimer's disease (AD) and other tauopathies' patients. That is why we consider that the application of tissue proteomics in primary olfactory structures is an ideal approach to explore early pathophysiological changes, detecting olfactory proteins that might be tested in CSF as potential biomarkers. In this chapter, we have applied peptide fractionation methods followed by tandem mass spectrometry (nanoLC-MS/MS), in silico analysis and semi-quantitative orthogonal techniques in the olfactory tract derived from AD subjects. After obtaining the differential OT proteome, we suggest that the use of different bioinformatic workflows might be a valuable workflow to discover potential disease biomarkers for AD and other tauopathies.

Key words Olfactory tract, Cerebrospinal fluid, Proteomics, Mass-spectrometry

1 Introduction

Olfactory impairment is a common occurrence in several neurodegenerative diseases including tauopathies such as Alzheimer's disease or frontotemporal dementia. Although often neglected in clinical practice, previous studies have reported olfactory deficits in 90% of AD patients whereas a moderate deficit has been described for frontotemporal dementia patients [1, 2]. Functionally, previous evidence suggests that the olfactory impairments occurring during AD progression are due to central sensorineural dysfunctions (smaller hippocampal volume, loss of noradrenergic neurons in the locus coeruleus and cholinergic neurons in the

Enrique Santamaría and Joaquín Fernández-Irigoyen (eds.), *Cerebrospinal Fluid (CSF) Proteomics: Methods and Protocols*, Methods in Molecular Biology, vol. 2914, https://doi.org/10.1007/978-1-0716-4462-1_19,

nucleus basalis of Meynert) [3–5]. On the other hand, it has also been demonstrated that both the human olfactory system and the brain areas with extensive connections to the olfactory system have pathological deposits of Aβ plaques and neurofibrillary tangles (NFT), formed by the poor processing of amyloid precursor protein and Tau hyperphosphorylation, respectively [1, 6]. Specifically, NFTs have been observed in the anterior olfactory nucleus [7] and other olfactory structures such as the olfactory bulb (OB) or the olfactory tract (OT) [1]. Interestingly, the deposition of the hyperphosphorylated form of the protein tau in both the OB and the OT two increases as the disease progresses and reflects what is happening in other affected brain regions. Importantly, according to Braak neuropathological staging based on the localization of NFTs, olfactory structures are already affected in the early stages of the disease, prior to the appearance of cognitive symptoms [8, 9]. Due to the demonstrated olfactory dysfunction in several neurological syndromes, our group has examined the molecular profiling of the early-affected OT region using high-throughput technologies, discovering incipient neurodegenerative changes in AD and other related tauopathies and novel sex-biased molecular mechanisms that enlighten the importance of studying the sex dimension in these disorders [10–12]. Considering the cellular complexity and protein heterogeneity present in the OB [13, 14], proteome-wide analysis based on high-resolution mass spectrometry (MS) [15] has become an attractive technology to characterize and quantify both OB and OT proteome, identifying protein mediators that may be used as potential therapeutic agents or even explored in biofluids [10, 16–18]. In fact, there is an urgent need to find early-stage biomarkers to define the disease staging and, thus, to establish a specific and accurate diagnosis for the neurodegenerative disease spectra. For this purpose, cerebrospinal fluid (CSF) represents the most valuable biological sample in terms of its close relation with the central nervous system [15, 19]. Interestingly, both CSF tau and phosphorylated tau are described as AD biomarkers and studies in large cohorts have confirmed the correlation of neuroimaging findings of Tau, CSF tau and cognitive function [20]. This scenario opens up doors for the finding of new disease biomarkers in CSF. In a biomarker discovery context, it is essential to identify the presence of brain tissue analytes in CSF. The enormous complexity, a great dynamic range of protein concentrations, and the protein heterogeneity originated by PTMs cause difficulties for nonbiased profiling of the human CSF proteome. To address these challenges, various proteomic workflows have been designed to systematically and comprehensively categorize CSF proteins, utilizing different complementary strategies. These include SDS-PAGE gel electrophoresis, an automated MudPIT system (an online 2D separation of SCX and RP-LC) combined with ESI-based IT or FT-ICR MS, immunoaffinity depletion chromatographic steps prior to MS analysis, combinatorial peptide ligand library technology associated with fractionation and nanoLC-MS/MS, and hydrazide-based glycopeptide capture before Orbitrap LC-MS/MS analysis [21–24].

On the other hand, in addition to the information available in extensive studies identifying CSF analytes [25], there are also online resources allowing the neuroscientific community to evaluate the potential secretability of biomarker candidates derived from high-throughput technologies applied in different brain structures, significantly contributing to decipher the human CSF proteomic landscape.

In this sense, we used a discovery platform combining neuropathological diagnosis, label-free quantitative olfactory proteomics, computational analysis, and biochemical approaches to detect and identify potential CSF biomarker candidates in tau-related pathologies, aiming to improve the current diagnostic assays.

2 Materials

2.1 Human Brain Collection

According to the Spanish Law 14/2007 of Biomedical Research, informed written consent forms should be obtained for research purposes from relatives of patients included in this study. Neuropathological assessment should be according to standardized neuropathological scoring/grading systems [26]. For this chapter, 34 postmortem olfactory tracts (OT) derived from controls with no known neurological history (n = 6F/11M) and AD (n = 4F/13M) subjects were analyzed.

2.2 Olfactory Tract Protein Extraction

1. Lysis buffer: 8 M urea, 50 mM DTT (Dithiothreitol) in 50 mM ammonium bicarbonate (ABC) (*see* **Note 1**).
2. Bradford.

2.3 Quantitative Proteomics by Label-Free nLC-MS/MS

1. Ammonium bicarbonate (ABC) solutions (25 mM and 100 mM).
2. Acetonitrile (ACN).
3. Reducing agent: 20 mM DTT in 50 mM ABC.
4. Alkylating agent: 30 mM iodoacetamide (IAA) in ABC 50 mM. Use fresh DTT and IAA in the reducing and alkylating buffers, respectively. Perform the reduction and alkylation steps as closely as possible.
5. Sequencing Grade Modified Trypsin (Promega). Store at −20 °C and keep it at this temperature or on ice until its use. The stock solution is lyophilized but can be dissolved in cold water or 25 mM ABC (*see* **Note 2**).
6. Formic acid (FA) Extraction solution: 2% ACN, 0.5% formic acid (FA).

7. Zip-Tip pipette tips p10, C18.
8. Solution A: 0.1% FA.
9. Solution B: 0.1% FA in ACN.
10. Solution C: 50% ACN, 0.1% FA.
11. Sample dissolution buffer: 2% ACN, 0.5% FA.
12. Buffer A: 0.1% FA.
13. Buffer B: 100% ACN, 0.1% FA.

2.4 Equipment

1. Potter-Elvehjem glass tissue homogenizer.
2. Thermo mixer.
3. Speed-vac concentrator.
4. Eksigent NanoLC-Ultra-2D HPLC.
5. TripleTOF 5600+ mass spectrometer (Sciex).
6. Analyst software (v.1.8.1-Sciex).
7. MaxQuant software (v.2.5).
8. Perseus software (v. 1.6.15.0).
9. Chemidoc MP Imaging System (Bio-rad).
10. Image Lab Software (v. 5.2; Bio-Rad).

3 Methods

We describe a comprehensive label-free approach to relative protein quantification including sample preparation, LC-MS/MS data acquisition, de-noising, mass and charge state estimation, chromatographic alignment, peptide quantification, secretability analysis, and tau-related biomarker discovery. Following experimental procedures in order to preliminary validate potential biomarkers can include CSF measurements by Western-blotting or Elisa. The present workflow was used for human OT and should be adaptable to other human or murine brain structures. Aiming to identify novel CSF biomarkers for AD, human samples from AD subjects together with control subjects with no neurological disease history were selected. Since this workflow implies the use of postmortem brain material, it is important to note that the success and comprehensiveness of the method will largely depend on the quality of biological material (both brain tissue and CSF). In this case, all human brains considered in this study had a postmortem interval lower than 22 h. The methodology described here uses 20–40 μg of olfactory tissue and ~100 μL of CSF, which allows the performance of the discovery phase by label-free proteomics at the tissue level (Fig. 1). To reduce experimental variability, we recommend

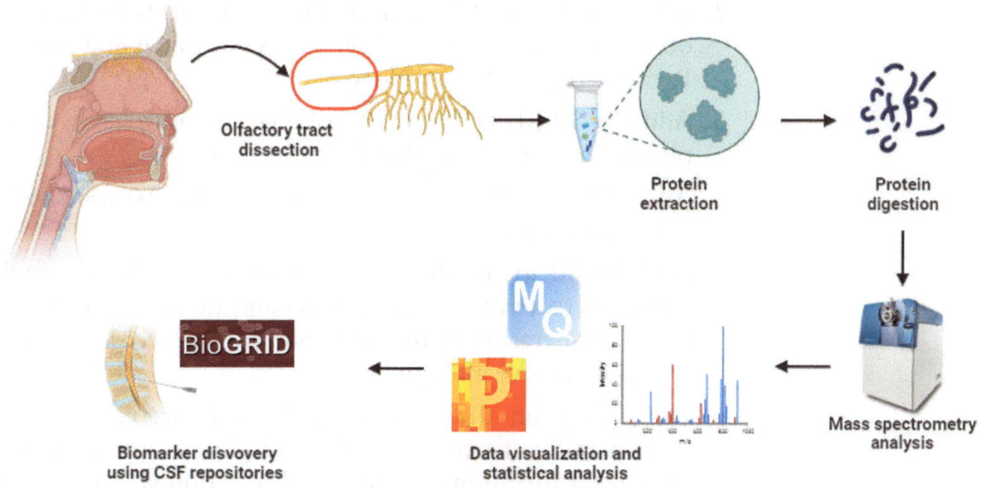

Fig. 1 Overview of the proteomic approach for tauopathy biomarker discovery via olfactory tract analysis. The strategy used by our group combines nano-liquid chromatography, mass-spectrometry, and computational analysis

preparing all samples that need to be analyzed on the same day under identical conditions.

3.1 Olfactory Tract Protein Extraction

1. Homogenize the OT tissue specimens with the appropriate amount of lysis buffer depending on the sample size (200 μL to process 50% of human OT structure, and 400 μL if the whole structure is processed. Sonicate briefly if necessary (*see* **Note 3**).
2. Ultracentrifuge at 100,000 × g for 1 h at 15 °C and carefully collect the supernatant in a LoBind tube (*see* **Notes 4** and **5**).
3. Measure protein concentration using a protein assay kit (*see* **Note 6**).

3.2 Quantitative Proteomics by Label-Free LC-MS/MS

1. Dilute 15–20 μg of protein extract in Laemmli sample buffer and load into a 0.75 mm thick polyacrylamide gel with a 4% stacking gel cast over a 12.5% resolving gel.
2. Stop the run as soon as the front enters 3 mm into the resolving gel with the aim of concentrating the whole proteome in the stacking/resolving gel interface.
3. Stain the bands with Coomassie brilliant blue and excised from the gel with a clean scalpel. Put the band in a LoBind tube (*see* **Note 7**).

3.3 In-gel Reduction, Alkylation, and Protein Digestion

1. Add 100 μL of 100 mM ABC/ACN (1:1, vol/vol) and incubate with vortexing for 30 min.

2. Add 500 μL of ACN and incubate at room temperature with occasional vortexing, until gel pieces become white and shrink. Remove the ACN. (*see* **Note 8**).
3. Add 30–50 μL of the DTT solution to completely cover gel pieces. Incubate 30 min at 56 °C.
4. Add 500 mL of acetonitrile, incubate for 10 min and then remove all liquid.
5. Add 30–50 μL of the IAA solution (the volume should be sufficient to cover the gel pieces) and incubate for 20 min at room temperature in the dark. Shrink gel pieces with acetonitrile and remove all liquid.
6. Depending on the volume of the gel matrix, add enough trypsin buffer to cover the dry gel pieces, and leave it in an ice bucket for 30 min. Check if all solution was absorbed and add more trypsin buffer, if necessary. Gel pieces should be completely covered with trypsin buffer.
7. Add 10–20 μL of ABC to cover the gel pieces and keep them wet during enzymatic cleavage.
8. Incubate samples overnight at 37 °C.
9. Add 30 μL 2% ACN and 0.5% FA to each tube and incubate for 15 min at 37 °C in a shaker.
10. Collect the supernatant to a loBind tube and repeat **step 9**, adding 25 μL 2% ACN, 0.5% FA. Add 25 μL de ACN 100%. Incubate 30 min. Collect the supernatant (*see* **Note 9**).
11. Dry down the supernatants in a vacuum centrifuge or speed-vac and resuspend it in the sample resuspension buffer (*see* **Note 10**).

3.4 Desalting

1. Purify and concentrate the peptide mixtures using C18 Zip Tip Solid Phase Extraction (Millipore).
2. Equilibrate the column with buffer B.
3. Wash the column with buffer A.
4. Load the sample in the column.
5. Wash the column with buffer A.
6. Elute the peptides with buffer C.
7. Dry down the sample in the speed-vac.

3.5 nLC-MS/MS (see Note 11)

1. For further LC MS/MS analysis, add 10–15 μL of 2% ACN, 0.5% FA into the tube, vortex and centrifuge for 15 min at 6.7 g (10,000 r.p.m.) at the bench-top centrifuge and withdraw the appropriate aliquot for further analysis. Dry down the rest in a vacuum centrifuge and store at −20 °C as a contingency.

2. Separate peptide mixtures by reverse phase chromatography using an Eksigent nanoLC ultra 2D pump fitted with a 75 μm ID column (Eksigent 0.075 × 250). Mobile phases are 100% water 0.1% FA (buffer A) and 100% ACN 0.1% FA (buffer B). Load the samples for desalting and concentration into a 0.5 cm length 100 μm ID precolumn packed with the same chemistry as the separating column.
3. Column gradient is developed in a 240 min two-step gradient from 5% B to 25% B in 210 min and 25% B to 40% B in 30 min (*see* **Note 12**). Equilibrate the column in 95% B for 5 min and 2% B for 15 min. During all processes, the pre-column has to be in line with the column and the flow maintained all along the gradient at 300 nL/min.
4. Eluting peptides from the column are analyzed using a 5600+ Triple-TOF system (Sciex). Information data acquisition is acquired upon a survey scan performed in a mass range from 350 m/z up to 1250 m/z in a scan time of 250 ms (*see* **Note 13**). The top 35 peaks are selected for fragmentation. The minimum accumulation time for MS/MS is set to 100 ms, giving a total cycle time of 3.8 s. Product ions are scanned in a mass range from 230 m/z up to 1500 m/z and excluded for further fragmentation during 15 s. (*see* **Note 14**).
5. MS/MS data acquisition is performed using Analyst 1.8.1 (Sciex) and spectra files are processed through MaxQuant Software (v.2.5) and searched against the concatenated target-decoy UniProt proteome reference Human database (Proteome ID: UP000005640). This software enables peptide and protein identification and quantification (*see* **Note 15**).
6. Perform false discovery rate analysis (<1% at protein level) using a non-linear fitting method [27]. (*see* **Note 16**).
7. Perform peptide quantification using the MaxQuant LC – MS software. Using the accurate mass measurements from full survey scans in the TOF detector and the observed retention times, runs are aligned to compensate for between-run variations in the nanoLC separation system. All LC-MS/MS runs are aligned to a reference run automatically chosen by the software, generating a master list of features considering m/z values and retention times.
8. Export protein quantifications from MaxQuant and import the matrix in Perseus software [28] to perform statistical analysis and data visualization (*see* **Note 17**).
9. Import the txt matrix to Perseus (*see* **Note 18**).
   (a) Sample name (=Protein spot values): Main column.
   (b) Label (=Protein names): Categorical column.
10. Transform the standardized values to Log2 (normalization).

11. Group the samples in the different groups among which you want to compare: Categorical annotations row.
12. Proteins identified by the site (identification based only on a modification), reverse proteins (identified by decoy database) and potential contaminants are filtered out using "rows based on categorical column."
13. Olfactory proteins should be quantified with at least two unique peptides. Use "Filter rows based on numerical column."
14. Perform the T-test. Select "Two-sample test" to obtain differentially expressed proteins between AD and control subjects. A p-value lower than 0.05, and an absolute fold change of <0.70 (down-regulation) or >1.3 (up-regulation) in linear scale are considered to be significantly differentially expressed between controls and AD subjects (*see* **Note 19**).

3.6 Potential Biomarker Discovery

1. Check the published MAPT interactome. Use the Biogrid tool (https://thebiogrid.org/) [29] and search for MAPT-related interactors to filter the list of potential biomarker candidates (Fig. 2) (*see* **Note 20**).

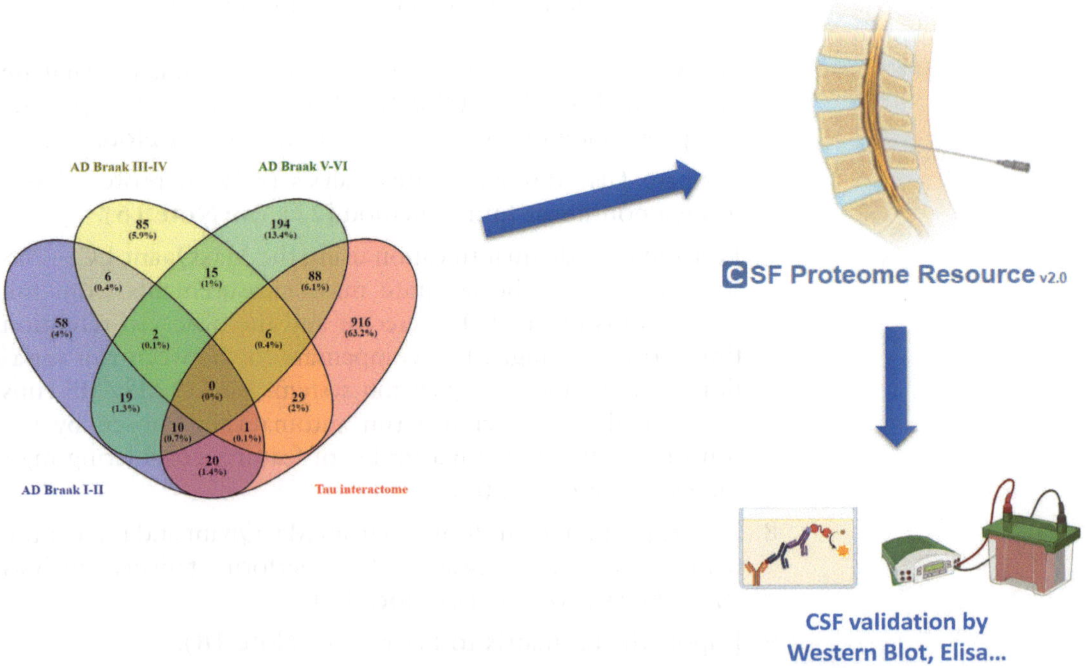

Fig. 2 Venn diagrams of proteins found in human OT by Braak staging and Tau interactome (by Biogrid). **a** Numbers represent the number of shared proteins in the respective overlapping areas. **b** Potential biomarkers can be explored in CSF Proteome Resource or **c** be analyzed by other immunodetection techniques

2. Check the potential secretability of olfactory proteins to the CSF. For that, use the CSF-PR 2.0 tool (http://probe.uib.no/csf-pr-2.0) [30] to search the protein candidates across 85 data sets containing quantitative information (*see* **Note 21**).
3. Proceed with subsequent preliminary validation techniques such as Western Blot or Elisa, among others, in CSF samples (*see* **Notes 22** and **23**).

4 Notes

1. Ensure that all the solutions containing urea are prepared freshly and do not heat above 37 °C to prevent protein carbamylation.
2. Modified trypsin is maximally active in the pH range of 7–9 and reversibly inactivated at pH 4. Dissolve the content of 20 μg vial in 1.5 mL of the buffer. Make shortly before use; discard unused volume. After thawing frozen aliquots, pH should be adjusted by adding 1.5 μL of 50 mM ABC shortly before use.
3. Alternative buffers/detergents may be used. Incompatible components with the trypsin action or nLC-MS/MS should be removed by buffer exchange or decrease their concentration.
4. Care should be taken when centrifuging to ensure items are fully contained within centrifuge buckets, that they are equally balanced, and that centrifugation speeds used do not result in damage. The temperature should be 15 °C because urea precipitates at 4 °C. In brain tissue samples, a white disc at the top of the supernatant is observed after the ultracentrifugation, mainly due to fat components. Try to take the supernatant without taking this layer. Using LoBind tubes helps minimize protein sample loss during sample preparation, especially when processing steps are extensive.
5. A precipitation step can be added to this workflow if needed. Protocols using methanol/chloroform, ice-cold acetone or commercial kits (i.e., 2D-clean up kit from Bio-rad or GE Healthcare) can be used. Protein precipitation allows changing the buffer and concentrates the sample.
6. Different quantification methods may be used depending on the equipment available in the laboratory. The best strategy with the higher sensitivity consists of quantifying at the amino acid level using the amino acid analysis technique. Other valid approaches are the Bradford assay, Qubit, etc.
7. Take special care to prevent massive keratin contamination of the samples. Wear gloves, perform all operations in a laminar

flow hood, and use a dedicated set of pipettes, tips, and tubes that should be stored in the hood in a dust-free environment.

8. Although the bulk of Coomassie staining should be removed, it is not necessary to destain the gel pieces completely. Samples are now ready for in-gel digestion. Alternatively, they can be stored at −20 °C for a few weeks.
9. Do not discard extracted gel pieces: if, for any reason, the digestion fails, it can be repeated with the same gel pieces.
10. Dried peptide extracts can be safely stored at −20 °C for a few months.
11. Different LC-MS/MS quantitative workflows may be used to detect differentially expressed proteins in the human brain such as iTRAQ, TMT, SWATH, etc. [31–34].
12. LC gradients and MS and MS/MS settings should be carefully optimized on each system and for each sample type to ensure optimal outcomes. For this reason, it is extremely recommended to try all these procedures and settings with known standards. These may be mixtures of 8–10 proteins, with some of them at the same level and others at a variable one. At the same time, this helps not only to know if the system is suitable but also whether the quantification accuracy is acceptable.
13. This system has a mass range from 5 to 40,000 m/z in Q1; the precursor selection range is 5–1250 m/z. In general, tryptic peptides generate precursors above 350 m/z, so we select this range to avoid as much noise as possible [35].
14. The MS conditions are likely to change due to instrument specificity.
15. Adequate the search settings to your sample preparation workflow. For label-free quantitative proteomics, other suitable software is available such as Proteome Discoverer (Thermo), etc.
16. Other statistical frameworks may be used [36, 37].
17. It is important to note that due to technical reasons, only the most abundant OT proteins were explored. Consequently, alterations other than those observed by our label-free workflow might also participate in the neurodegenerative process at the level of the OT. For that, different experimental approaches such as subcellular fractionation or laser microdissection coupled with mass-spectrometry could spatially resolve the molecular heterogeneity present in the tissue.
18. Other data processing software may be used (R scripts, Proteome discoverer).
19. Fold change is commonly used in quantitative proteomic analysis where proteins differing by more than an arbitrary cut-off

value in abundance are considered to be differentially expressed. This cut-off is arbitrary and can be modified.

20. Performing a Venn diagram comparing the secretable olfactory proteins in CSF with MAPT interactors is recommended as a straight-handed workflow to filter potential disease biomarkers. As mentioned, comparison with other CSF proteomic studies may complement the analysis.
21. Other data derived from recent CSF proteomic studies [24, 38, 39] may also be used to complement the secretability analysis.
22. The most common procedure for collection of CSF is the lumbar puncture (LP), which consists of a needle insertion into the spinal canal. With relative frequency LP procedure is complicated by the contamination of CSF with blood cells as a result of needle trauma. The presence of blood cells and the concentration affected by some substances due to blood contamination in CSF can confound the analysis and the corresponding interpretation. Although, it is difficult to judge the degree of contamination by visual inspection alone, discard reddish samples. Flow cytometry may be used to specifically count and separate the red blood cells.
23. Depending on the sample availability, complementary approaches should be used to validate potential CSF markers. For example, ELISA (enzyme-linked immunosorbent assays) or multiple reaction monitoring mass spectrometry (MRM-MS) are suitable devices to detect and quantify the protein/s of interest in large sample collections.

Acknowledgments

The Clinical Neuroproteomics Unit is member of the Spanish Olfactory Network—ROE (grant RED2022-134081-T from the Spanish Ministry of Science & Innovation) and is supported by grants PID2023-152593OB-I00 funded by MCIU/AEI/10.13039/501100011033 / FEDER, UE to ES and JFI and 0011-1411-2023-000028 (from Government of Navarra—the Department of Economic and Business Development-S4). PC-C is supported by a predoctoral fellowship from the Public University of Navarra (UPNA). ML-M is supported by a postdoctoral fellowship from Miguel Servet Foundation-Navarrabiomed. EA-C is supported by "Programa MRR Investigo 2023" in the framework of the European Union recovery and resilience facility.

References

1. Attems J, Walker L, Jellinger KA (2014) Olfactory bulb involvement in neurodegenerative diseases. Acta Neuropathol 127:459. https://doi.org/10.1007/s00401-014-1261-7
2. Doty RL (2017) Olfactory dysfunction in neurodegenerative diseases: is there a common pathological substrate? Lancet Neurol 16:478. https://doi.org/10.1016/S1474-4422(17)30123-0
3. Kjelvik G, Saltvedt I, White LR, Stenumgård P, Sletvold O, Engedal K, Skåtun K, Lyngvær AK, Steffenach HA, Håberg AK (2014) The brain structural and cognitive basis of odor identification deficits in mild cognitive impairment and Alzheimer's disease. BMC Neurol 14:168
4. Wilson RS, Arnold SE, Schneider JA, Boyle PA, Buchman AS, Bennett DA (2009) Olfactory impairment in presymptomatic Alzheimer's disease. Ann N Y Acad Sci 1170:730–735
5. Wang J, Eslinger PJ, Doty RL et al (2010) Olfactory deficit detected by fMRI in early Alzheimer's disease. Brain Res 1357:184–194
6. Christen-Zaech S, Kraftsik R, Pillevuit O, Kiraly M, Martins R, Khalili K, Miklossy J (2003) Early olfactory involvement in Alzheimer's disease. Can J Neurol Sci 30:20–25
7. Drummond E, Nayak S, Pires G, Ueberheide B, Wisniewski T (2018) Isolation of amyloid plaques and neurofibrillary tangles from archived alzheimer's disease tissue using laser-capture microdissection for downstream proteomics. Methods Mol Biol. https://doi.org/10.1007/978-1-4939-7558-7_18
8. Attems J, Lintner F, Jellinger KA (2005) Olfactory involvement in aging and Alzheimer's disease: an autopsy study. J Alzheimers Dis 7:149–180
9. McKeith IG (2016) Clinical diagnostic criteria for dementia with lewy bodies. In: Dementia with lewy bodies: clinical and biological aspects. https://doi.org/10.1007/978-4-431-55948-1
10. Lachén-Montes M, González-Morales A, Zelaya MV, Pérez-Valderrama E, Ausín K, Ferrer I, Fernández-Irigoyen J, Santamaría E (2017) Olfactory bulb neuroproteomics reveals a chronological perturbation of survival routes and a disruption of prohibitin complex during Alzheimer's disease progression. Sci Rep 7:9115. https://doi.org/10.1038/s41598-017-09481-x
11. Lachén-Montes M, González-Morales A, Schvartz D, Zelaya MV, Ausin K, Fernández-Irigoyen J, Sánchez JC, Santamaría E (2019) The olfactory bulb proteotype differs across frontotemporal dementia spectrum. J Proteome 201:37. https://doi.org/10.1016/j.jprot.2019.04.011
12. Cartas-Cejudo P, Lachén-Montes M, Ferrer I, Fernández-Irigoyen J, Santamaría E (2023) Sex-divergent effects on the NAD+-dependent deacetylase sirtuin signaling across the olfactory–entorhinal–amygdaloid axis in Alzheimer's and Parkinson's diseases. Biol Sex Differ 14:5. https://doi.org/10.1186/s13293-023-00487-x
13. Fernandez-Irigoyen J, Labarga A, Zabaleta A, de Morentin XM, Perez-Valderrama E, Zelaya MV, Santamaria E (2015) Toward defining the anatomo-proteomic puzzle of the human brain: an integrative analysis. Proteom Clin Appl 9:796. https://doi.org/10.1002/prca.201400127
14. Aebersold R, Mann M (2016) Mass-spectrometric exploration of proteome structure and function. Nature 537:347. https://doi.org/10.1038/nature19949
15. Hühmer AF, Biringer RG, Amato H, Fonteh AN, Harrington MG (2006) Protein analysis in human cerebrospinal fluid: physiological aspects, current progress and future challenges. Dis Markers 22:3–26
16. Lachen-Montes M, Zelaya MV, Segura V, Fernández-Irigoyen J, Santamaría E (2017) Progressive modulation of the human olfactory bulb transcriptome during Alzheimer's disease evolution: novel insights into the olfactory signaling across proteinopathies. Oncotarget 10.18632/oncotarget.18193
17. Nagayama S, Homma R, Imamura F (2014) Neuronal organization of olfactory bulb circuits. Front Neural Circuit 8:98
18. Lachén-Montes M, Íñigo-Marco I, Cartas-Cejudo P, Fernández-Irigoyen J, Santamaría E (2021) Olfactory bulb proteomics reveals widespread proteostatic disturbances in mixed dementia and guides for potential serum biomarkers to discriminate Alzheimer disease and mixed dementia phenotypes. J Pers Med 11. https://doi.org/10.3390/jpm11060503
19. Li X, Li T-Q, Andreasen N, Wiberg MK, Westman E, Wahlund L-O (2014) The association between biomarkers in cerebrospinal fluid and structural changes in the brain in patients with Alzheimer's disease. J Intern Med 275:418–427
20. Shoji M (2019) Cerebrospinal fluid and plasma Tau as a biomarker for brain tauopathy. Adv Exp Med Biol 1184:393–405

21. Schutzer SE, Liu T, Natelson BH et al (2010) Establishing the proteome of normal human cerebrospinal fluid. PLoS One 5:e10980
22. Mouton-Barbosa E, Roux-Dalvai F, Bouyssié D et al (2010) In-depth exploration of cerebrospinal fluid by combining peptide ligand library treatment and label-free protein quantification. Mol Cell Proteom 9:1006–1021
23. Guldbrandsen A, Vethe H, Farag Y, Oveland E, Garberg H, Berle M, Myhr K-M, Opsahl JA, Barsnes H, Berven FS (2014) In-depth characterization of the cerebrospinal fluid (CSF) proteome displayed through the CSF proteome resource (CSF-PR). Mol Cell Proteom 13: 3152–3163
24. Macron C, Lane L, Núñez Galindo A, Dayon L (2018) Deep dive on the proteome of human cerebrospinal fluid: a valuable data resource for biomarker discovery and missing protein identification. J Proteome Res 17:4113–4126
25. Schilde LM, Steinbach S, Serschnitzki B, Maass F, Bähr M, Lingor P, Marcus K, May C (2020) Human cerebrospinal fluid data for use as spectral library, for biomarker research. Data Brief 32:106048
26. Andrews GL, Simons BL, Young JB, Hawkridge AM, Muddiman DC (2011) Performance characteristics of a new hybrid quadrupole time-of-flight tandem mass spectrometer (TripleTOF 5600). Anal Chem 83: 5442–5446
27. Tang WH, Shilov IV, Seymour SL (2008) Nonlinear fitting method for determining local false discovery rates from decoy database searches. J Proteome Res 7:3661–3667
28. Tyanova S, Temu T, Sinitcyn P, Carlson A, Hein MY, Geiger T, Mann M, Cox J (2016) The Perseus computational platform for comprehensive analysis of (prote)omics data. Nat Methods 13:731. https://doi.org/10.1038/nmeth.3901
29. Oughtred R, Rust J, Chang C et al (2021) The BioGRID database: a comprehensive biomedical resource of curated protein, genetic, and chemical interactions. Protein Sci 30:187. https://doi.org/10.1002/pro.3978
30. Guldbrandsen A, Farag Y, Kroksveen AC, Oveland E, Lereim RR, Opsahl JA, Myhr K-M, Berven FS, Barsnes H (2017) CSF-PR 2.0: an interactive literature guide to quantitative cerebrospinal fluid mass spectrometry data from neurodegenerative disorders. Mol Cell Proteom 16:300–309
31. Ping L, Kundinger SR, Duong DM, Yin L, Gearing M, Lah JJ, Levey AI, Seyfried NT (2020) Global quantitative analysis of the human brain proteome and phosphoproteome in Alzheimer's disease. Sci Data 7:315
32. Shao S, Guo T, Koh CC, Gillessen S, Joerger M, Jochum W, Aebersold R (2015) Minimal sample requirement for highly multiplexed protein quantification in cell lines and tissues by PCT-SWATH mass spectrometry. Proteomics 15:3711–3721
33. Chang RYK, Etheridge N, Nouwens AS, Dodd PR (2015) SWATH analysis of the synaptic proteome in Alzheimer's disease. Neurochem Int 87:1–12
34. Navarro P, Vázquez J (2009) A refined method to calculate false discovery rates for peptide identification using decoy databases. J Proteome Res 8:1792–1796
35. Shilov IV, Seymour SL, Patel AA, Loboda A, Tang WH, Keating SP, Hunter CL, Nuwaysir LM, Schaeffer DA (2007) The Paragon Algorithm, a next generation search engine that uses sequence temperature values and feature probabilities to identify peptides from tandem mass spectra. Mol Cell Proteom 6:1638–1655
36. Nesvizhskii AI (2010) A survey of computational methods and error rate estimation procedures for peptide and protein identification in shotgun proteomics. J Proteome 73:2092–2123
37. Lehnert S, Jesse S, Rist W et al (2012) iTRAQ and multiple reaction monitoring as proteomic tools for biomarker search in cerebrospinal fluid of patients with Parkinson's disease dementia. Exp Neurol 234:499–505
38. van Zalm PW, Ahmed S, Fatou B, Schreiber R, Barnaby O, Boxer A, Zetterberg H, Steen JA, Steen H (2023) Meta-analysis of published cerebrospinal fluid proteomics data identifies and validates metabolic enzyme panel as Alzheimer's disease biomarkers. Cell Rep Med 4: 101005
39. Johnson ECB, Bian S, Haque RU et al (2023) Cerebrospinal fluid proteomics define the natural history of autosomal dominant Alzheimer's disease. Nat Med 29:1979–1988

Chapter 20

Inspecting Biological Deregulation, Putative Markers, and Therapeutic Targets for Neurodegenerative Diseases Through an Integrative Bioinformatics Analysis of the Human Cerebrospinal Fluid Proteome: A Tutorial

Fábio Trindade, Rita Nogueira-Ferreira, Paulo Bastos, Francisco Amado, Rita Ferreira, and Rui Vitorino

Abstract

Cerebrospinal fluid (CSF) is a source of valuable information concerning brain disorders. The technical advances of high throughput omics platforms to analyze body fluids can generate a huge amount of data, whose translation of the biological meaning can be a challenge. Several bioinformatics tools have emerged to help handle this data from a systems biology perspective. Herein, we describe a step-by-step tutorial for CSF proteome data analysis in the set of neurodegenerative diseases using: (i) ShinyGO webtool to perform functional enrichment analysis envisioning the characterization of the biological pathways and processes deregulated in neurodegenerative diseases including Alzheimer's and Parkinson's diseases; (ii) Cytoscape to map disease-specific proteins based on evidence from proteomics; (iii) DisGeNET to identify the proteins more strongly and more specifically associated with neurodegenerative diseases to date; (iv) STRING to identify putative therapeutic targets through a combined protein-protein interaction and network topological analyses. This step-by-step guide might help researchers to better characterize disease pathogenesis and to identify putative disease biomarkers and therapeutic targets.

Key words CSF proteome, Neurodegenerative diseases, Gene ontology enrichment analysis, Candidate biomarkers, Protein-protein interaction, Therapeutic targets

1 Introduction

Cerebrospinal fluid (CSF) is an important source of potential biomarkers because of its proximity to the brain parenchyma. CSF is also generated from brain interstitial fluid, thus providing a more direct reflection of biochemical changes within the central nervous system (CNS) [1, 2]. There has been a growing interest in the application of CSF proteomics to search for CNS diseases' specific biomarkers. Classically, the proteomic approaches rely on mass spectrometry (MS), combining gel-free shotgun MS approaches

Enrique Santamaría and Joaquín Fernández-Irigoyen (eds.), *Cerebrospinal Fluid (CSF) Proteomics: Methods and Protocols*, Methods in Molecular Biology, vol. 2914, https://doi.org/10.1007/978-1-0716-4462-1_20,

with the depletion of high-abundance proteins for the detection and quantification of low-abundance proteins [2–4]. The latest advances in high-throughput MS have driven the identification of thousands of proteins in CSF [5]. Nevertheless, as a result, the new challenge of digging the biological meaning out of big data has emerged in CSF proteome research. As such, bioinformatics is now viewed as a cornerstone of life and health sciences because it accelerates the integration, translation, and interpretation of big data into meaningful discoveries.

Several bioinformatics tools (marketed and freely available) have been released to help researchers deal with large lists of proteins, envisioning the identification of the molecular processes modulated by a given pathophysiological condition and the identification of putative biomarkers and therapeutic targets. Marketed tools are widely popular as they tend to offer more user-friendly environments, whilst open-source programs are more flexible in terms of the possibilities to modify existing algorithms [6, 7]. Since different programs might produce dissimilar and (in some cases) contradictory results, the selection of a specific software/algorithm is crucial and should take into account the following variables: (i) MS methodology applied (depletion or enrichment procedures, labelling for quantitative purposes, normalization, protein concentration) [7]; (ii) data format support (e.g. excel or notepad); (iii) interface provided (source code vs user-friendly interface) [8].

Herein, we exemplify following a step-by-step guide on the applicability of the freely available bioinformatics tools ShinyGO [9], Cytoscape [10], DisGeNET [11, 12], and STRING [13, 14] for the integration of a CSF protein dataset. ShinyGO will be used for gene ontology and functional enrichment analyses [9]. Besides ShinyGO, there are other web-based tools like EnrichR, DAVID, GOrilla, Panther, g:Profiler, which have a different coverage of organisms and source different databases. However, ShinyGO was selected due to its intuitive environment, especially helpful in a first approach for researchers not familiarized with bioinformatics tools. We will then use Cytoscape to visualize the associations/specificity between the proteins retrieved from literature search and neurodegenerative diseases. This is often the first step towards the identification of proteins with putative candidacy for biomarkers. The Cytoscape platform is an open-source software for complex network analysis and visualization, hosting a multitude of plugins for different types of analysis. There are several plugins available, including for gene ontology enrichment and cluster-based analyses, such as ClueGO and CluePedia, which were addressed in the previous edition of this book [15]. In parallel with a user-made network of protein-disease associations, we will source DisGeNET, also within the Cytoscape environment. The DisGeNET is a database of gene-disease associations (GDA) retrieved from expert-curated repositories, GWAS catalogues, animal models and text-mining of scientific literature [12]. DisGeNET delivers a GDA score, which is a measure of the strength of the association, integrating the number and types of sources (curation degree, type of model organism, among others). It also provides disease specificity indexes, a measure of the specificity of a given protein towards one or more conditions. Therefore, DisGeNET not only stores GDA data but also helps prioritize targets for biomarker experiments by computing and grading the degree of evidence through easily interpreted scores. Finally, also in the Cytoscape environment, we will source the STRING database, containing annotated functional interactions between proteins [13], to draw a protein-protein interaction network. We will also perform topological analysis, to identify protein hubs and bottlenecks. Network hubs are evidenced through the node degree, i.e., the number of direct interacting patterns, while bottlenecks are evidenced through betweenness centrality, i.e., a measure of the number of shortest paths that pass through the node (the gene/protein) to fulfil the entire network. Protein hubs and bottlenecks regulate the flow of biological information and are, therefore, appealing targets for modulation from a pharmacological perspective [16]. The applicability of these bioinformatic tools is exemplified below.

2 Materials

The choice of a bioinformatic tool depends on the biological or clinical question to be investigated. For example, we compare CSF proteomes between distinct neurodegenerative diseases: Alzheimer's disease (AD), Parkinson's disease (PD), Dementia with Lewy Bodies (DLB), Frontotemporal dementia (FTD) and Huntington disease (HD), to decipher specifically deregulated biological processes, hypothetical disease-specific markers and putative therapeutic targets.

2.1 Datasets

See Table 1 (created based on [2]).

2.2 Freely Available Bioinformatic Web Tools, Databases, and Applications

1. UniProtKB (http://www.uniprot.org/).
2. ShinyGO (version 0.80, http://bioinformatics.sdstate.edu/go/).
3. Cytoscape (version 3.10.2, http://www.cytoscape.org/), and the following plugins:
4. DisGeNET (version 7.3.0, plugin available at https://apps.cytoscape.org/apps/disgenetapp, DisGeNET source database: https://www.disgenet.org/).

Table 1
List of CSF proteins modulated by the neurodegenerative diseases Alzheimer's disease (AD), Parkinson's disease (PD), dementia with Lewy bodies (DLB), frontotemporal dementia (FTD), and Huntington's disease (HD)

Accession	Gene	Protein	Disease	Disease type	Change
P24593	IGFBP5	Insulin-like growth factor-binding protein 5	Alzheimer's disease	Dementia and primarily neurodegenerative	+
Q14257	RCN2	Reticulocalbin-2	Alzheimer's disease	Dementia and primarily neurodegenerative	+
P01344	IGF2	Insulin-like growth factor II	Alzheimer's disease	Dementia and primarily neurodegenerative	+
Q9BXI6	TBC1D10A	TBC1 domain family member 10A	Alzheimer's disease	Dementia and primarily neurodegenerative	+
Q14767	LTBP2	Latent-transforming growth factor beta-binding protein 2	Alzheimer's disease	Dementia and primarily neurodegenerative	+
P41271	NBL1	Neuroblastoma suppressor of tumorigenicity 1	Alzheimer's disease	Dementia and primarily neurodegenerative	+
Q9H4L7	SMARCAD1	SWI/SNF-related matrix-associated actin-dependent regulator of chromatin subfamily A containing DEAD/H box 1	Alzheimer's disease	Dementia and primarily neurodegenerative	+
P02763	ORM1	Alpha-1-acid glycoprotein 1	Alzheimer's disease	Dementia and primarily neurodegenerative	+
P02652	APOA2	Apolipoprotein A-II	Alzheimer's disease	Dementia and primarily neurodegenerative	+
P05090	APOD	Apolipoprotein D	Alzheimer's disease	Dementia and primarily neurodegenerative	+
P02749	APOH	Beta-2-glycoprotein 1	Alzheimer's disease	Dementia and primarily neurodegenerative	+
P08123	COL1A2	Collagen alpha-2(I) chain	Alzheimer's disease	Dementia and primarily neurodegenerative	+

(continued)

Table 1
(continued)

Accession	Gene	Protein	Disease	Disease type	Change
Q8NBJ4	GOLM1	Golgi membrane protein 1	Alzheimer's disease	Dementia and primarily neurodegenerative	+
Q9HAV4	XPO5	Exportin-5	Alzheimer's disease	Dementia and primarily neurodegenerative	+
Q99523	SORT1	Sortilin	Alzheimer's disease	Dementia and primarily neurodegenerative	+
A0A024QZN8	HERC4	Hect domain and RLD 4	Alzheimer's disease	Dementia and primarily neurodegenerative	+
Q92876	KLK6	Kallikrein-6	Alzheimer's disease	Dementia and primarily neurodegenerative	+
O14776	TCERG1	Transcription elongation regulator 1	Alzheimer's disease	Dementia and primarily neurodegenerative	+
O43405	COCH	Cochlin	Alzheimer's disease	Dementia and primarily neurodegenerative	+
P08493	MGP	Matrix Gla protein	Alzheimer's disease	Dementia and primarily neurodegenerative	+
Q9HCB6	SPON1	Spondin-1	Alzheimer's disease	Dementia and primarily neurodegenerative	+
P02760	AMBP	Protein AMBP	Alzheimer's disease	Dementia and primarily neurodegenerative	+
Q9NRR1	CYTL1	Cytokine-like protein 1 (protein C17)	Alzheimer's disease	Dementia and primarily neurodegenerative	+
P02675	FGB	Fibrinogen beta chain	Alzheimer's disease	Dementia and primarily neurodegenerative	+
P03989	HLA-B	HLA class I histocompatibility antigen B-27 alpha chain	Alzheimer's disease	Dementia and primarily neurodegenerative	+

(continued)

Table 1
(continued)

Accession	Gene	Protein	Disease	Disease type	Change
P13747	HLA-E	HLA class I histocompatibility antigen alpha chain E	Alzheimer's disease	Dementia and primarily neurodegenerative	+
P00738	HP	Haptoglobin	Alzheimer's disease	Dementia and primarily neurodegenerative	+
P19827	ITIH1	Inter-alpha-trypsin inhibitor heavy chain H1	Alzheimer's disease	Dementia and primarily neurodegenerative	+
P02766	TTR	Transthyretin	Alzheimer's disease	Dementia and primarily neurodegenerative	+
P02753	RBP4	Retinol-binding protein 4	Alzheimer's disease	Dementia and primarily neurodegenerative	+
P61769	B2M	Beta-2-microglobulin	Alzheimer's disease	Dementia and primarily neurodegenerative	+
P61769	B2M	Beta-2-microglobulin	Alzheimer's disease	Dementia and primarily neurodegenerative	+
P01034	CST3	Cystatin C	Alzheimer's disease	Dementia and primarily neurodegenerative	+
P01009	SERPINA1	α-1 antitrypsin	Alzheimer's disease	Dementia and primarily neurodegenerative	+
P10636	MAPT	Microtubule-associated protein tau	Alzheimer's disease	Dementia and primarily neurodegenerative	+
P10909	CLU	Clusterin	Alzheimer's disease	Dementia and primarily neurodegenerative	+
P02649	APOE	Apolipoprotein E	Alzheimer's disease	Dementia and primarily neurodegenerative	+
P05067	APP	Amyloid beta A4 protein	Alzheimer's disease	Dementia and primarily neurodegenerative	+

(continued)

Table 1
(continued)

Accession	Gene	Protein	Disease	Disease type	Change
P01024	C3	Complement C3	Alzheimer's disease	Dementia and primarily neurodegenerative	+
P07602	PSAP	Prosaposin	Alzheimer's disease	Dementia and primarily neurodegenerative	+
P07998	RNASE1	Ribonuclease pancreatic	Alzheimer's disease	Dementia and primarily neurodegenerative	+
P36222	CHI3L1	Chitinase-3-like protein 1	Alzheimer's disease	Dementia and primarily neurodegenerative	+
P41222	PTGDS	Prostaglandin-H2 D-isomerase	Alzheimer's disease	Dementia and primarily neurodegenerative	+
P01034	CST3	Cystatin C	Alzheimer's disease	Dementia and primarily neurodegenerative	+
P10599	TXN	Thioredoxin	Alzheimer's disease	Dementia and primarily neurodegenerative	+
P61769	B2M	Beta-2-microglobulin	Alzheimer's disease	Dementia and primarily neurodegenerative	+
P60880	SNAP25	Synaptosomal-associated protein 25	Alzheimer's disease	Dementia and primarily neurodegenerative	+
P10636	MAPT	Microtubule-associated protein tau	Alzheimer's disease	Dementia and primarily neurodegenerative	+
P43652	AFM	Afamin precursor	Alzheimer's disease	Dementia and primarily neurodegenerative	+
P39060	COL18A1	α1-Type XVIII collagen isoform 3 precursor	Alzheimer's disease	Dementia and primarily neurodegenerative	+
P04217	A1BG	α1B-Glycoprotein precursor	Alzheimer's disease	Dementia and primarily neurodegenerative	+

(continued)

Table 1 (continued)

Accession	Gene	Protein	Disease	Disease type	Change
P05067	APP	Amyloid β A4 protein isoform g	Alzheimer's disease	Dementia and primarily neurodegenerative	+
P43251	BTD	Biotinidase precursor	Alzheimer's disease	Dementia and primarily neurodegenerative	+
P42857	NSG1	Brain neuron cytoplasmic protein 1	Alzheimer's disease	Dementia and primarily neurodegenerative	+
P14209	CD99	CD99 antigen isoform b precursor	Alzheimer's disease	Dementia and primarily neurodegenerative	+
P00450	CP	Ceruloplasmin precursor	Alzheimer's disease	Dementia and primarily neurodegenerative	+
Q01459	CTBS	Di-N-acetyl-chitobiase	Alzheimer's disease	Dementia and primarily neurodegenerative	+
P06681	C2	Complement component 2 isoform 2 preproprotein	Alzheimer's disease	Dementia and primarily neurodegenerative	+
P01031	C5	Complement component 5 preproprotein	Alzheimer's disease	Dementia and primarily neurodegenerative	+
P13671	C6	Complement component 6 precursor	Alzheimer's disease	Dementia and primarily neurodegenerative	+
P07358	C8B	Complement component 8, β polypeptide preproprotein	Alzheimer's disease	Dementia and primarily neurodegenerative	+
P00751	CFB	Complement factor B preproprotein	Alzheimer's disease	Dementia and primarily neurodegenerative	+
P09104	ENO2	Enolase 2	Alzheimer's disease	Dementia and primarily neurodegenerative	+
P04075	ALDOA	Fructose-bisphosphate aldolase A	Alzheimer's disease	Dementia and primarily neurodegenerative	+

(continued)

Table 1 (continued)

Accession	Gene	Protein	Disease	Disease type	Change
P48058	GRIA4	Glutamate receptor, ionotrophic, AMPA 4 isoform 3 precursor	Alzheimer's disease	Dementia and primarily neurodegenerative	+
P24592	IGFBP6	Insulinlike growth factor binding protein 6	Alzheimer's disease	Dementia and primarily neurodegenerative	+
P19827	ITIH1	Inter-α (globulin) inhibitor H1	Alzheimer's disease	Dementia and primarily neurodegenerative	+
P32004	L1CAM	L1 cell adhesion molecule isoform 3 precursor	Alzheimer's disease	Dementia and primarily neurodegenerative	+
P02750	LRG1	Leucine-rich α2-glycoprotein 1	Alzheimer's disease	Dementia and primarily neurodegenerative	+
P43121	MCAM	Melanoma cell adhesion molecule	Alzheimer's disease	Dementia and primarily neurodegenerative	+
Q7Z3B1	NEGR1	Neuronal growth regulator 1	Alzheimer's disease	Dementia and primarily neurodegenerative	+
P20774	OGN	Osteoglycin preproprotein	Alzheimer's disease	Dementia and primarily neurodegenerative	+
Q96PD5	PGLYRP2	Peptidoglycan recognition protein 2 precursor	Alzheimer's disease	Dementia and primarily neurodegenerative	+
P00747	PLG	Plasminogen	Alzheimer's disease	Dementia and primarily neurodegenerative	+
Q15223	PVRL1	Poliovirus receptor–related 1 isoform 1	Alzheimer's disease	Dementia and primarily neurodegenerative	+
P30101	PDIA3	Protein disulfide-isomerase A3 precursor	Alzheimer's disease	Dementia and primarily neurodegenerative	+
Q16849	PTPRN	Protein tyrosine phosphatase, receptor type, N precursor	Alzheimer's disease	Dementia and primarily neurodegenerative	+

(continued)

Table 1 (continued)

Accession	Gene	Protein	Disease	Disease type	Change
P14618	PKM	Pyruvate kinase, muscle isoform M1	Alzheimer's disease	Dementia and primarily neurodegenerative	+
P10451	SPP1	Secreted phosphoprotein 1 isoform a	Alzheimer's disease	Dementia and primarily neurodegenerative	+
P10451	SPP1	Secreted phosphoprotein 1 isoform b	Alzheimer's disease	Dementia and primarily neurodegenerative	+
Q9BYH1	SEZ6L	Seizure-related 6 homologue (mouse)–like precursor	Alzheimer's disease	Dementia and primarily neurodegenerative	+
P01011	SERPINA3	Serpin peptidase inhibitor, clade A, member 3 precursor	Alzheimer's disease	Dementia and primarily neurodegenerative	+
Q6EMK4	VASN	Slitlike 2	Alzheimer's disease	Dementia and primarily neurodegenerative	+
Q9HCB6	SPON1	Spondin 1, extracellular matrix protein	Alzheimer's disease	Dementia and primarily neurodegenerative	+
P08294	SOD3	Superoxide dismutase 3, extracellular precursor	Alzheimer's disease	Dementia and primarily neurodegenerative	+
P02787	TF	Transferrin	Alzheimer's disease	Dementia and primarily neurodegenerative	+
O15240	VGF	VGF nerve growth factor–inducible precursor	Alzheimer's disease	Dementia and primarily neurodegenerative	+
P41222	PTGDS	Prostaglandin-H2 D-isomerase	Alzheimer's disease	Dementia and primarily neurodegenerative	+
P02647	APOA1	Apolipoprotein A-I	Alzheimer's disease	Dementia and primarily neurodegenerative	+
P10645	CHGA	Chromogranin-A	Alzheimer's disease	Dementia and primarily neurodegenerative	+

(continued)

Table 1 (continued)

Accession	Gene	Protein	Disease	Disease type	Change
P02766	TTR	Transthyretin	Alzheimer's disease	Dementia and primarily neurodegenerative	+
P01023	A2M	Alpha-2-macroglobulin	Alzheimer's disease	Dementia and primarily neurodegenerative	+
P02749	APOH	Beta-2-glycoprotein 1	Alzheimer's disease	Dementia and primarily neurodegenerative	+
P36222	CHI3L1	Chitinase-3-like protein 1	Alzheimer's disease	Dementia and primarily neurodegenerative	+
P0C0L5	C4B	Complement C4-B	Alzheimer's disease	Dementia and primarily neurodegenerative	+
P23142	FBLN1	Fibulin 1	Alzheimer's disease	Dementia and primarily neurodegenerative	+
P0C0L5	C4B	C4B1	Alzheimer's disease	Dementia and primarily neurodegenerative	+
P02749	APOH	β2-glycoprotein	Alzheimer's disease	Dementia and primarily neurodegenerative	+
P01019	AGT	Angiotensinogen	Alzheimer's disease	Dementia and primarily neurodegenerative	+
P36955	SERPINF1	PEDF	Alzheimer's disease	Dementia and primarily neurodegenerative	+
P00738	HP	Haptoglobin	Alzheimer's disease	Dementia and primarily neurodegenerative	+
P02766	TTR	TTR dimer	Alzheimer's disease	Dementia and primarily neurodegenerative	+

5. stringApp for Cytoscape (version 2.1.1, plugin available at https://apps.cytoscape.org/apps/stringapp, STRING source database: https://string-db.org/).

3 Methods

3.1 Step-by-Step Analysis of the Biological Pathways and Processes Associated with the CSF Proteomes Across Neurodegenerative Diseases Using ShinyGO

For this tutorial, the analysis of the biological pathways and processes using ShinyGO will be restricted to Alzheimer's disease, Dementia with Lewy Bodies, and Parkinson's disease, given the scarcity of data available for the remainder diseases (Table 1). We will perform separate analyses for the down- and up-regulated proteins.

1. For the identification of the deregulated biological pathways more representative for each disease, go to ShinyGO 0.80 (http://bioinformatics.sdstate.edu/go/).
2. Maintain the species as "Human."
3. Paste the list of the down-regulated proteins identified in CSF in Alzheimer's disease, using the corresponding gene name (Table 1, column 2).
4. Keep the Background gene list. By default, ShinyGO compares our gene list with a background of all protein-coding genes in the genome (*see* **Note 1**).
5. Keep the "FDR cutoff" as 0.05, the "number of pathways to show" as 20, change the "Pathway size. Min." to 3 and keep the "Max." as 5000. However, you can change these parameters as desired (*see* **Note 2**).
6. Uncheck the option "Abbreviated pathways," so that the full name of the pathways is displayed. Keep only a check on "Remove redundancy." This option eliminates similar pathways that share 95% of their genes and 50% of the words in their names, keeping only the pathway with the highest significance.
7. Click on "Submit."
8. On the right, click on the tab "Chart" to see a graphical representation of the analysis (*see* **Note 3**). By default, a graph with the KEGG pathways is shown.
9. On the left, in the item "Pathway database," select "Curated. Reactome" to obtain a graph with the pathways from the Reactome database enriched with the down-regulated proteins in CSF in Alzheimer's disease.
10. You can change the parameters below the graph as desired. For instance, we changed some of the parameters as follows:
    - 10.1. Select 3 on "Aspect Ratio."
    - 10.2. Decrease the "Font Size" to 10.
    - 10.3. Change the "Color:Low" to yellow.
    - 10.4. Change the "Plot theme:" to linedraw.

11. Click on "Download Plot" to download the graph. Choose the dimensions desired for the graph. For instance, select 12 inches "Width (in)" and 4 inches "Height (in)." Then click on PNG to save the graph in the PNG format. The graph can also be saved in PDF and SVG formats.
12. On the tab "Enrichment," you can see the results of the enrichment analysis in a table. Only pathways within specified size limits are considered. After the analysis, pathways are filtered by a user-defined FDR cutoff. Significant pathways are then ranked based on FDR, Fold Enrichment, or other metrics. When the "Sort by average ranks (FDR & Fold Enrichment)" option is selected, pathways are sorted by the average of their ranks based on both FDR and Fold Enrichment. By selecting "Select by FDR, sort by Fold Enrichment," users first filter pathways by FDR and then sort those pathways by Fold Enrichment (*see* **Note 4**).
13. To save the pathways shown in the table, click on "Top pathways shown above" and a file in CSV format will be downloaded. If you want to save all the pathways, click on "Results on All Pathways."
14. To visualize the biological pathways associated with the proteins up-regulated in CSF in Alzheimer's disease, follow the previous steps with the respective list of the up-regulated proteins (Table 1).
15. Follow the steps described above to visualize the biological pathways associated with the down- and up-regulated proteins identified in the CSF in Dementia with Lewy Bodies and Parkinson's disease, pasting the corresponding gene lists (Fig. 1).
16. The analysis of the biological processes using ShinyGO can be performed with the proteins up- and down-regulated identified in CSF in Alzheimer's disease, Dementia with Lewy Bodies, and Parkinson's disease in a similar way to the one described for the biological pathways. After obtaining the graph in the "Chart" tab, select on the left, on "Pathway database," the option "GO Biological Process." Then, change the "Font size" to 10, the "Plot theme:" to line-draw and download the plot as PNG (Width (in): 8, Height (in): 4) (Fig. 2).

In the case of Alzheimer's disease, "HDL remodeling" appears as a relevant pathway associated with the proteins down-regulated in the CSF, followed by "Chylomicron assembly" and "Chylomicron remodeling." The pathways associated with the up-regulated proteins include "Terminal pathway of complement," and also "Chylomicron assembly" and "Chylomicron remodeling." These two pathways commonly associated with down- and up-regulated

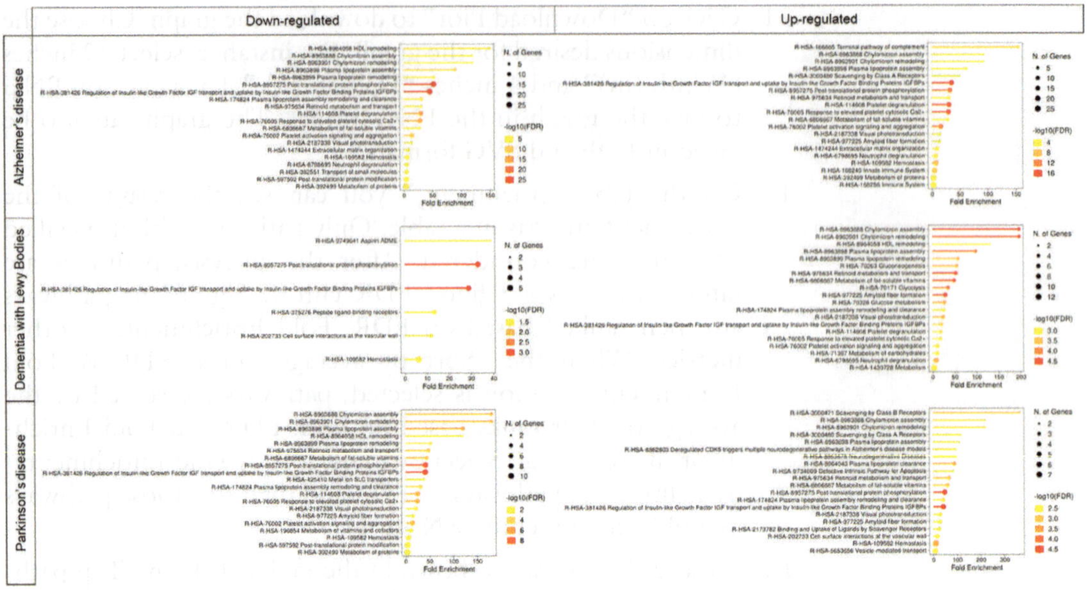

Fig. 1 Biological pathways associated with the proteins down- and up-regulated in CSF in Alzheimer's disease, Dementia with Lewy Bodies, and Parkinson's disease, obtained using ShinyGO. Database sourced on June 20th, 2024

proteins highlight inconsistencies between the studies regarding the levels of some proteins. This is also observed in the case of Parkinson's disease. On the contrary, other pathways ("Aspirin ADME,""Post-translational protein phosphorylation," and "Hemostasis") emerged as being exclusively associated with the down-regulated proteins in Dementia with Lewy Bodies. This most probably reflects the lower level of evidence rather than pointing to effectively downregulated pathways in this condition. In any case, "Chylomicron assembly," "Chylomicron remodeling," and "HDL remodeling" pathways surge associated with the up-regulated proteins. The identification of common deregulated pathways in these three neurodegenerative diseases suggests they may share similar pathophysiological processes.

Some biological processes are common to the diseases in study; nonetheless, there are also disease-specific deregulated biological pathways and processes. For instance, "Positive regulation of amyloid fibril formation" was the most relevant up-regulated biological process assigned to Alzheimer's disease. The processes "Astrocyte activation involved in immune response" and "Negative regulation of long-term synaptic potentiation" were also exclusively up-regulated in Parkinson's disease.

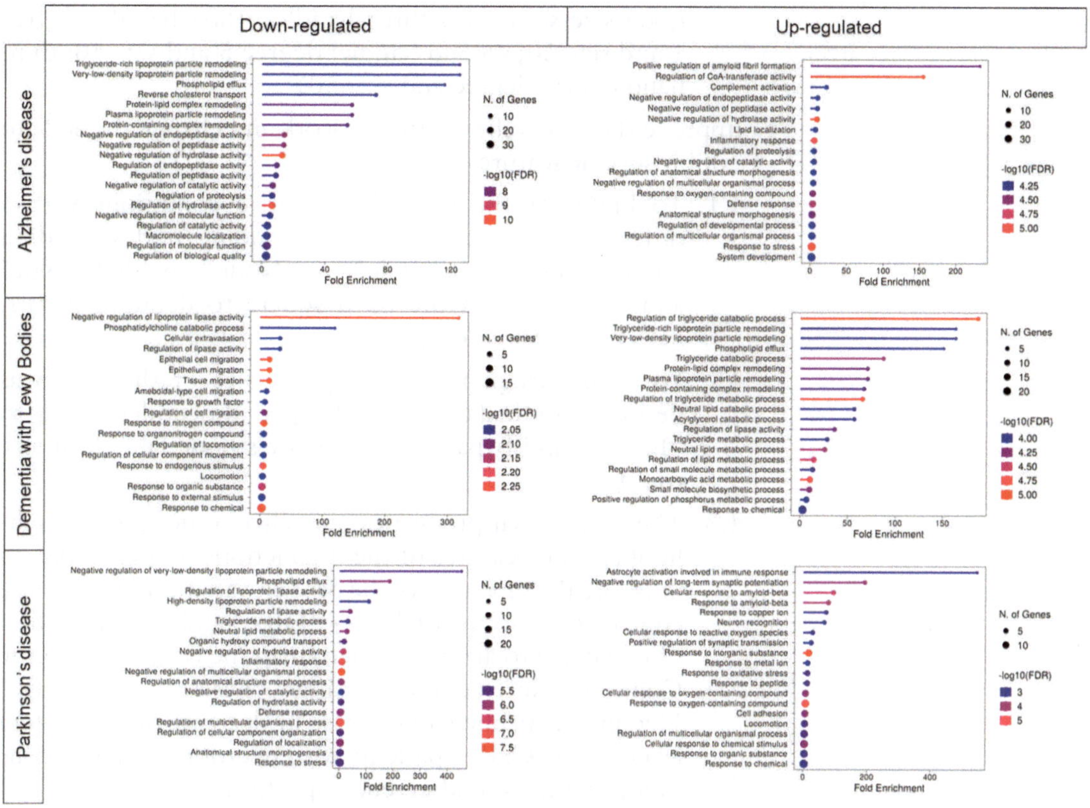

Fig. 2 Biological processes associated with the proteins down- and up-regulated in CSF in Alzheimer's disease, Dementia with Lewy Bodies, and Parkinson's disease, obtained using ShinyGO. Database sourced on June 20th, 2024

3.2 Step-by-Step Analysis of the Associations Between CSF Proteins and Neurodegenerative Diseases Using Cytoscape and Comparison with the DisGeNET Database

1. Open Cytoscape v3.10.2.
2. Select "File" → "Import" →"Network from File" → Select "Table 1." A new window pops up.
   2.1. In the Gene column select the green dot (Source Node).
   2.2. In the Disease column select "◉" (Target Node).
   2.3. In the Change column select "►"(Interaction type; *see* **Note 5**). Click "OK."
3. Select "Tools" → "Analyze Network."
   3.1. Keep the default option "Analyze as Directed Graph?" blank (*see* **Note 6**).
   3.2. A new table is shown in a new tab with the main network topological characteristics, including the number of nodes and edges.
   3.3. In the "Style" tab, tick the option "Lock node width and height," map node size to the Column "Degree" and select "Continuous Mapping." We defined a continuous

node size scale with three handles: (first handle: degree 1 and size 20; second handle: degree 5 and size 40; third handle: max degree and size 50).

4. Improve the readability of the network and change the style to exhibit specific features.

   4.1. To highlight the neurodegenerative conditions with a different color, select all corresponding nodes, holding control, and go to "Style" → "Node" → "Fill Color" and right-click on the third square to define a bypass. Select the desired color.

   4.2. You can change the visual properties of the nodes as desired. For instance, we changed the nodes shape to "Ellipse" and bypassed the disease's node label font size to 20.

   4.3. The edge visual properties may also be defined to highlight specific features of the interactions. For example, to illustrate the direction of the protein variation, one can use edge colors. Go to "Style" → "Edge" and tick the option "Edge color to arrows." Then, map the edge color to the Column "Interaction," using "Discrete Mapping." Herein, we mapped an increase or decrease of the protein levels in the CSF of patients with neurodegenerative conditions with red and green, respectively.

   4.4. For a better visualization of the network (rotation, scaling) you can use the Layout tools available on the "Layout" menu.

5. Go to "File" → "Export" →"Network to Image... ." Select the desired file format, name it (e.g., Fig. 3) and find the proper location. Files should be preferentially saved as *.png*. Maximum resolution is achieved at maximum zoom and with "inches" selected (*see* **Note** 7). To save the session go to "File" → "Save As" and select the desired directory.

   This analysis allows the identification of common and unique proteins *per* disease. In the resulting network, a snapshot of the relevance of each protein *per* condition is attained. Larger nodes represent proteins associated with more conditions and/or with a higher number of evidences. By marking the proteins that are increased or decreased in CSF, the edge colors are useful to evidence the agreement between different studies. For instance, the protein CST3 (Cystatin-C) is associated simultaneously with Parkinson's disease and Alzheimer's disease. However, for the latter, there is no agreement between the studies: two studies show that CST3 is increased (red edges) in the CSF of Alzheimer's disease patients, and two studies report otherwise (green edges). From this analysis, a first filter to identify potential disease biomarkers can be

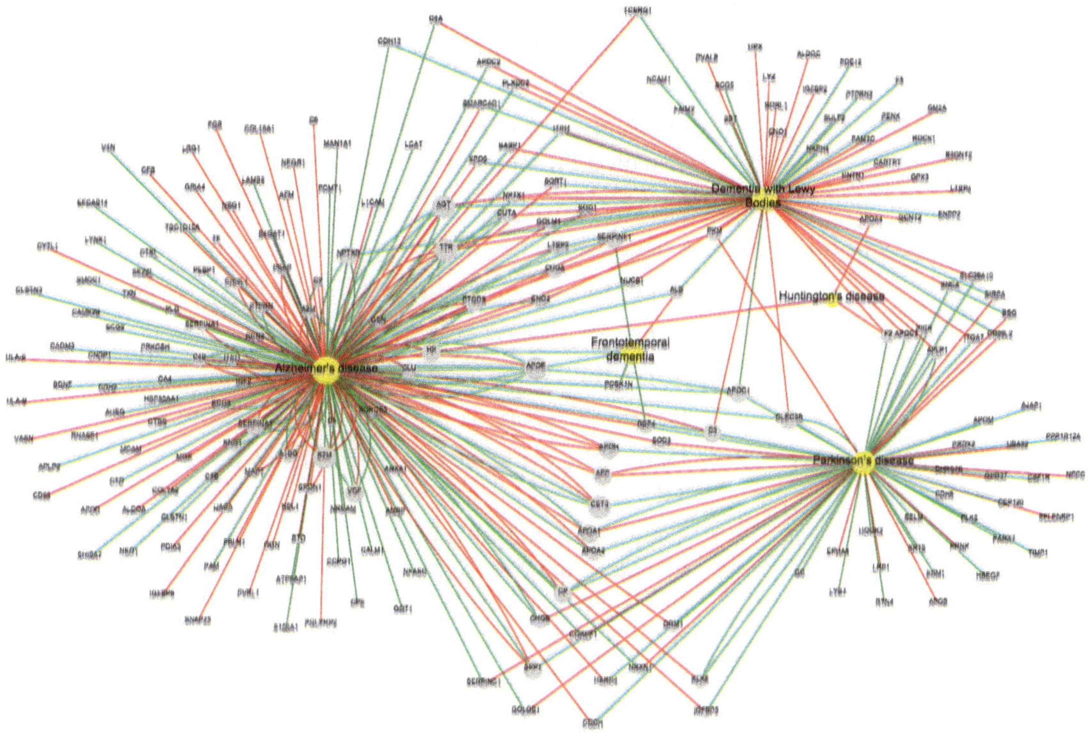

Fig. 3 Network depicting the association between CSF proteins and the selected neurodegenerative diseases. Proteins' node size is proportional to the number of studies supporting the association. Red and green edges represent an increase and decrease in the CSF protein levels, respectively. Proteins are identified through the respective gene name and the extended protein name can be found in Table 1

applied. Ideally, proteins uniquely associated with one condition, with a higher number of evidence and showing a consistent variation, should be regarded as preferable candidates. For instance, the protein APOH (apolipoprotein H) would be one of the best candidate biomarkers among the proteins depicted in Fig. 3. APOH was found upregulated in the CSF of Alzheimer's disease patients in three independent studies and downregulated in the CSF of Parkinson's disease in another study. Still, one should mind that many other considerations are needed in the pipeline towards the definition of potential biomarkers (sample size of the discovery cohort, study design, effect size, performance of quantification platform, associations with other non-neurodegenerative conditions, among others). The putative candidate biomarkers identified through this strategy do not account for potential associations with other non-neurodegenerative conditions and do not integrate other types of evidence, such as GWAS, animal models, and text-mining, among others. DisGeNET analysis integrates all these evidences and, in addition, provides a score for the association strength as well as a disease specificity index, as exemplified next.

6. Open Cytoscape v.3.10.2.
7. Select "Apps" → "DisGeNET" → "Start DisGeNET."
   7.1. The first time running the app, you need to download the DisGeNET database. Select the desired directory and download it. It can take several minutes.
   7.2. On the DisGeNET tab, select "Gene Disease Network."
   7.3. Select source "ALL."
   7.4. Select the association type "Biomarker" (*see* **Note 8**).
   7.5. Select the disease class "Any."
   7.6. Select the evidence level "Any."
   7.7. Define the score from 0.5 to 1.0 (*see* **Note 9**).
   7.8. Define the evidence index (EI) from 0.8 to 1.0 (*see* **Note 10**).
   7.9. Type the name of the disease, for instance, "Alzheimer's disease" or the respective Concept Unique Identifier (CUI), which can be found on the DisGeNET database website. The CUI for Alzheimer's Disease is C0002395 (*see* **Note 11**).
   7.10. Click on "Create Network" (*see* **Note 12**).
   7.11. Repeat the steps for the remaining conditions: Parkinson's disease (CUI: C0030567), Dementia with Lewy Bodies (CUI: C0752347), Frontotemporal dementia (CUI: C0338451), and Huntington's disease (CUI: C0020179).
8. Merge Networks.
   8.1. Go to "Tools" → "Merge" → "Networks."
   8.2. Select "Union" and all "Available Networks."
   8.3. Click "Merge."
9. Style the network to emphasize DisGeNET data.
   9.1. By default, the node labels correspond to geneId (number). To depict both gene names and disease names, go to "Style" → "Node" → "Label," select the column "styleName" and keep the default "Passthrough Mapping."
   9.2. DisGeNET networks depict, also by default, each evidence of a gene-disease association by a different edge. To simplify the network and improve visualization, go to "Edit" →" Remove Duplicate Edges." This step cannot be undone.
   9.3. The score for each association can be shown through edge's visual properties. Go to "Style" →" Edge" → "Width." Select the column "Score" and define a "Continuous Mapping." We defined a continuous edge width scale with two handles (first handle: GDA score 0.5

and width 1; second handle: GDA score 1.0 and width 10).

9.4. You can perform additional changes to edge visualization properties. For example, if you wish to change the color of the edges, go to "Style" → "Edges" and tick the option "Edge color to arrows." Then, choose the desired color on "Color."

9.5. Node style can also be modified to highlight the diseases and other DisGeNET scores. If you want to highlight the diseases with a different color, first select all relevant nodes, holding the Control key to allow multiple selections. Then, go to "Style" → "Node" → "Fill Color" and right-click on the third square to "Set Bypass." Then, define the preferred color.

9.6. The disease specificity index (DSI) can also be evidenced by node size. Go to "Style" → "Node" → "Size." Then, select the column "DSI" and use a Continuous Mapping. We defined a continuous node size scale with two handles (first handle: min DSI and size 10; second handle: max DSI, node size 40).

10. Repeat **step 5** to save the session and export the network (Fig. 4).

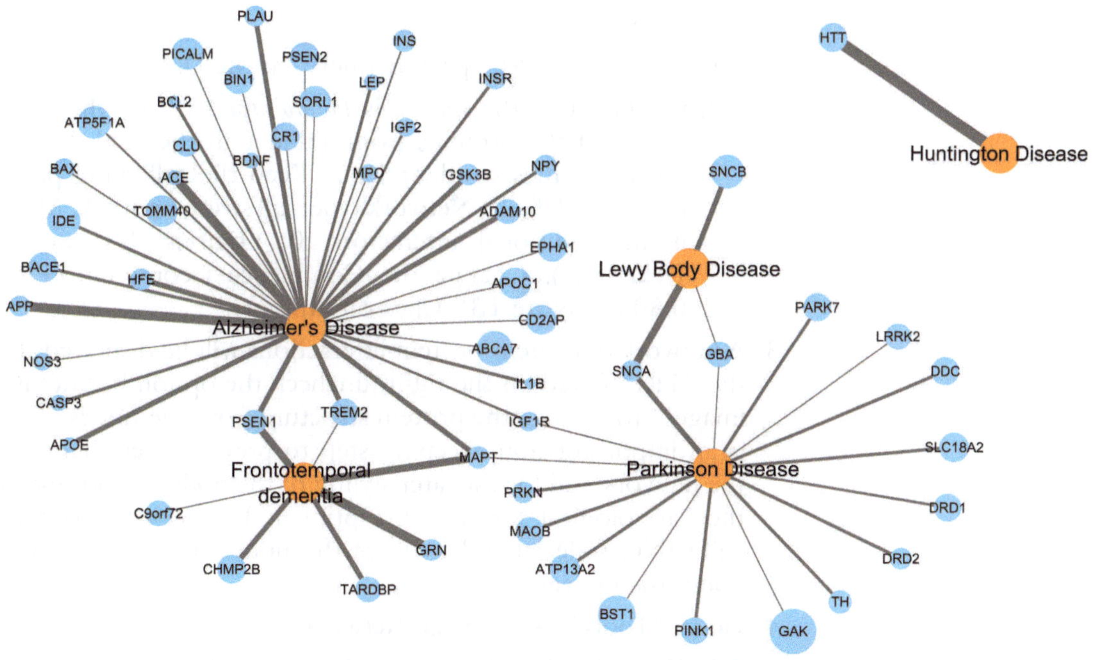

Fig. 4 Network showing the known associations between genes/proteins and the selected neurodegenerative diseases, according to DisGeNET. Proteins' node size is proportional to the disease specificity index (DSI), while the edge's width is proportional to the gene-disease association score. Proteins are identified through the respective gene name. Database sourced on June 20th, 2024

This analysis makes clear the difference between the strength of association and the specificity of a given protein to a disease. For instance, despite the strong association between the amyloid-beta precursor protein (APP) and Alzheimer's disease or alpha-synuclein (SNCA) with Parkinson's disease or Lewy Body disease (thicker edges), there are other proteins that are more specific to these conditions. The mitochondrial import receptor subunit TOM40 homolog (TOMM40), the cyclin-G-associated kinase (GAK) and the protein beta-synuclein (SNCB) are more specifically associated with Alzheimer's, Parkinson's and Lewy Body diseases, respectively, as shown by the bigger node size (reflecting a higher DSI). This might reflect either disease-specific deregulated pathways or the lack of evidence for other pathologies. The proteome data, as shown in Fig. 3, can be integrated and fed to the DisGeNET database, providing complementary information on the strongest candidates for biomarker purposes.

3.3 Step-by-Step Protein-Protein Interaction Analysis and Identification of Putative Therapeutic Targets Using STRING

To analyze protein-protein interactions and possible therapeutic targets, the STRING app from Cytoscape was used. For this analysis, we only considered the up- and down-regulated proteins found in CSF in Alzheimer's disease, Dementia with Lewy Bodies, and Parkinson's disease.

1. Open Cytoscape v.3.10.2.
2. Go to "File" → "Import" → "Network from Public Databases."
   2.1. Select "STRING: protein query" as the Data Source.
   2.2. Confirm that the species is *Homo sapiens.* Paste the gene names of the proteins from Table 1 in the field "Enter protein names or identifiers." Keep the following parameters as default (Network type: full string network, Maximum additional interactors: 0, Options: Use Smart Delimiters). Change the Confidence (score) cutoff to 0.80 (*see* **Note 13**). Click on "Import."
3. A network of protein-protein interactions will be displayed. In the STRING tab on the right, uncheck the option "Structure images" to remove the protein structure from the nodes. This is an important housekeeping step to prevent overwhelming the network and for the later styling of the nodes according to their topological properties. Optional: Uncheck the option "STRING style labels" to layer the nodes' labels above the nodes themselves.
4. Go to "Tools" → "Analyze Network."
   4.1. Keep the default option "Analyze as Directed Graph?" unchecked (*see* **Note 6**).

4.2. A new tab will appear on the right (Analyzer) where you can see the network's topological characteristics.

5. Next, we will go to change node style according to node degree and betweenness centrality, to identify protein hubs and bottlenecks, respectively. Go to the tab "Style" on the left.

6. Check the option "Lock node width and height." This will allow the mapping of the nodes' properties according to their "Size."

   6.1. Drop the size feature down. Click on "Column" and select "Degree." Click on "Mapping Type" and select "Continuous Mapping." The nodes in the network will be defined with different sizes, depending on the node degree (the higher the size, the higher the node degree). Double-click the graph to change the minimum and maximum size of the nodes. In this case, we used the minimum size of 30.0 and the maximum of 80.0 to visualize the differences in the nodes' degrees.

7. Then, drop the transparency feature down. Click on "Column" and select "BetweennessCentrality." Click on "Mapping Type" and select "Continuous Mapping." Double-click the graph to change the minimum and maximum Transparency of the nodes. In this case, we defined the minimum as 150.0 and the maximum as 255.0, to distinguish the nodes with a higher betweenness centrality (that corresponds to the nodes with higher opacity) from the nodes with a lower betweenness centrality (more transparent). Select the option "Cytoscape design" on the STRING tab on the right to see the differences in node transparency.

8. You can perform additional changes to node visualization properties. For example, if you wish to change the label font size of the nodes, go to "Style" → "Node." Then, choose the desired size on "Label Font Size." Like in **step 4.4**. from Subheading 3.2, you can use the Layout tools available on the "Layout" menu, for a better visualization of the network.

9. When the edge widths make the visualization of the nodes difficult, you can change the Edge style on the Style option. In this case, we changed the Transparency default value to 100.

10. A disease-specific node coloring might also be useful to facilitate the interpretation of the network. In this case, we changed the color of the nodes according to the identification of the respective proteins in the CSF of Alzheimer's disease, Dementia with Lewy Bodies, Parkinson's disease, and more than one of these diseases. For this, you can select the respective nodes and bypass their color under "Style." To avoid selecting each node one by one, you can:

10.1. Export the node table to a .csv file.

10.2. Import data to a spreadsheet and obtain the nodes under the "name" column corresponding to the "display name" for the proteins identified in the CSF of Alzheimer's disease, Dementia with Lewy Bodies, Parkinson's disease, and in more than one of these diseases (*see* **Note 14**).

10.3. Copy and paste the different groups of proteins to distinct .txt files and name them accordingly.

10.4. Back on Cytoscape, go to "Select" → "Nodes" → "From ID list file," and select, for instance, the file corresponding to the proteins common to all diseases. The nodes corresponding to these proteins should automatically pop up as yellow.

10.5. To change their color, go to "Style" → Node tab and make a bypass to "Fill color" (right-click on the third square and select "Set Bypass..."). Then, choose the desired color.

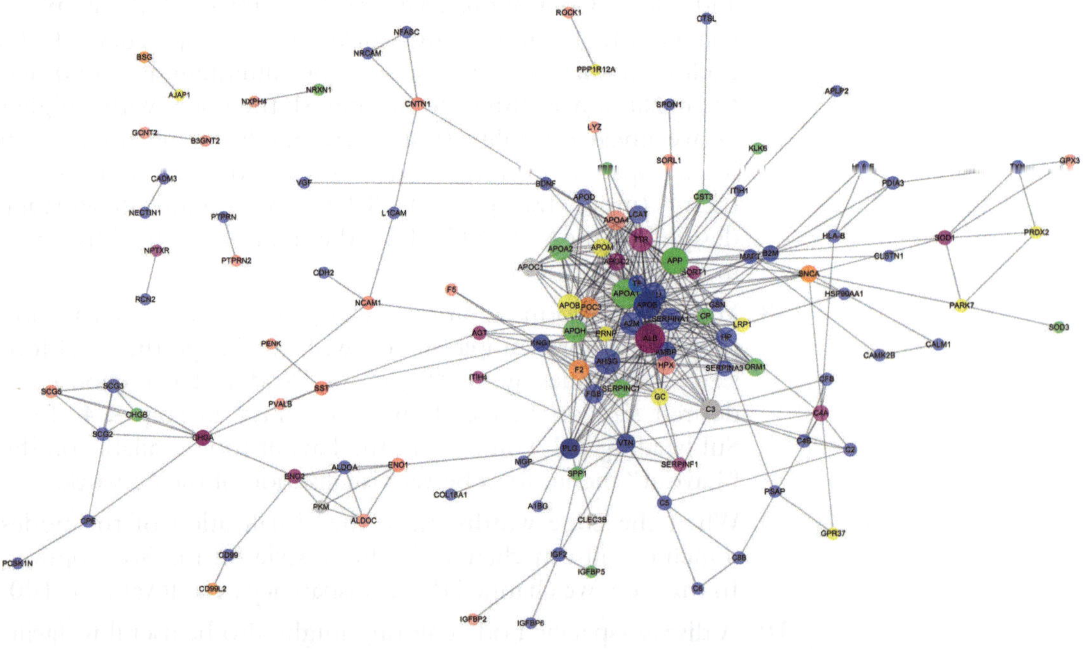

Fig. 5 STRING's protein-protein interaction network obtained using Cytoscape. Nodes in blue: proteins deregulated only in Alzheimer's disease (AD), yellow: proteins deregulated only in Parkinson's disease (PD), rose: proteins deregulated only in Dementia with Lewy Bodies (DLB), green: proteins deregulated in both AD and PD, orange: proteins deregulated in both PD and DLB, purple: proteins deregulated in both AD and DLB, and grey: proteins deregulated in AD, PD, and DLB. Proteins are identified by their gene name. Node size is proportional to its degree, representing hub proteins as larger nodes. Node opacity is proportional to its betweenness centrality, highlighting bottleneck proteins as opaquer nodes. Nodes corresponding to non-interacting proteins were removed to improve readability. Database sourced on July 17th, 2024

Herein, we set grey to define the proteins common to all diseases.

10.6. Repeat 9.4. and 9.5., selecting the different files and choosing a different color. Herein, we set blue to define proteins identified only in Alzheimer's disease, yellow to proteins identified only in Parkinson's disease, rose to proteins identified only in Dementia with Lewy Bodies, green to proteins present in both AD and PD, orange to proteins present in both PD and DLB and purple to proteins present in both AD and DLB (Fig. 5).

11. Save the image as described above (Subheading 3.2, **step 5**).
12. To obtain a graph with the relationship between the Degree and Betweenness Centrality of the nodes, click on the "Betweenness by Degree" option in the Analyzer tab on the right (Fig. 6).

Not surprisingly, amyloid-beta precursor protein (APP), identified in the CSF of Alzheimer's disease and Parkinson's disease, was found to be the protein with a higher degree and betweenness centrality (Fig. 6). Also, apolipoprotein A-I (APOA1), common to these two conditions, was highlighted as having greater hub/-bottleneck properties. The same is true for the protein plasminogen (PLG), associated with Alzheimer's disease alone. The proteins albumin (ALB), apolipoprotein E (APOE), and alpha-2-HS-glycoprotein (AHSG) present a high degree, but a lower betweenness

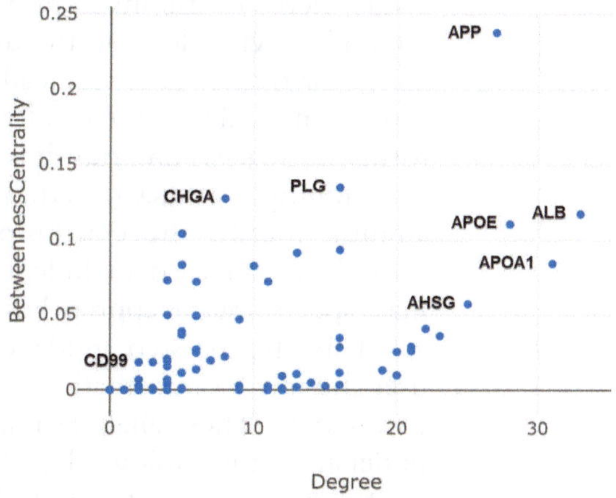

Fig. 6 Node degree versus betweenness centrality. Distribution of the proteins according to their degree and centrality highlights APP as the most interconnected (hub) and central (bottleneck) protein. AHSG: alpha-2-HS-glycoprotein, ALB: albumin, APOA1: apolipoprotein A-I, APOE: apolipoprotein E, APP: amyloid-beta precursor protein, CD99: CD99 antigen, CHGA: chromogranin-A, PLG: plasminogen. Database sourced on July 17th, 2024

centrality than PLG. In comparison, for instance, the protein chromogranin-A (CHGA) presents a lower degree and a higher betweenness centrality. On the other extreme are proteins like the CD99 which present both a low degree and betweenness centrality (Fig. 6). The proteins APP, APOA1, APOE and PLG emerge as the ones with greater potential for consideration as therapeutic targets for neurodegenerative disease. However, these results should be interpreted with caution, because some proteins like ALB and apolipoproteins circulate at a higher concentration; therefore, they can be more often identified as false-positive interactions. On a final note, careful integration of the protein expression/levels data, interactions, disease associations, and participation in specific pathways and biological processes should guide the researcher on the decision of the most promising therapeutic targets.

Variability in proteomic studies is a common challenge that arises from multiple sources. Differences in CSF collection methods, sample processing, and patient populations can lead to significant variability in the identification and quantification of proteins. Different mass spectrometry platforms, data acquisition settings, and pre-processing methods (data cleaning, imputation, and normalization) can also lead to different proteomic outputs. Similarly, in bioinformatics analysis, different algorithms and parameters can affect the reliability and reproducibility of the results. Although each tool has mechanisms to control for false positives such as adjustment for multiple testing in ShinyGO, or for the volume and quality of evidence, such as confidence scores in STRING and scores for the strength and specificity of the association in DisGeNET, there are still inherent limitations. Furthermore, the results are highly dependent on the quality and extent of the input data, and poor-quality data can lead to misleading conclusions. Therefore, to maximize the reliability and biological validity of the bioinformatics analysis findings, it is recommended to integrate datasets from multiple studies to reduce the impact of bias from individual studies. Ideally, the conclusions of the bioinformatics analyses should be validated in independent experiments using different mass spectrometric approaches or other independent techniques.

This step-by-step guide exemplifies the usefulness of some user-friendly bioinformatics tools in the integrated analysis of proteome data. These analyses might be of great interest to: i) better understand the biological pathways and processes modulated by neurological diseases through the analysis of CSF with tools such as ShinyGO; ii) quickly identify putative biomarkers with Cytoscape and DisGeNET; iii) uncover potential therapeutic targets using STRING.

4 Notes

1. For enrichment analysis, a background gene list is needed. Usually, this list includes all protein-coding genes in the genome. However, the utilization of a customized background gene list with the inclusion of protein-coding genes from a specific tissue sample, in this case, brain tissue, could give more biologically meaningful enrichment results.
2. Enrichment analysis P-values are calculated using the hypergeometric test. To correct for multiple testing, a False Discovery Rate (FDR) is calculated using the Benjamini-Hochberg method.
3. Several tabs are presented. For instance, in the tab "Enrichment," a table with the enrichment analysis is shown. In the tab "Chart," a graphical representation of the analysis can be found. By default, a graph with the KEGG pathways is shown, but other ontologies can be selected within the "Pathway database" dropdown menu. In the tab "Tree," a hierarchical clustering tree summarizing the correlation between significant pathways listed in the Enrichment tab is obtained. In the tab "Network," an interactive network is shown, also representing the relationship between enriched pathways. In the tab "KEGG," it is possible to explore in detail a specific KEGG pathway (only possible for KEGG enrichment analysis). In the tab "Genes," a mapping table of the input genes can be found. In addition, in the tab "STRING," genes are sent to the STRING-db website for enrichment analysis and a protein-protein network is obtained. This is not explored further, as the Cytoscape-based STRING analysis is explained in **step 3.3**.
4. In the GO Enrichment Results (table from the Enrichment tab):
   - The P-value reflects the statistical significance of the enrichment. Lower values suggest a lower likelihood of the result occurring by chance. FDR q-values are P-values adjusted for multiple testing to reduce type I errors.
   - The Fold Enrichment measures the effect size of the enrichment.
   - The Pathway Genes refers to the total number of genes in a pathway or GO term.
   - The nGenes means the number of genes in the pathway that overlap with the input gene list.
5. The selection of gene, "►" in the Change column and "◉" in the Disease column defines the type of interaction with a focus on the disease, the target node. With these definitions, the gene name can be visualized and the multiple associations to the conditions in analysis are depicted.

6. To study associations between proteins and diseases no directionality is required, i.e., the statements 'condition A is associated to protein B' and 'protein B is associated to condition A' mean the same. Still, when working with protein-protein interactions, e.g., when studying signaling pathways, directionality is sometimes required. In other words, in a reaction cascade A → B → C (where A activates B and this activates C) is not biologically equal to C → B → A. In such cases, make sure to "Treat the network as directed."
7. Maximum resolution is achieved at 600 dpi.
8. According to the DisGeNET ontology, "Biomarker" is any of the following categories: "Genomic Alterations," "Altered Expression," and "Post-translational Modification." Selecting "Biomarker" only excludes "Therapeutic," which is not the goal of this specific analysis.
9. Refers to the gene-disease association score, a measure of the strength of the association.
10. The evidence index is a measure of the degree of agreement between the available evidences. The higher the number of contradicting evidences, the lower the score.
11. Using CUI is recommended for reproducibility purposes (slight variations of disease names and disease subtypes may lead to different results).
12. This command might be hidden depending on your screen configurations. If that is the case, just pop out the DisGeNET plugin window and stretch down until the command is visible.
13. In STRING, this score is an indicator of confidence, meaning the likelihood of a protein-protein interaction to be true, given the available evidence. This score ranks from 0 to 1, with 1 being the highest possible confidence. Choosing a higher score can be useful to clean overcrowded networks, i.e., with a high number of interactions. The user should fine-tune this score, according to the level of evidence available and the stringency required for the desired downstream analysis.
14. To obtain the proteins identified only in one specific condition and common to different conditions, you can use tools such as Venny, available at https://bioinfogp.cnb.csic.es/tools/venny/.

Acknowledgments

Acknowledgments are due to "Fundação para a Ciência e a Tecnologia"—FCT, European Union, QREN, FEDER and COMPETE for the financial support for the UnIC (UIDB/IC/00051/

2020 and UIDP/00051/2020), LAQV-REQUIMTE (UIDB/50006/2020, https://doi.org/10.54499/UIDP/50006/2020; LA/P/0008/2020, https://doi.org/10.54499/LA/P/0008/2020; UIDP/50006/2020, https://doi.org/10.54499/UIDP/50006/2020), and RISE—Research Network-From the Lab to the Community (LA/P/0053/2020). Rita Nogueira-Ferreira acknowledges FCT for the research contract CEECIND/03935/2021 under the CEEC Individual 2021 (https://doi.org/10.54499/2021.03935.CEECIND/CP1685/CT0001).

References

1. Jeromin A, Bowser R (2017) Biomarkers in neurodegenerative diseases. Adv Neurobiol 15:491–528. https://doi.org/10.1007/978-3-319-57193-5_20
2. Bastos P, Ferreira R, Manadas B, Moreira PI, Vitorino R (2017) Insights into the human brain proteome: disclosing the biological meaning of protein networks in cerebrospinal fluid. Crit Rev Clin Lab Sci 54:185–204. https://doi.org/10.1080/10408363.2017.1299682
3. Macron C, Lane L, Núñez Galindo A, Dayon L (2018) Deep dive on the proteome of human cerebrospinal fluid: a valuable data resource for biomarker discovery and missing protein identification. J Proteome Res 17:4113–4126. https://doi.org/10.1021/acs.jproteome.8b00300
4. Kroksveen AC, Guldbrandsen A, Vaudel M, Lereim RR, Barsnes H, Myhr K-M, Torkildsen Ø, Berven FS (2017) In-depth cerebrospinal fluid quantitative proteome and deglycoproteome analysis: presenting a comprehensive picture of pathways and processes affected by multiple sclerosis. J Proteome Res 16:179–194. https://doi.org/10.1021/acs.jproteome.6b00659
5. Guldbrandsen A, Farag Y, Kroksveen AC, Oveland E, Lereim RR, Opsahl JA, Myhr K-M, Berven FS, Barsnes H (2017) CSF-PR 2.0: an interactive literature guide to quantitative cerebrospinal fluid mass spectrometry data from neurodegenerative disorders. Mol Cell Proteom 16:300–309. https://doi.org/10.1074/mcp.O116.064477
6. Havugimana PC, Hu P, Emili A (2017) Protein complexes, big data, machine learning and integrative proteomics: lessons learned over a decade of systematic analysis of protein interaction networks. Expert Rev Proteomics 14:845–855. https://doi.org/10.1080/14789450.2017.1374179
7. Calderon-Gonzalez KG, Hernandez-Monge J, Herrera-Aguirre ME, Luna-Arias JP (2016) Bioinformatics tools for proteomics data interpretation. Adv Exp Med Biol 919:281–341. https://doi.org/10.1007/978-3-319-41448-5_16
8. Bindea G, Galon J, Mlecnik B (2013) CluePedia Cytoscape plugin: pathway insights using integrated experimental and in silico data. Bioinformatics (Oxford, England) 29:661–663. https://doi.org/10.1093/bioinformatics/btt019
9. Ge SX, Jung D, Yao R (2020) ShinyGO: a graphical gene-set enrichment tool for animals and plants. Bioinformatics 36:2628–2629. https://doi.org/10.1093/bioinformatics/btz931
10. Shannon P, Markiel A, Ozier O, Baliga NS, Wang JT, Ramage D, Amin N, Schwikowski B, Ideker T (2003) Cytoscape: a software environment for integrated models of biomolecular interaction networks. Genome Res 13:2498–2504. https://doi.org/10.1101/gr.1239303
11. Bauer-Mehren A, Rautschka M, Sanz F, Furlong LI (2010) DisGeNET: a Cytoscape plugin to visualize, integrate, search and analyze gene-disease networks. Bioinformatics 26:2924–2926. https://doi.org/10.1093/bioinformatics/btq538
12. Piñero J, Ramírez-Anguita JM, Saüch-Pitarch J, Ronzano F, Centeno E, Sanz F, Furlong LI (2020) The DisGeNET knowledge platform for disease genomics: 2019 update. Nucleic Acids Res 48:D845–d855. https://doi.org/10.1093/nar/gkz1021
13. Szklarczyk D, Kirsch R, Koutrouli M, Nastou K, Mehryary F, Hachilif R, Gable AL, Fang T, Doncheva NT, Pyysalo S, Bork P, Jensen LJ, von Mering C (2023) The STRING database in 2023: protein-protein association networks and functional enrichment analyses

for any sequenced genome of interest. Nucleic Acids Res 51:D638–d646. https://doi.org/10.1093/nar/gkac1000

14. von Mering C, Huynen M, Jaeggi D, Schmidt S, Bork P, Snel B (2003) STRING: a database of predicted functional associations between proteins. Nucleic Acids Res 31:258–261. https://doi.org/10.1093/nar/gkg034

15. Trindade F, Nogueira-Ferreira R, Bastos P, Amado F, Ferreira R, Vitorino R (2019) Bioinformatics to tackle the biological meaning of human cerebrospinal fluid proteome. Methods Mol Biol 2044:393–553. https://doi.org/10.1007/978-1-4939-9706-0_26

16. Murakami Y, Tripathi LP, Prathipati P, Mizuguchi K (2017) Network analysis and in silico prediction of protein-protein interactions with applications in drug discovery. Curr Opin Struct Biol 44:134–142. https://doi.org/10.1016/j.sbi.2017.02.005

Chapter 21

Cerebrospinal Fluid Metabolomics and Proteomics Integration in Neurological Syndromes

Haitao Sun, Shilan Chen, and Jingjing Kong

Abstract

The integration of multi-omics data has increasingly been recognized as an effective approach to addressing complex problems and advancing precision medicine. Metabolomics and proteomics are closely related, and their integration provides complementary insights and enables cross-validation of experimental results. The integration of proteomics and metabolomics in cerebrospinal fluid is expected to reconstruct the complex biological networks underlying nervous system diseases, enhance understanding of molecular mechanisms, and aid in disease classification and prognosis prediction. However, integrating multi-omics data still faces numerous challenges, limiting the application of combined proteomic and metabolomic analyses in neurological diseases. Based on the advantages of integrated proteomics and metabolomics, this chapter introduces, for the first time, common strategies for the integrated analyses of omics data. Furthermore, we review advances in cerebrospinal fluid proteomics and metabolomics for neurological syndromes, highlighting current challenges and future directions.

Key words Proteomics, Metabolomics, Integrated analyses, Neurological syndromes, Cerebrospinal fluid

1 Introduction

Proteins are the downstream products in the central dogma process and act as primary agents in biological processes [1, 2]. Generalized proteomics studies proteins comprehensively within a biological system, encompassing protein identification, quantification, post-translational modifications, and protein-protein interactions [3, 4]. Meanwhile, the characterization of protein post-translational modifications, such as phosphorylation and ubiquitination, significantly differs from others in terms of research objectives, data acquisition, data processing, and data analyses. Therefore, the identification of protein post-translational modifications is commonly referred to as "modification omics" alone. The demanding technique and high cost of modification omics hinder its application in research; thus, this chapter focuses on narrow

Enrique Santamaría and Joaquín Fernández-Irigoyen (eds.), *Cerebrospinal Fluid (CSF) Proteomics: Methods and Protocols*, Methods in Molecular Biology, vol. 2914, https://doi.org/10.1007/978-1-0716-4462-1_21,

proteomics, which excludes the study of post-translational modification. Proteomic analyses can unveil alterations in cellular state and function associated with specific diseases. Cerebrospinal fluid (CSF) offers unique benefits for the diagnosis and research of neurological disorders [5, 6], and CSF proteomics has been extensively employed in the investigation of various neurological syndromes [7–10]. However, the functional state of proteins is regulated not only by genes and transcription but also by downstream metabolites, temperature, and other environmental factors [11, 12]. Proteomics alone cannot fully capture the functional state and phenotype of a biological system [13, 14].

Metabolomics, an emerging discipline developed after proteomics, has unique advantages compared to genomics and proteomics. Metabolomics is utilized to identify and quantify metabolites, the substrates and products of enzymes (proteins) in cells, tissues, or body fluids [15, 16]. In addition to being influenced by enzymes, metabolic processes are highly responsive to variations in the external environment and the physicochemical properties of the human body [15, 17]. Moreover, numerous essential cellular processes, such as intercellular communication and energy transfer, take place at the metabolite level. Therefore, metabolites, refltecting genetic and environmental influences, serve as direct indicators of biochemical reactions [17–20]. Metabolites reflect the functional and physiological state of organisms, making metabolomics the omics approach most closely linked to phenotyping [21, 22]. Additionally, minor alterations in gene and protein levels can be magnified through protein cascades during the metabolite generation process. The capture of alteration is more feasible at the metabolite level than at the protein level.

In reality, findings from a single dataset often lack reproducibility due to biases stemming from various sources, such as subject inclusion criteria, sample processing, and data analysis strategies [17, 23]. Consequently, the integration of multi-omics data has increasingly been recognized as an effective approach for tackling complex issues and advancing precision medicine [24]. Metabolomics and proteomics are closely related, and the integration of these two disciplines can mutually validate findings from diverse perspectives, reducing experimental errors caused by confounding factors. Additionally, proteins and metabolites form complex networks of biochemical reactions by interacting with each other [25]. Therefore, the integration of proteomics and metabolomics can also offer complementary insights, revealing patterns that may not be discernible when relying on a single data source [20, 26]. Such strategy, also known as horizontal data integration, enables a multi-dimension understanding of biological systems. Multi-omics data integration holds significant promise for capturing the integrity and complexity of biological networks, providing reliable insights into diseases mechanisms, classification, and prognostic [27–30].

Methods for co-extracting proteins and metabolites from a single CSF sample, along with proteomics workflows, are detailed in other chapters of this book [31]. Based on the acquired data, this chapter will further introduces strategies for integrating proteomic and metabolomic data. Additionally, we review current research on CSF proteomics and metabolomics, summarizing existing challenges and potential future directions in omics data integration. Notably, despite significant advancements in omics techniques in recent decades, the scarcity of CSF samples and the high cost of sequencing have perpetuated the predominance of single-omics approaches in scientific research [20]. The integration of omics data is an emerging field. To the best of our knowledge, this chapter is among the first to propose strategies for integrating proteomics and metabolomics.

2 Strategies for Integrating Proteomics and Metabolomics

Multi-omics generally refers to a collection of high-throughput screening techniques, including genomics, transcriptomics, proteomics, and metabolomics [32], where any combination can address scientific issue. Most studies that combine multiple omics will integrate the omics data from different dimensions to achieve maximum utility. Strategies for integrating omics data vary depending on the combination. Strategies for integrating proteomic and metabolomic data include function-based association analyses, correlation-based association analyses, and machine learning (Fig 1).

2.1 Function-Based Association Analyses

After qualitative and quantitative analyses of proteins and metabolites, differentially expressed proteins and metabolites are usually subjected to pathway enrichment analyses, which facilitates the interpretation of data at a functional level. A fundamental step in function-based association analyses is the identification of co-enriched pathways. Co-enriched pathways are those that undergo significant changes in the specific disease state and include differentially expressed proteins and metabolites. These shared enrichment pathways are considered to mutually support each other in two dimensions and warrant further investigation as disease-specific pathways.

In contemporary research, function-based association analyses are commonly conducted in two formations. One of the formations involves conducting pathway enrichment analyses separately on proteomics and metabolomics initially. Then Venn analyses are conducted to acquire co-enriched pathways [33]. The exhibition of integrated results usually begins with an overview. Venn diagrams, bar charts, or bubble charts are commonly employed to illustrate all co-enriched pathways [34]. Subsequently, researchers

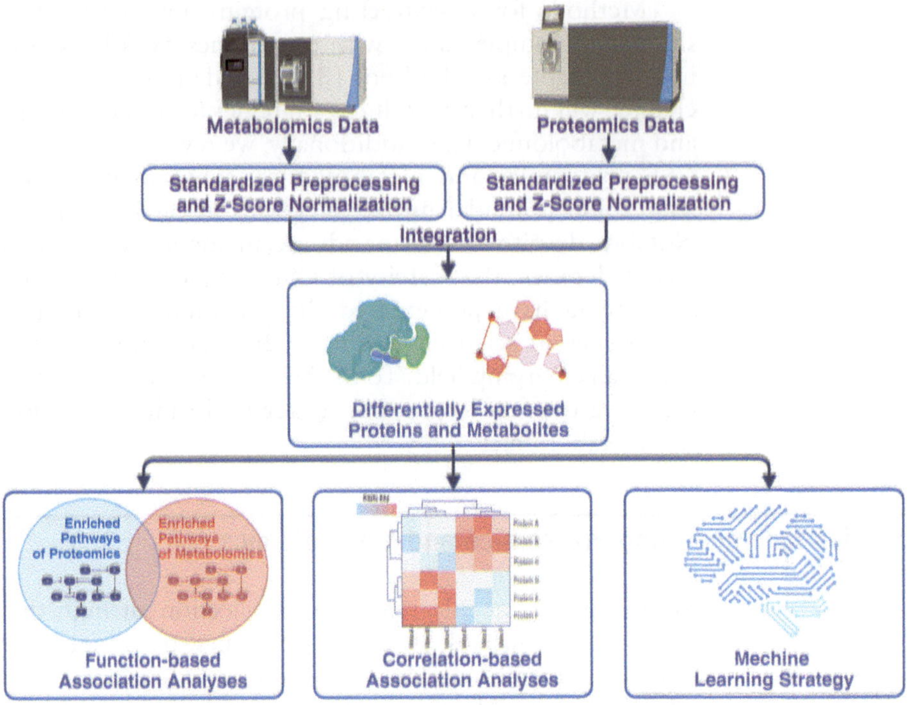

Fig 1 Workflow and strategies for integrating proteomics and metabolomics. The first step following the qualitative and quantitative analyses of proteins and metabolites is standardized data preprocessing. Data from both proteomics and metabolomics can be combined after Z-Score normalization. Statistical analyses approaches, such as Student's t-test, are employed for identifying differentially expressed proteins and metabolites. Subsequently, the data derived from proteomics and metabolomics can be collectively interpreted using function-based association analyses, correlation-based association analyses, and machine learning

will identify and prioritize the co-enriched pathways of interest for further investigation. Metabolic network maps, with the KEGG map being the most common, are used to display the relationships between differentially expressed proteins and metabolites within pathways [35]. To better visualize metabolic network disruptions, some researchers also integrated heat maps into the network maps to label the expression levels of differentially expressed metabolites or proteins.

Another formation of function-based association analyses involves conducting pathway enrichment using just one kind of omics data (metabolomics or proteomics). Subsequently, researchers systematically evaluate enriched pathways according to existing literature or their interests. Once the selection is complete, the results from another omics data will be directly mapped onto the selected pathway [36]. This strategy carries certain risks, as data from the other omics may not align with the selected pathway. Comprehensive mastery of research advancements is crucial for selecting appropriate pathways and improving success rates. As this approach utilizes pathway enrichment analyses for a single omics type, key pathways are often visualized directly through metabolic network maps [36]. In addition, proteins and metabolites involved in key pathways can also be highlighted directly in the heat maps of differential proteins or metabolites [37].

Function-based association analyses are preferred for integrating proteomics and metabolomics due to their ability to directly identify functional pathways. However, this strategy appears to be less commonly used in reality compared to the correlation-based association analyses strategy, which will be introduced next. One reason may be its heavy reliance on the quality of database annotations. Currently, there exist substantial disparities among various databases, and the attainment of effective data integration remains elusive [15]. To identify co-enriched pathways, researchers may need to extract and manually compile information from multiple databases, thereby significantly increasing the workload and introducing potential biases into the final results [38]. In addition, the integration is based on the data acquired separately from proteomics and metabolomics. Any variables that can influence the results of single-omic may similarly influence the outcomes of integrated analyses. Moreover, the intricate nature of biological processes may fundamentally underlie this phenomenon. Ideally, multidimensional data for the same disease should be consistent. Nevertheless, previous research has encountered difficulties in identifying a statistically significant common pathway, regardless of the omics types used [39–42]. Variations in regulatory levels across different omics, feedback mechanisms within biological networks, and the impact of environmental factors may contribute to this phenomenon [43]. For instance, genetic upregulation of an enzyme in CSF may be offset by microenvironmental pH inhibiting its activity. In such condition, corresponding changes in proteins might not be reflected in metabolomic data. Furthermore, if CSF samples for proteomic and metabolomic analyses are not collected simultaneously from the same patients, disease progression and individual heterogeneity may increase data discordance. This scenario may pose greater challenges in generating positive findings in the integration of proteomic and metabolomic data. In general, while omics technology has emerged as a potent instrument for exploring the life sciences, unidimensional data interpretation poses a substantial risk of being misled by technological artifacts. Given potential errors at each research stage, researchers should interpret integrated omics outcomes cautiously, as ostensibly negative outcomes may, in fact, contribute to the progression of the disease.

2.2 Correlation-Based Association Analyses

Correlation-based association analyses employ statistical methods to calculate the combinations of variables that exhibit concordant or discordant changes based on the quantitative results of proteomics and metabolomics. Among these combinations, those exhibiting strong correlations often be prioritized for further investigation. Correlation-based association analyses can be categorized into unidimensional and multidimensional analyses based on the quantitative relationship between variables. Unidimensional analyses, also referred to as univariate statistical analyses, calculate the statistical association between one variable (e.g., a protein) and another variable (e.g., a metabolite). The correlation algorithms commonly used for unidimensional analyses encompass Pearson correlation, Spearman correlation and Kendall correlation [44–46]. Correlation between proteins and metabolites can be visualized using correlation clustering heatmaps, correlation network diagrams, and Sankey diagrams. Among these, correlation clustering heatmap is the most common form of visualization, providing an intuitive representation of the correlation between differentially expressed proteins and metabolites. Protein-metabolite correlation network diagrams and Sankey diagrams are employed to illustrate potential interactions between differentially expressed proteins and metabolites.

Another type of correlation-based association analyses is multidimensional analyses, also referred to as multivariate statistical analyses. Multidimensional analyses can serve as a complementary approach when the outcomes obtained from the aforementioned strategies are deemed unsatisfactory. Compared to unidimensional analyses, multidimensional analyses focus on the overall association between two omics datasets. It is often employed to unveil the extent of correlation between two omics datasets, or to identify the variables in one omic that have the most significant influence on another omic. Statistical models used in multidimensional analyses include bidirectional orthogonal partial least squares (O2PLS), the Mantel test, and canonical correlation analyses (CCA) [47]. O2PLS, the most commonly used statistical model, adheres to the "orthogonal" principle of OPLS. By constructing an integrated model of the protein expression matrix and metabolic expression matrix, O2PLS can predict potential correlated datasets within both matrices. The results of O2PLS analyses are usually presented in loading plots. According to the load value, the first two dimensions are typically chosen for integration and plotting, illustrating proteins and metabolites with the strongest correlations.

The protein-metabolite relationships obtained from correlation-based association analyses represent statistical correlations and can reveal additional insights. Therefore, correlation-based association analyses are often used as a supplementary approach to function-based association analyses. Nevertheless, a high correlation between proteins and metabolites does not necessarily imply a direct biological interaction in authentic biological processes. Consequently, after obtaining the correlation insights, researchers should ultimately return to the functional level to interpret the findings.

In pathological states, complex relationships between proteins and metabolites may exist that go beyond our current comprehension. Existing metabolic pathways might undergo restructuring, while novel ones may be established. Hence, when investigating uncharted regulatory pathways, it is more likely to uncover disease-specific insights through correlation-based association analyses than through function-based association analyses. However, metabolomics and proteomics data only offer a snapshot of information. Neither function-based association analyses nor correlation-based association analyses can clarify the causal relationship between molecular changes and disease. Consequently, it is imperative for researchers to further optimize cellular experiments and animal studies or employ Mendelian randomization to draw more robust conclusions and establish definitive causality after identifying key clues [45, 48].

2.3 Data Processing and Statistical Analyses

Following standardized preprocessing, the proteomics and metabolomics data can be further utilized for subsequent data analyses. By applying Z-score normalization, the proteomics and metabolomics data can be combined. A combination of univariate analyses (e.g., T-test) and multivariate analyses (e.g., OPLS-DA) is subsequently employed for the screening of differentially expressed proteins and metabolites. The p-value from the t-test is utilized for assessing the significance of disparities between groups, while the variable importance in projection (VIP) values from OPLS-DA quantifies the ability of biological molecule expression patterns to classify and discriminate samples across distinct groups. In general, a biological molecule is considered differentially expressed only if its $VIP>1$ and p-value <0.05.

The integrated analyses of proteomics and metabolomics prioritize function-based association analyses. In studies integrating proteomics and metabolomics, the Kyoto Encyclopedia of Genes and Genomes (KEGG) database (https://www.kegg.jp/kegg/) is widely employed for function-based association analyses. Utilizing KEGG IDs of differentially expressed molecules for pathway enrichment analyses and employing the hypergeometric test (with a p-value <0.05 as a common criterion) enables the identification of enriched pathways. Correlation-based association analyses can serve as a complement to function-based association analyses. The ratio of the average expression level of proteins or metabolites, often referred to as Fold Change, is a common index in omics calculations. Data items with larger absolute values of log2 (Fold Change) will be used to calculate the correlation coefficient. When comparing differences across multiple groups, data items are filtered based on p-values obtained from single-omic analyses. The correlation coefficient ranges from −1 to 1. A protein-metabolite pair showing a strong positive correlation (approaching 1) or a strong negative correlation (approaching −1) warrants further investigation.

2.4 Machine Learning Strategy

Machine learning strategies are widely used in various stages of omics research. Machine learning can be categorized into traditional machine learning and deep learning based on their development timelines. Deep learning, which emerged after traditional machine learning, requires a substantial amount of accessible, multi-feature and highly structured data. In current research, developing and training deep learning algorithms is both time-consuming and resource-intensive. Therefore, traditional machine learning is more commonly applied in omics data interpretation, as reflected in published scholarly literature. Based on the characteristics of the models, traditional machine learning can be broadly divided into supervised learning and unsupervised learning. Unsupervised learning models encompass techniques such as dimensionality reduction (e.g., principal component analyses) and clustering (e.g., K-means algorithm and hierarchical clustering). These techniques are commonly utilized to portray the relationships among samples and assess intra-group consistency. Based on model complexity, supervised learning models can be categorized into single and ensemble learning models. The Least Absolute Shrinkage and Selection Operator (Lasso) from single models and Random Forest form ensemble learning models are commonly employed for constructing biomarker panels. In this context, we focus on the application of LASSO and Random Forests in the integrated analyses of proteomics and metabolomics for generating biomarker panels.

Lasso regression is an extension of linear regression. This machine learning strategy imposes penalties on parameters, driving some of them towards zero to effectively eliminate redundant features. As a result, it identifies the most influential features that significantly impact the dependent variable. The Lasso regression model can be implemented using the R package "glmnet" [49]. In the "glmnet" package, the parameter lambda represents the strength of global regularization. A higher lambda value corresponds to a stronger penalty, resulting in fewer retained variables. The cross-validation curve for Lasso regression aids in selecting the optimal regularization parameter. The point at which the curve reaches its minimum, denoted as lambda.min, corresponds to the model with the lowest average cross-validation error. The Lasso regression cross-validation curve also includes another point, lambda.lse, where the cross-validation error falls within one standard deviation of the minimum error. The model corresponding to lambda.lse retains fewer variables and avoids overfitting, making it often regarded as the optimal model. Lasso regression is characterized by its simplicity and computational efficiency, enabling the development of concise and highly generalizable models. Nevertheless, this approach may fail to accurately identify features in the presence of multicollinearity, leading to a decline in model performance.

Random Forest is an ensemble learning strategy that can be implemented using the "random Forest" package in the R programming language. Random Forest constructs and aggregates multiple decision trees to classify and predict data, characterized by high accuracy, strong resistance to overfitting, and straightforward implementation. Random Forest can be employed to assess the significance of proteins and metabolites for classification. Proteins or metabolites with higher importance scores can be further utilized to construct a Random Forest model. When constructing a model, data are randomly shuffled and cross-validated over multiple rounds. The dataset is divided into a training set and a validation set in a 1:1 or 2:1 ratio through random selection [46, 50]. The training set is used to construct the model, while the validation set is used to assess the predictive performance of the model.

3 CSF Metabolomic and Proteomic Combination in Neurological Syndromes

The molecular landscape of diseases is exceedingly intricate. Relying solely on single-dimensional data is insufficient for capturing comprehensive molecular information, which also risks compromising the accuracy of the information [51]. The rapid advancement and widespread adoption of omics technologies have led to increasing recognition that the integration of multi-omics data holds significant potential to address this issue. Although the prevailing trend in current research has traditionally focused on utilizing a single omic approach, recent years have witnessed a significant shift, with an increasing number of studies adopting integrated multi-omics strategies [7–10, 52, 53]. However, given the maturity of genomics, transcriptomics, and proteomics technologies, multi-omics studies in the field of neurological diseases predominant emphasize the integration of these three well-established omics techniques [54–57].

The majority of studies integrating CSF metabolomic and proteomic focus on the normal physiological properties of CSF, Alzheimer's disease, psychosis, and multiple sclerosis. The prerequisite for evaluating any abnormal state is the clear definition of a healthy baseline. To establish a database capable of characterizing the healthy state, Lilley et al. and Myint et al. integrated metabolomics and proteomics approaches to define the baseline molecule profile of CSF. Interestingly, Myint et al. found that, in contrast to metabolites, protein levels in CSF exhibit significant variability among individuals [58]. In addition to analyzing CSF samples, Lilley et al. simultaneously conducted proteomic and metabolomic analyses of plasma, identifying common pathways between serum and CSF in healthy individuals [42]. The combination of proteomics and metabolomics has been widely applied in Alzheimer's disease research, utilizing diverse biological samples such as CSF, serum, and brain tissue [45, 59, 60]. In a study analyzing CSF samples, Kamil et al. described the association between lipid and protein in CSF from patients with Alzheimer's disease [61]. Moreover, in December 2023, Daniel et al. reported, for the first time, the pathophysiological processes in CSF from patients with different subtypes of Alzheimer's disease, based on the ATN classification framework proposed by the National Institute on Aging and Alzheimer's Association (NIA-AA) in 2018 [62]. Apart from Alzheimer's disease, CSF proteomics and metabolomics have also been used to explore first-episode psychosis (FEP) and the initial prodromal state of psychosis (IPS) [63–65]. In addition to aiding in the comprehension of the multidimensional changes associated with psychosis, Jeffrey et al.'s study also re-emphasizes the importance of longitudinal, multi-timepoint sample collection and dynamic monitoring in clinical research [63]. Besides, CSF proteomics and metabolomics have also been used to identify biomarkers for multiple sclerosis with higher sensitivity and specificity [51, 66, 67]. These three studies collectively established robust biomarker panels composed of proteins and metabolites, highlighting the significant effectiveness of integrating proteomics and metabolomics in enhancing the precision of disease diagnosis [51, 66, 67].

Other studies on neurological syndromes that integrating CSF proteomic and metabolomic include research on neurologic post-acute sequelae of SARS-CoV-2 infection (neuro-PASC) [46], spinal muscular atrophy (SMA) [68], anxiety [69], medulloblastoma [70], leptomeningeal metastasis (LM) [71], seizures [72], iron deficiency [73], and ischemic stroke [74]. The objectives, species under investigation, and findings of these studies are summarized and outlined in Table 1.

4 Challenges and Future Direction

CSF, which surrounds the brain and contains numerous dynamic biomolecules, is invaluable for diagnosing and researching neurological diseases. However, existing proteomic and metabolomic studies on neurological disorders rely on plasma or urine [55, 75–78]. This preference may stem from the invasive nature of CSF collection. Additionally, CSF collection is prone to blood contamination, and the stringent requirements for preprocessing CSF may also limit its usage. Certain pathological conditions can affect the permeability of the blood-brain barrier, potentially enabling the use of blood-based biomarkers in clinical practice [17]. However, predicting alterations in blood-brain barrier permeability remains

Table 1
CSF Metabolomics and proteomics combination in neurological syndromes

Disease	Research objective	Species	Research findings	Reference
Neurologic post-acute sequelae of SARS-CoV-2 infection (neuro-PASC)	Exploring the molecular changes and mechanism of neuro-PASC.	Human	Disorders in sphingolipid metabolism and attenuated inflammatory responses may contribute to the occurrence of neuro-PASC, whereas activation of steroid hormone biosynthesis may play a role in the repair process.	[46]
Spinal muscular atrophy (SMA)	Identifying SMA type 3 patients' response to nusinersen treatment and discovering predictive markers of treatment response.	Human	The reduction in SEMA6A levels, along with the increase in COL1A2 and GRIA4, might reflect the therapeutic efficacy of nusinersen. Additionally, the treatment induced changes in the utilization of certain amino acids, while the levels of most other metabolites remained largely unaffected.	[68]
Anxiety	Investigating the longitudinal effects of anxiety on cognitive impairment in non-dementia elderly and exploring the underlying biological processes through multi-omics approaches.	Human	Anxiety is a risk factor for cognitive progression in non-dementia elderly, and axon/synapse damage associated with energy metabolism imbalance may underlie this phenomenon.	[69]
Medulloblastoma (MB)	Exploring the microenvironment of MB tumor.	Human	The formation of 12,13-DiHOME and increased beta-oxidation induce a metabolic shift, potentially facilitate drug resistance and the acquisition of stem cell properties in MB cells.	[70]
Leptomeningeal metastasis (LM)	Exploration of molecular differences between ventricular and lumbar CSF samples.	Human	Metabolite and protein profiles differed between paired lumbar and ventricular CSF	[71]

(continued)

Table 1 (continued)

Disease	Research objective	Species	Research findings	Reference
			samples, with LM-associated molecular changes being more pronounced in the lumbar compartment.	
Seizures	Investigation of the function of NaCT (SLC13A5), a plasma membrane sodium-dependent citrate transporter, in seizures.	SlC13A5 deletion mouse model	SLC13A5 is involved in brain citrate regulation, and its dysfunction can result in seizures.	[72]
Iron deficiency	Identification of biomarkers of ID-induced metabolic dysfunction in serum and CSF.	Rhesus macaques	Metabolomic and proteomic analyses revealed disruption in synaptogenesis during the anemic period.	[73]
Ischemic stroke	Establishment of clinical indicators for Buyang Huanwu decoction (BHD) efficacy in stroke treatment and prognosis.	Male ICR mice	BHD may be effective in the treatment and prevention of stroke-related neurodegenerative disease.	[74]

challenging. A combined approach, using CSF for initial biomarker discovery and blood for validation, could enhance the reliability of biomarker panels. Besides, the cost of sequencing and other factors pose challenges in conducting multidimensional analyses on the same sample. Nevertheless, conducting multidimensional analyses on the same sample can reduce sample variability, improve data consistency, and provide deeper insights into disease mechanisms [79–81]. Hence, future studies should prioritize the integrated multi-omics analyses of single samples.

Many existing studies rely solely on proteomics or metabolomics, resulting in largely under-integrated and under-utilized data. Integrating data from multiple databases at the individual sample level can maximize benefits by leveraging available data resources. Nonetheless, substantial difficulties and challenges persist in the assessment and integration of existing data. On the one hand, the variability in inclusion and exclusion criteria across studies, as well as the differences in proteomic and metabolomic methodologies between laboratories, hinder reproducibility and complicate data integration [17]. Furthermore, disparities in standard data formats across various omics data impose greater demands on researchers' computational skills [82, 83]. Therefore, to reduce research heterogeneity and facilitate the re-mining and utilization of omics data, it is crucial to establish standardized operating procedures for omics methods [41]. Moreover, developing user-friendly integrated data tools for clinical investigators is a promising future direction.

Additionally, although proteomics encompasses the study of post-translational modification (PTM) of proteins, the current application of proteomics is primarily confined to the qualitative and quantitative characterization of proteins. Some metabolites can regulate PTM, thereby exerting influence on enzyme activity, which in turn can govern metabolic reactions [19, 20]. Therefore, the integration of PTM-based proteomics and metabolomics is expected to yield a comprehensive understanding of the PTM-enzyme-metabolite cycle in specific diseases. In addition to employing proteomics for qualitative and quantitative analyses of proteins, additional data on post-translational modifications of proteins should also be integrated.

5 Conclusion

Despite the clear advantages of multidimensional data integration in scientific research, the integration of proteomics and metabolomics is still in its early stages and encounters various technical and non-technical challenges. In this chapter, strategies for integrated analyses of proteomics and metabolomics are reviewed, including the analyses of the advantages and disadvantages of each strategy. We then review the advancements in integrating CSF proteomics and metabolomics for studying neurological syndromes and summarize factors that may hinder omics integration in CSF. Based on current challenges in data integration, we analyze study design, sample collection, technology application, and data integration, and propose potential future research directions. The superiority of integrating CSF proteomics and metabolomics analyses in studying neurological diseases has garnered increasing attention from research teams. Such research is expected to significantly advance the field of neuroscience and deepen our understanding of neurological syndromes in the future.

Acknowledgments

This work was supported by Special Funds for the Cultivation of Guangdong College Students' Scientific and Technological Innovation (project number pdjh2024a088).

References

1. Li GW, Xie XS (2011) Central dogma at the single-molecule level in living cells. Nature 475(7356):308–315. https://doi.org/10.1038/nature10315
2. Ding Z, Wang N, Ji N, Chen ZS (2022) Proteomics technologies for cancer liquid biopsies. Mol Cancer 21(1):53. https://doi.org/10.1186/s12943-022-01526-8
3. Fountoulakis M (2004) Application of proteomics technologies in the investigation of the brain. Mass Spectrom Rev 23(4):231–258. https://doi.org/10.1002/mas.10075
4. Oppermann FS, Gnad F, Olsen JV, Hornberger R, Greff Z, Kéri G, Mann M, Daub H (2009) Large-scale proteomics analysis of the human kinome. Mol Cell Proteomics 8(7):1751–1764. https://doi.org/10.1074/mcp.M800588-MCP200
5. Conly JM, Ronald AR (1983) Cerebrospinal fluid as a diagnostic body fluid. Am J Med 75(1b):102–108. https://doi.org/10.1016/0002-9343(83)90080-3
6. Wyart C, Carbo-Tano M, Cantaut-Belarif Y, Orts-Del'Immagine A, Böhm UL (2023) Cerebrospinal fluid-contacting neurons: multimodal cells with diverse roles in the CNS. Nat Rev Neurosci 24(9):540–556. https://doi.org/10.1038/s41583-023-00723-8
7. Sirko S, Schichor C, Della Vecchia P, Metzger F, Sonsalla G, Simon T, Bürkle M, Kalpazidou S, Ninkovic J, Masserdotti G, Sauniere JF, Iacobelli V, Iacobelli S, Delbridge C, Hauck SM, Tonn JC, Götz M (2023) Injury-specific factors in the cerebrospinal fluid regulate astrocyte plasticity in the human brain. Nat Med 29(12):3149–3161. https://doi.org/10.1038/s41591-023-02644-6
8. Pham L, Wright DK, O'Brien WT, Bain J, Huang C, Sun M, Casillas-Espinosa PM, Shah AD, Schittenhelm RB, Sobey CG, Brady RD, O'Brien TJ, Mychasiuk R, Shultz SR, McDonald SJ (2021) Behavioral, axonal, and proteomic alterations following repeated mild traumatic brain injury: Novel insights using a clinically relevant rat model. Neurobiol Dis 148:105151. https://doi.org/10.1016/j.nbd.2020.105151
9. Peppercorn K, Kleffmann T, Hughes SM, Tate WP (2023) Secreted amyloid precursor protein alpha (sAPPα) regulates the cellular proteome and secretome of mouse primary astrocytes. Int J Mol Sci 24(8). https://doi.org/10.3390/ijms24087165
10. Sun L, Bhawal R, Xu H, Chen H, Anderson ET, Haroutunian V, Cross AC, Zhang S, Gibson GE (2021) The human brain acetylome reveals that decreased acetylation of mitochondrial proteins associates with Alzheimer's disease. J Neurochem 158(2):282–296. https://doi.org/10.1111/jnc.15377
11. Rinschen MM, Ivanisevic J, Giera M, Siuzdak G (2019) Identification of bioactive metabolites using activity metabolomics. Nat Rev Mol Cell Biol 20(6):353–367. https://doi.org/10.1038/s41580-019-0108-4
12. Guijas C, Montenegro-Burke JR, Warth B, Spilker ME, Siuzdak G (2018) Metabolomics activity screening for identifying metabolites that modulate phenotype. Nat Biotechnol 36(4):316–320. https://doi.org/10.1038/nbt.4101
13. Metallo CM, Vander Heiden MG (2013) Understanding metabolic regulation and its influence on cell physiology. Mol Cell 49(3):388–398. https://doi.org/10.1016/j.molcel.2013.01.018
14. Rabinowitz JD, Silhavy TJ (2013) Systems biology: metabolite turns master regulator. Nature 500(7462):283–284. https://doi.org/10.1038/nature12544
15. Cambiaghi A, Ferrario M, Masseroli M (2017) Analysis of metabolomic data: tools, current strategies and future challenges for omics data integration. Brief Bioinform 18(3):498–510. https://doi.org/10.1093/bib/bbw031
16. Xin C, Guan X, Wang L, Liu J (2023) Integrative multi-omics research in cerebral palsy: current progress and future prospects. Neurochem Res 48(5):1269–1279. https://doi.org/10.1007/s11064-022-03839-y
17. Schumacher-Schuh A, Bieger A, Borelli WV, Portley MK, Awad PS, Bandres-Ciga S (2021) Advances in proteomic and metabolomic profiling of neurodegenerative diseases. Front Neurol 12:792227. https://doi.org/10.3389/fneur.2021.792227
18. Laíns I, Gantner M, Murinello S, Lasky-Su JA, Miller JW, Friedlander M, Husain D (2019) Metabolomics in the study of retinal health and disease. Prog Retin Eye Res 69:57–79. https://doi.org/10.1016/j.preteyeres.2018.11.002
19. Fahrmann JF, Grapov DD, Wanichthanarak K, DeFelice BC, Salemi MR, Rom WN, Gandara DR, Phinney BS, Fiehn O, Pass H, Miyamoto S (2017) Integrated metabolomics and proteomics highlight altered nicotinamide- and polyamine pathways in lung adenocarcinoma. Carcinogenesis 38(3):271–280. https://doi.org/10.1093/carcin/bgw205

20. Nalbantoglu S, Karadag A (2021) Metabolomics bridging proteomics along metabolites/oncometabolites and protein modifications: Paving the way toward integrative multiomics. J Pharm Biomed Anal 199:114031. https://doi.org/10.1016/j.jpba.2021.114031
21. Nicholson JK, Lindon JC (2008) Systems biology: metabonomics. Nature 455(7216):1054–1056. https://doi.org/10.1038/4551054a
22. Patti GJ, Yanes O, Siuzdak G (2012) Innovation: Metabolomics: the apogee of the omics trilogy. Nat Rev Mol Cell Biol 13(4):263–269. https://doi.org/10.1038/nrm3314
23. Kitchen RR, Rozowsky JS, Gerstein MB, Nairn AC (2014) Decoding neuroproteomics: integrating the genome, translatome and functional anatomy. Nat Neurosci 17(11):1491–1499. https://doi.org/10.1038/nn.3829
24. Yu XT, Zeng T (2018) Integrative analysis of omics big data. Methods Mol Biol 1754:109–135. https://doi.org/10.1007/978-1-4939-7717-8_7
25. Venegas-Molina J, Molina-Hidalgo FJ, Clicque E, Goossens A (2021) Why and how to dig into plant metabolite-protein interactions. Trends Plant Sci 26(5):472–483. https://doi.org/10.1016/j.tplants.2020.12.008
26. Zhao J, Feng Q, Wei WQ (2022) Integration of omics and phenotypic data for precision medicine. Methods Mol Biol 2486:19–35. https://doi.org/10.1007/978-1-0716-2265-0_2
27. Hieke S, Benner A, Schlenl RF, Schumacher M, Bullinger L, Binder H (2016) Integrating multiple molecular sources into a clinical risk prediction signature by extracting complementary information. BMC Bioinformatics 17(1):327. https://doi.org/10.1186/s12859-016-1183-6
28. Li J, Xiong M, Fu XH, Fan Y, Dong C, Sun X, Zheng F, Wang SW, Liu L, Xu M, Wang C, Ping J, Che S, Wang Q, Yang K, Zuo Y, Lu X, Zheng Z, Lan T, Wang S, Ma S, Sun S, Zhang B, Chen CS, Cheng KY, Ye J, Qu J, Xue Y, Yang YG, Zhang F, Zhang W, Liu GH (2023) Determining a multimodal aging clock in a cohort of Chinese women. Med 4(11):825–848.e813. https://doi.org/10.1016/j.medj.2023.06.010
29. Badhwar A, McFall GP, Sapkota S, Black SE, Chertkow H, Duchesne S, Masellis M, Li L, Dixon RA, Bellec P (2020) A multiomics approach to heterogeneity in Alzheimer's disease: focused review and roadmap. Brain 143(5):1315–1331. https://doi.org/10.1093/brain/awz384
30. Schwarz E, Torrey EF, Guest PC, Bahn S (2012) Biomarker discovery in human cerebrospinal fluid: the need for integrative metabolome and proteome databases. Genome Med 4(4):39. https://doi.org/10.1186/gm338
31. Hörmann P, Barkovits K, Marcus K, Hiller K (2019) Co-extraction for metabolomics and proteomics from a single CSF sample. Methods Mol Biol 2044:337–342. https://doi.org/10.1007/978-1-4939-9706-0_21
32. Chen C, Wang J, Pan D, Wang X, Xu Y, Yan J, Wang L, Yang X, Yang M, Liu GP (2023) Applications of multi-omics analysis in human diseases. MedComm 4(4):e315. https://doi.org/10.1002/mco2.315
33. Li H, Li X, Wu Q, Wang X, Qin Z, Wang Y, He Y, Wu Q, Li L, Chen H (2022) Plasma proteomic and metabolomic characterization of COVID-19 survivors 6 months after discharge. Cell Death Dis 13(3):235. https://doi.org/10.1038/s41419-022-04674-3
34. Zhang Y, Chen C, Cui Y, Du Q, Tang W, Yang W, Kou G, Tang W, Chen H, Gong R (2023) Potential regulatory genes of light induced anthocyanin accumulation in sweet cherry identified by combining transcriptome and metabolome analysis. Front Plant Sci 14:1238624. https://doi.org/10.3389/fpls.2023.1238624
35. Li L, Zhu S, Shu W, Guo Y, Guan Y, Zeng J, Wang H, Han L, Zhang J, Liu X, Li C, Hou X, Gao M, Ge J, Ren C, Zhang H, Schedl T, Guo X, Chen M, Wang Q (2020) Characterization of metabolic patterns in mouse oocytes during meiotic maturation. Mol Cell 80(3):525–540.e529. https://doi.org/10.1016/j.molcel.2020.09.022
36. Chen YM, Zheng Y, Yu Y, Wang Y, Huang Q, Qian F, Sun L, Song ZG, Chen Z, Feng J, An Y, Yang J, Su Z, Sun S, Dai F, Chen Q, Lu Q, Li P, Ling Y, Yang Z, Tang H, Shi L, Jin L, Holmes EC, Ding C, Zhu TY, Zhang YZ (2020) Blood molecular markers associated with COVID-19 immunopathology and multi-organ damage. EMBO J 39(24):e105896. https://doi.org/10.15252/embj.2020105896
37. Shen B, Yi X, Sun Y, Bi X, Du J, Zhang C, Quan S, Zhang F, Sun R, Qian L, Ge W, Liu W, Liang S, Chen H, Zhang Y, Li J, Xu J, He Z, Chen B, Wang J, Yan H, Zheng Y, Wang D, Zhu J, Kong Z, Kang Z, Liang X, Ding X, Ruan G, Xiang N, Cai X, Gao H, Li L, Li S, Xiao Q, Lu T, Zhu Y, Liu H, Chen H, Guo T (2020) Proteomic and metabolomic

characterization of COVID-19 patient sera. Cell 182(1):59–72.e15. https://doi.org/10.1016/j.cell.2020.05.032

38. Xia J, Wishart DS (2010) MSEA: a web-based tool to identify biologically meaningful patterns in quantitative metabolomic data. Nucleic Acids Res 38(Web Server issue):W71–W77. https://doi.org/10.1093/nar/gkq329
39. Greenbaum D, Colangelo C, Williams K, Gerstein M (2003) Comparing protein abundance and mRNA expression levels on a genomic scale. Genome Biol 4(9):117. https://doi.org/10.1186/gb-2003-4-9-117
40. Lundberg E, Fagerberg L, Klevebring D, Matic I, Geiger T, Cox J, Algenäs C, Lundeberg J, Mann M, Uhlen M (2010) Defining the transcriptome and proteome in three functionally different human cell lines. Mol Syst Biol 6:450. https://doi.org/10.1038/msb.2010.106
41. Blasco H, Patin F, Madji Hounoum B, Gordon PH, Vourc'h P, Andres CR, Corcia P (2016) Metabolomics in amyotrophic lateral sclerosis: how far can it take us? Eur J Neurol 23(3):447–454. https://doi.org/10.1111/ene.12956
42. Lilley LM, Sanche S, Moore SC, Salemi MR, Vu D, Iyer S, Hengartner NW, Mukundan H (2022) Methods to capture proteomic and metabolomic signatures from cerebrospinal fluid and serum of healthy individuals. Sci Rep 12(1):13339. https://doi.org/10.1038/s41598-022-16598-1
43. Graw S, Chappell K, Washam CL, Gies A, Bird J, Robeson MS 2nd, Byrum SD (2021) Multi-omics data integration considerations and study design for biological systems and disease. Mol Omics 17(2):170–185. https://doi.org/10.1039/d0mo00041h
44. Kuan PF, Yang X, Kotov R, Clouston S, Bromet E, Luft BJ (2022) Metabolomics analysis of post-traumatic stress disorder symptoms in World Trade Center responders. Transl Psychiatry 12(1):174. https://doi.org/10.1038/s41398-022-01940-y
45. Xu J, Bankov G, Kim M, Wretlind A, Lord J, Green R, Hodges A, Hye A, Aarsland D, Velayudhan L, Dobson RJB, Proitsi P, Legido-Quigley C (2020) Integrated lipidomics and proteomics network analysis highlights lipid and immunity pathways associated with Alzheimer's disease. Transl Neurodegener 9(1):36. https://doi.org/10.1186/s40035-020-00215-0
46. Chen S, Liang J, Chen D, Huang Q, Sun K, Zhong Y, Lin B, Kong J, Sun J, Gong C, Wang J, Gao Y, Zhang Q, Sun H (2024) Cerebrospinal fluid metabolomic and proteomic characterization of neurologic post-acute sequelae of SARS-CoV-2 infection. Brain Behav Immun 115:209–222. https://doi.org/10.1016/j.bbi.2023.10.016
47. Du Y, Qiu R, Chen L, Chen Y, Zhong Z, Li P, Fan F, Cheng Y (2023) Identification of serum exosomal metabolomic and proteomic profiles for remote ischemic preconditioning. J Transl Med 21(1):241. https://doi.org/10.1186/s12967-023-04070-1
48. Benson MD, Eisman AS, Tahir UA, Katz DH, Deng S, Ngo D, Robbins JM, Hofmann A, Shi X, Zheng S, Keyes M, Yu Z, Gao Y, Farrell L, Shen D, Chen ZZ, Cruz DE, Sims M, Correa A, Tracy RP, Durda P, Taylor KD, Liu Y, Johnson WC, Guo X, Yao J, Chen YI, Manichaikul AW, Jain D, Yang Q, Bouchard C, Sarzynski MA, Rich SS, Rotter JI, Wang TJ, Wilson JG, Clish CB, Sarkar IN, Natarajan P, Gerszten RE (2023) Protein-metabolite association studies identify novel proteomic determinants of metabolite levels in human plasma. Cell Metab 35(9):1646–1660.e1643. https://doi.org/10.1016/j.cmet.2023.07.012
49. Xu RH, Wei W, Krawczyk M, Wang W, Luo H, Flagg K, Yi S, Shi W, Quan Q, Li K, Zheng L, Zhang H, Caughey BA, Zhao Q, Hou J, Zhang R, Xu Y, Cai H, Li G, Hou R, Zhong Z, Lin D, Fu X, Zhu J, Duan Y, Yu M, Ying B, Zhang W, Wang J, Zhang E, Zhang C, Li O, Guo R, Carter H, Zhu JK, Hao X, Zhang K (2017) Circulating tumour DNA methylation markers for diagnosis and prognosis of hepatocellular carcinoma. Nat Mater 16(11):1155–1161. https://doi.org/10.1038/nmat4997
50. Sun K, Zhang X, Li X, Li X, Su S, Luo Y, Tian H, Zeng M, Wang C, Xie Y, Zhang N, Cao Y, Zhu Z, Ni Q, Liu W, Xia F, He X, Shi Z, Duan C, Sun H (2023) Plasma metabolic signatures for intracranial aneurysm and its rupture identified by pseudotargeted metabolomics. Clin Chim Acta 538:36–45. https://doi.org/10.1016/j.cca.2022.11.002
51. Herman S, Khoonsari PE, Tolf A, Steinmetz J, Zetterberg H, Åkerfeldt T, Jakobsson PJ, Larsson A, Spjuth O, Burman J, Kultima K (2018) Integration of magnetic resonance imaging and protein and metabolite CSF measurements to enable early diagnosis of secondary progressive multiple sclerosis. Theranostics 8(16):4477–4490. https://doi.org/10.7150/thno.26249
52. Locasale JW, Melman T, Song S, Yang X, Swanson KD, Cantley LC, Wong ET, Asara JM (2012) Metabolomics of human cerebrospinal fluid identifies signatures of malignant glioma.

Mol Cell Proteomics 11(6):M111.014688. https://doi.org/10.1074/mcp.M111.014688

53. Blasco H, Corcia P, Pradat PF, Bocca C, Gordon PH, Veyrat-Durebex C, Mavel S, Nadal-Desbarats L, Moreau C, Devos D, Andres CR, Emond P (2013) Metabolomics in cerebrospinal fluid of patients with amyotrophic lateral sclerosis: an untargeted approach via high-resolution mass spectrometry. J Proteome Res 12(8):3746–3754. https://doi.org/10.1021/pr400376e

54. Panyard DJ, Reus LM, Ali M, Liu J, Deming YK, Lu Q, Kollmorgen G, Carboni M, Wild N, Visser PJ, Bertram L, Zetterberg H, Blennow K, Gobom J, Western D, Sung YJ, Carlsson CM, Johnson SC, Asthana S, Cruchaga C, Tijms BM, Engelman CD, Snyder MP (2024) Post-GWAS multiomic functional investigation of the TNIP1 locus in Alzheimer's disease highlights a potential role for GPX3. Alzheimers Dement 20(7):5044–5053. https://doi.org/10.1002/alz.13848

55. Li C, Li X, He K, Wu Y, Xie X, Yang J, Zhang F, Yue Y, Hao H, Zhao S, Li X, Tian G (2023) Discovery of the mechanisms of acupuncture in the treatment of migraine based on functional magnetic resonance imaging and omics. Front Med 17(5):993–1005. https://doi.org/10.1007/s11684-023-0989-7

56. Hesham D, On J, Alshahaby N, Amer N, Magdeldin S, Okada M, Tsukamoto Y, Hiraishi T, Imai C, Okuda S, Wakai T, Kakita A, Oishi M, El-Naggar S, Natsumeda M (2024) Multi-omics analyses of choroid plexus carcinoma cell lines reveal potential targetable pathways and alterations. J Neuro-Oncol 166(1):27–38. https://doi.org/10.1007/s11060-023-04484-3

57. Chen PC, Han X, Shaw TI, Fu Y, Sun H, Niu M, Wang Z, Jiao Y, Teubner BJW, Eddins D, Beloate LN, Bai B, Mertz J, Li Y, Cho JH, Wang X, Wu Z, Liu D, Poudel S, Yuan ZF, Mancieri A, Low J, Lee HM, Patton MH, Earls LR, Stewart E, Vogel P, Hui Y, Wan S, Bennett DA, Serrano GE, Beach TG, Dyer MA, Smeyne RJ, Moldoveanu T, Chen T, Wu G, Zakharenko SS, Yu G, Peng J (2022) Alzheimer's disease-associated U1 snRNP splicing dysfunction causes neuronal hyperexcitability and cognitive impairment. Nat Aging 2(10):923–940. https://doi.org/10.1038/s43587-022-00290-0

58. Stoop MP, Coulier L, Rosenling T, Shi S, Smolinska AM, Buydens L, Ampt K, Stingl C, Dane A, Muilwijk B, Luitwieler RL, Sillevis Smitt PA, Hintzen RQ, Bischoff R, Wijmenga SS, Hankemeier T, van Gool AJ, Luider TM (2010) Quantitative proteomics and metabolomics analysis of normal human cerebrospinal fluid samples. Mol Cell Proteomics 9(9):2063–2075. https://doi.org/10.1074/mcp.M900877-MCP200

59. Kodam P, Sai Swaroop R, Pradhan SS, Sivaramakrishnan V, Vadrevu R (2023) Integrated multi-omics analysis of Alzheimer's disease shows molecular signatures associated with disease progression and potential therapeutic targets. Sci Rep 13(1):3695. https://doi.org/10.1038/s41598-023-30892-6

60. Mast N, Butts M, Pikuleva IA (2024) Unbiased insights into the multiplicity of the CYP46A1 brain effects in 5XFAD mice treated with low dose-efavirenz. J Lipid Res 65(6):100555. https://doi.org/10.1016/j.jlr.2024.100555

61. Borkowski K, Seyfried NT, Arnold M, Lah JJ, Levey AI, Hales CM, Dammer EB, Blach C, Louie G, Kaddurah-Daouk R, Newman JW (2023) Integration of plasma and CSF metabolomics with CSF proteomic reveals novel associations between lipid mediators and central nervous system vascular and energy metabolism. Sci Rep 13(1):13752. https://doi.org/10.1038/s41598-023-39737-8

62. Panyard DJ, McKetney J, Deming YK, Morrow AR, Ennis GE, Jonaitis EM, Van Hulle CA, Yang C, Sung YJ, Ali M, Kollmorgen G, Suridjan I, Bayfield A, Bendlin BB, Zetterberg H, Blennow K, Cruchaga C, Carlsson CM, Johnson SC, Asthana S, Coon JJ, Engelman CD (2023) Large-scale proteome and metabolome analysis of CSF implicates altered glucose and carbon metabolism and succinylcarnitine in Alzheimer's disease. Alzheimers Dement 19(12):5447–5470. https://doi.org/10.1002/alz.13130

63. Huang JT, Leweke FM, Tsang TM, Koethe D, Kranaster L, Gerth CW, Gross S, Schreiber D, Ruhrmann S, Schultze-Lutter F, Klosterkötter J, Holmes E, Bahn S (2007) CSF metabolic and proteomic profiles in patients prodromal for psychosis. PLoS One 2(8):e756. https://doi.org/10.1371/journal.pone.0000756

64. Shang P, Ho AM, Tufvesson-Alm M, Lindberg DR, Grant CW, Orhan F, Eren F, Bhat M, Engberg G, Schwieler L, Fatouros-Bergman H, Imbeault S, Iverson RM, Dasari S, Piehl F, Cervenka S, Sellgren CM, Erhardt S, Choi DS (2022) Identification of cerebrospinal fluid and serum metabolomic biomarkers in first episode psychosis patients. Transl Psychiatry 12(1):229. https://doi.org/10.1038/s41398-022-02000-1

65. Haroon H, Ho AM, Gupta VK, Dasari S, Sellgren CM, Cervenka S, Engberg G, Eren F, Erhardt S, Sung J, Choi DS (2024) Cerebrospinal fluid proteomic signatures are associated with symptom severity of first-episode psychosis. J Psychiatr Res 171:306–315. https://doi.org/10.1016/j.jpsychires.2024.02.002
66. Probert F, Yeo T, Zhou Y, Sealey M, Arora S, Palace J, Claridge TDW, Hillenbrand R, Oechtering J, Kuhle J, Leppert D, Anthony DC (2021) Determination of CSF GFAP, CCN5, and vWF levels enhances the diagnostic accuracy of clinically defined MS from Non-MS patients with CSF oligoclonal bands. Front Immunol 12:811351. https://doi.org/10.3389/fimmu.2021.811351
67. Blanchet L, Smolinska A, Attali A, Stoop MP, Ampt KA, van Aken H, Suidgeest E, Tuinstra T, Wijmenga SS, Luider T, Buydens LM (2011) Fusion of metabolomics and proteomics data for biomarkers discovery: case study on the experimental autoimmune encephalomyelitis. BMC Bioinformatics 12:254. https://doi.org/10.1186/1471-2105-12-254
68. Faravelli I, Gagliardi D, Abati E, Meneri M, Ongaro J, Magri F, Parente V, Petrozzi L, Ricci G, Farè F, Garrone G, Fontana M, Caruso D, Siciliano G, Comi GP, Govoni A, Corti S, Ottoboni L (2023) Multi-omics profiling of CSF from spinal muscular atrophy type 3 patients after nusinersen treatment: a 2-year follow-up multicenter retrospective study. Cell Mol Life Sci 80(8):241. https://doi.org/10.1007/s00018-023-04885-7
69. Sun L, Li W, Qiu Q, Hu Y, Yang Z, Xiao S (2023) Anxiety adds the risk of cognitive progression and is associated with axon/synapse degeneration among cognitively unimpaired older adults. EBioMedicine 94:104703. https://doi.org/10.1016/j.ebiom.2023.104703
70. Reichl B, Niederstaetter L, Boegl T, Neuditschko B, Bileck A, Gojo J, Buchberger W, Peyrl A, Gerner C (2020) Determination of a tumor-promoting microenvironment in recurrent medulloblastoma: a multi-omics study of cerebrospinal fluid. Cancers (Basel) 12(6). https://doi.org/10.3390/cancers12061350
71. Kwon JW, Im JH, Lee KY, Yoo BC, Lee JH, Kim KH, Kim JH, Shin SH, Yoo H, Gwak HS (2022) Different metabolomic and proteomic profiles of cerebrospinal fluid in ventricular and lumbar compartments in relation to leptomeningeal metastases. Metabolites 12(1). https://doi.org/10.3390/metabo12010080
72. Henke C, Töllner K, van Dijk RM, Miljanovic N, Cordes T, Twele F, Bröer S, Ziesak V, Rohde M, Hauck SM, Vogel C, Welzel L, Schumann T, Willmes DM, Kurzbach A, El-Agroudy NN, Bornstein SR, Schneider SA, Jordan J, Potschka H, Metallo CM, Köhling R, Birkenfeld AL, Löscher W (2020) Disruption of the sodium-dependent citrate transporter SLC13A5 in mice causes alterations in brain citrate levels and neuronal network excitability in the hippocampus. Neurobiol Dis 143:105018. https://doi.org/10.1016/j.nbd.2020.105018
73. Sandri BJ, Kim J, Lubach GR, Lock EF, Guerrero C, Higgins L, Markowski TW, Kling PJ, Georgieff MK, Coe CL, Rao RB (2022) Tandem mass tag proteomic and untargeted metabolomic profiling reveals altered serum and CSF biochemical datasets in iron deficient monkeys. Data Brief 45:108591. https://doi.org/10.1016/j.dib.2022.108591
74. Hsu WH, Shen YC, Shiao YJ, Kuo CH, Lu CK, Lin TY, Ku WC, Lin YL (2019) Combined proteomic and metabolomic analyses of cerebrospinal fluid from mice with ischemic stroke reveals the effects of a Buyang Huanwu decoction in neurodegenerative disease. PLoS One 14(1):e0209184. https://doi.org/10.1371/journal.pone.0209184
75. Hu L, Dong MX, Huang YL, Lu CQ, Qian Q, Zhang CC, Xu XM, Liu Y, Chen GH, Wei YD (2020) Integrated metabolomics and proteomics analysis reveals plasma lipid metabolic disturbance in patients with Parkinson's disease. Front Mol Neurosci 13:80. https://doi.org/10.3389/fnmol.2020.00080
76. Xu X, Yang Q, Liu Z, Zhang R, Yu H, Wang M, Chen S, Xu G, Shao Y, Le W (2023) Integrative analysis of metabolomics and proteomics unravels purine metabolism dysregulation in the SOD1(G93A) mouse model of amyotrophic lateral sclerosis. Neurobiol Dis 181:106110. https://doi.org/10.1016/j.nbd.2023.106110
77. Babu H, Sperk M, Ambikan AT, Rachel G, Viswanathan VK, Tripathy SP, Nowak P, Hanna LE, Neogi U (2019) Plasma metabolic signature and abnormalities in HIV-infected individuals on long-term successful antiretroviral therapy. Metabolites 9(10). https://doi.org/10.3390/metabo9100210
78. Chen C, Wang J, Yang C, Yu H, Zhang B, Yang X, Xiong B, Xie Y, Li S, Zhang Z, Zhu F, Liu J, Liu GP, Yang X (2023) Multiomics analysis of human peripheral blood reveals marked molecular profiling changes caused by one night of sleep deprivation. MedComm

4(3):e252. https://doi.org/10.1002/mco2.252

79. Kreitmaier P, Katsoula G, Zeggini E (2023) Insights from multi-omics integration in complex disease primary tissues. Trends Genet 39(1):46–58. https://doi.org/10.1016/j.tig.2022.08.005
80. Shin A, Kashyap PC (2023) Multi-omics for biomarker approaches in the diagnostic evaluation and management of abdominal pain and irritable bowel syndrome: what lies ahead. Gut Microbes 15(1):2195792. https://doi.org/10.1080/19490976.2023.2195792
81. Gegner HM, Naake T, Aljakouch K, Dugourd A, Kliewer G, Müller T, Schilling D, Schneider MA, Kunze-Rohrbach N, Grünewald TGP, Hell R, Saez-Rodriguez J, Huber W, Poschet G, Krijgsveld J (2024) A single-sample workflow for joint metabolomic and proteomic analysis of clinical specimens. Clin Proteomics 21(1):49. https://doi.org/10.1186/s12014-024-09501-9
82. Gomez-Cabrero D, Abugessaisa I, Maier D, Teschendorff A, Merkenschlager M, Gisel A, Ballestar E, Bongcam-Rudloff E, Conesa A, Tegnér J (2014) Data integration in the era of omics: current and future challenges. BMC Syst Biol 8(Suppl 2):I1. https://doi.org/10.1186/1752-0509-8-s2-i1
83. Goble C, Stevens R (2008) State of the nation in data integration for bioinformatics. J Biomed Inform 41(5):687–693. https://doi.org/10.1016/j.jbi.2008.01.008

Chapter 22

Drug Repositioning Based on Cerebrospinal Fluid Proteomes Using Connectivity Map Framework

Adriana Cortés, Silvia Romero-Murillo, Elena Anaya-Cubero, Paz Cartas-Cejudo, Leire Extramiana, Mercedes Lachén-Montes, Joaquín Fernández-Irigoyen, and Enrique Santamaría

Abstract

Selecting a fluid near an affected organ can improve the likelihood of identifying a biomarker panel from pathological tissue. Cerebrospinal fluid (CSF), in close contact with the brain, is a valuable source of biomarkers for neurological disorders due to the inaccessibility of brain tissue. Moreover, the altered CSF proteome identified in neurological diseases can facilitate the repurposing of drugs already used for other therapeutic purposes. In this context, Connectivity Map (CMap) is a valuable tool as it provides information on compounds and gene modifications that can be utilized to reverse specific pathological signatures. Analyzing CSF differential proteomics through the CMap framework offers an efficient and cost-effective approach to identifying potential novel therapies for neurodegenerative diseases.

Key words Mass spectrometry, LC-MS/MS, Proteomics, Connectivity map, CSF, Drug repurposing, Label-Free

1 Introduction

Over the past decade, drug-repurposing methods have proven crucial in identifying novel therapies for neurodegenerative diseases, offering a promising approach to accelerate the development of effective treatments by utilizing existing medications [1]. Repurposing hypotheses can be formulated by integrating machine learning computational methods with genotype data, as well as large-scale transcriptomics and proteomics data [2]. One of the most widely accepted tools for this purpose is Connectivity Map (CMap) [3, 4]. CMap operates on the principle of signature reversion, wherein compounds, drugs, or specific genetic manipulations can potentially reverse omics pathological profiles in silico, pointing out potential therapeutic options. These workflows are based on transcriptional signatures from experiments conducted in human

Enrique Santamaría and Joaquín Fernández-Irigoyen (eds.), *Cerebrospinal Fluid (CSF) Proteomics: Methods and Protocols*, Methods in Molecular Biology, vol. 2914, https://doi.org/10.1007/978-1-0716-4462-1_22,

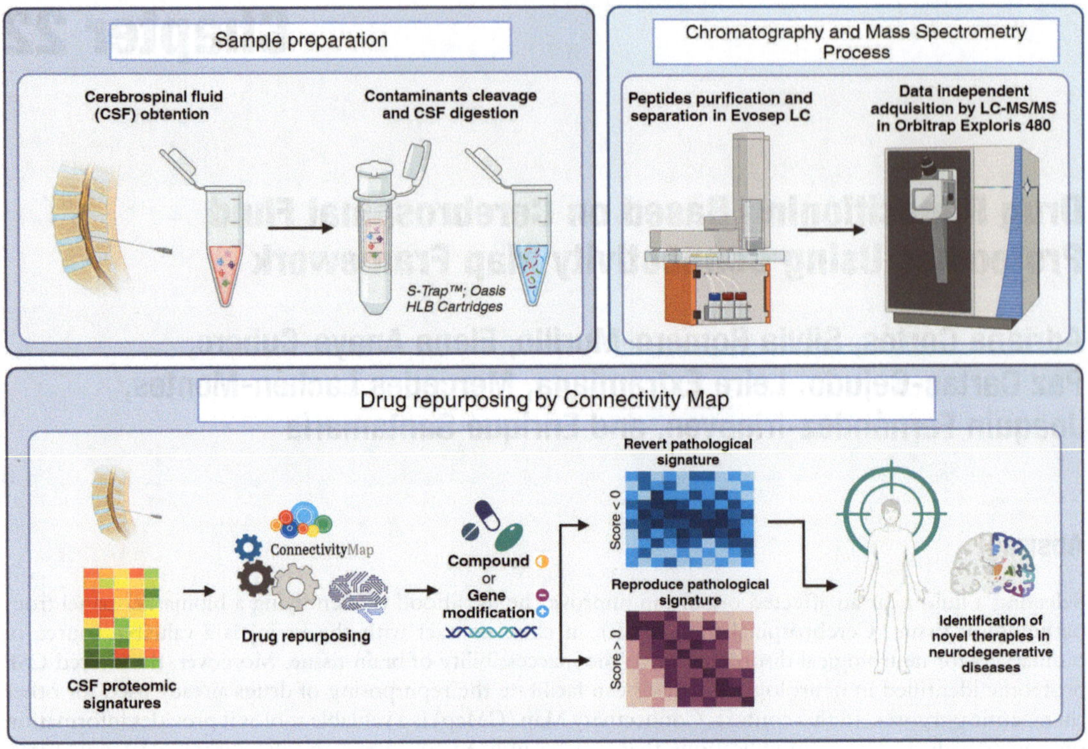

Fig. 1 Experimental workflow to combine CSF proteomics and drug repositioning

cell lines under specific perturbations, such as drugs or genetic manipulations, with a database containing approximately 1.5 million signatures from cells exposed to over 5000 compounds and 3000 genetic perturbations [3]. Selecting a fluid in close proximity to an affected organ can enhance the probability of identifying a biomarker panel originating from pathological tissue. In this context, cerebrospinal fluid (CSF), due to its intimate contact with the brain, serves as a pertinent reservoir of potentially clinically relevant biomarkers for neurological disorders, given the inaccessibility of brain tissue [5]. CSF proteomic assays can reflect brain pathology and enable the study of molecular mechanisms that occur before the onset of neurodegeneration or throughout the progression of the disease [6, 7]. Said that, the altered proteome identified in CSF samples from patients with neurological diseases can facilitate the repurposing of drugs already used for other therapeutic purposes [8]. CMap offers a pioneering approach to identify candidates for potential effective therapies based on the altered CSF proteomes from patients with neurological diseases.

In this chapter, we describe the Cmap-based drug repurposing workflow using CSF proteome pathological signatures obtained by Direct Data-Independent Acquisition (dDIA) mass spectrometry

(Fig. 1). Thus, we exemplify following a step-by-step guide on the applicability of Cmap to differential CSF proteome datasets.

2 Materials

2.1 S-TRAP™ Digestion (Up to 100 µg of Protein)

1. S-Trap™ micro spin column (1–100 µg protein loading capacity) or S-Trap™ 96-well plate (100–250 µg protein loading capacity) (Protifi).
2. Denaturing buffer: 10% SDS, 100 mM TEAB pH 8.5 (triethylammonium bicarbonate, Sigma).
3. 120 mM TCEP Bond-Breaker™ solution (neutral pH) in 100 mM TEAB (*see* **Note 1**).
4. Cysteine-blocking buffer: 500 mM MMTS (methyl methanethiosulfonate) in isopropanol.
5. 27.5% ortho-phosphoric for S-Trap™ micro spin column. Dilute ortho-phosphoric acid 85% with water.
6. Binding buffer: 90% methanol, 100 m TEAB (triethylammonium bicarbonate).
7. Trypsin solution: dissolve 2 µg of MS-grade trypsin (Promega) in 20 µL of 100 mM TEAB.
8. Elution solution 1: 50 mM TEAB.
9. Elution solution 3: 50% acetonitrile (ACN), 50 mM TEAB pH 8.5.

2.2 Desalting Proteolytic/Tryptic Peptides After Digestion Using Oasis HLB Cartridges

1. Oasis HLB 96-well µElution Plate, 2 mg Sorbent per Well, 30 µm (Waters).
2. Extraction Plate Manifold for Oasis 96-Well Plates (Waters).
3. ACN.
4. TFA 0.1% in water.
5. 70% ACN in water.

2.3 Chromatography and Mass Spectrometry

1. Mobile Phase A: 0.1% formic acid in water.
2. Mobile Phase B: 0.1% formic acid in ACN.
3. Liquid chromatography (Evosep):
   (a) Evosep One LC.
   (b) Analytical column. Performance Evosep column E1112 (C18, 1.9 µm beads, 15 cm × 75 µm).
   (c) E1111 Silica emitter (10 µm).
4. High-resolution mass spectrometry instrument (e.g., Orbitrap Exploris-480 mass spectrometer).

2.4 Other Materials

1. Low Protein Binding Microcentrifuge Tubes, 2.0 mL.
2. Low Protein Binding Microcentrifuge Tubes, 1.5 mL.
3. pH control strips.
4. Ultracentrifuge with adjustable rotor speed up to 100,000× *g*.
5. Microcentrifuge with adjustable rotor speed up to 7000× *g*.
6. Speed-Vacuum centrifuge.

3 Methods

3.1 Sample Preparation

3.1.1 CSF Extraction

1. The samples are obtained by lumbar puncture (LP) following ethically approved protocols, using a dry clean tube without preservatives, immediately frozen, and stored at −80 °C until use (*see* **Note 2**).
2. Collect samples via lumbar or ventricular puncture according to ethically approved protocols. Place the samples on ice immediately after collection.
3. Centrifuge the CSF samples at 2500× *g* for 10 min at 4 °C to remove cellular components and blood traces (*see* **Note 3**).

3.1.2 S-TRAP™ Micro Spin Column Digestion

1. Mix 50 μL of CSF with 50 μL denaturing buffer in a 1.5 mL microcentrifuge tube (*see* **Note 4).**
2. Add 4.4 μL of TCEP 120 mM and incubate at 55 °C for 15 min in a thermoblock with gentle mixing (final concentration 5 mM).
3. Add 4.4 μL of 500 mM cysteine-blocking reagent MMTS (final concentration 20 mM) and incubate for 10 min at room temperature in a thermoblock with gentle mixing.
4. Add 11 μL of 27.5% ortho-phosphoric acid solution (final concentration 2.5%) and mix (*see* **Note 5**).
5. Add 719 μL (dilution 1:7 vol/vol) of binding buffer. Colloidal protein particulate is instantaneously formed in this step.
6. Place the S-Trap filter in a 2-mL tube for flow through. Load 200 μL of the whole sample (colloidal and non-colloidal part) into the column and centrifuge at 4000× *g* 30″.
7. Repeat the step as many times as required to filter the sample volume totality (*see* **Note 6**). At this point, protein is trapped by matrix of the spin column.
8. Add 150 μL of binding buffer to wash the spin column. Repeat three times (4000 g, 1 min).
9. Place the filter onto a clean 1.5-mL tube for protein digestion. Add n μg of MS-grade trypsin in 20 μL of a 50 mM TEAB

solution into the top of the column in a ratio 1:10 (w/w) and allow digestion at 47 °C for 1 h.

10. After 1 h, add 50 μL more of Elution solution 1 (50 mM TEAB)—to avoid sample drying out—and return the plate to the incubator for another 1 h.
11. Elute proteolytic peptides:
    (a) Add 40 μL of elution solution 1 to the S-Trap and centrifuge (4000× *g*, 1 min).
    (b) Add 40 μL of elution solution 2 to the S-Trap and centrifuge (4000× *g*, 1 min).
12. Pool eluted peptides and dry them down in a vacuum concentrator system. After resuspension with a suitable solution, the sample is ready for reverse-phase chromatography. Additional C18 clean-up may be required with samples with high contaminant content.

3.1.3 S-Trap™ 96-Well Plate Digestion

1. Mix 25 μL of CSF with 25 μL denaturing buffer in a 1.5 mL microcentrifuge tube (final volume should not exceed 50 μL).
2. Add 2.2 μL of TCEP 120 mM and incubate at 95 °C for 10 min in a thermoblock with gentle mixing (final concentration 5 mM). Let cool down the sample.
3. Add 2.2 μL of 500 mM cysteine-blocking reagent MMTS (final concentration 20 mM) and incubate for 10 min at room temperature in a thermoblock with gentle mixing.
4. Add 5.5 μL of 27.5% ortho-phosphoric acid solution (final concentration 2.5%) and mix.
5. Add 360 μL (dilution 1:7 vol/vol) of binding buffer.
6. Place S-Trap™ 96-well plate in a 2-mL plate for waste flow through. Load the sample into the S-Trap™ 96-well plate and centrifuge at 2000 RCF 2′. Since all wells exhibit flow behavior, a vacuum manifold may also be used, applying 5 Hg (2.5 Psi) of negative pressure to filter and wash the sample.
7. Add 200 μL of binding buffer to wash S-Trap plate. Repeat three times (2000 RCF, 2′).
8. Place S-Trap™ 96-well plate in a new clean 2-mL plate to recover the peptides. Add n μg of MS-grade trypsin in 125 μL of a 50 mM TEAB solution into the top of the column in a ratio 1:10 (w/w). Make sure the digestion solution completely covers the protein capture matrix (*see* **Note** 7).
9. Loosely cover the digestion plate to limit evaporative loss and allow digestion at 47 °C for 2 h. Do not seal the plate as pressure build-up will force the solution prematurely through the trapping matrix.

10. After 1 h, add 70 μL of Elution solution 1 (50 mM TEAB)—to avoid sample drying out—and return the plate to the incubator for another 1 h.
11. Remove the plate from the incubator, add 75 μL of Elution solution 1, and centrifuge the plate at 2000 RCF min or until all solution has passed through.
12. Elute proteolytic peptides:
    (a) Add 80 μL of elution solution 1 to the S-Trap and centrifuge (2000× *g*, 2 min).
    (b) Add 80 μL of elution solution 2 to the S-Trap and centrifuge (2000× *g*, 2 min).
13. Pool eluted peptides and dry them down in a vacuum concentrator system. After resuspension with a suitable solution, the sample is ready for reverse-phase chromatography. Additional C18 clean-up may be required (e.g., Oasis HLB Cartridges) with samples with high contaminant content.

3.1.4 Desalting Tryptic Peptides Using Oasis HLB Cartridges (Additional Clean-Up)

1. Resuspend dry-down tryptic peptides from S-Trap™ digestion with 500 μL 0.1% TFA.
2. Place Oasis HLB 96-well μElution Plate onto the extraction plate manifold with the collection-wash plate inside (*see* **Note 8**).
3. Add 200 μL of ACN to conditionate the HLB plate.
4. Add 200 μL of TFA 0.1% to equilibrate the HLB plate.
5. Load the acidified tryptic peptides onto the HLB plate.
6. Wash the bound peptides with 800 μL of TFA 0.1%.
7. Wash the bound peptides with 200 μL of water to remove salts and buffers.
8. Replace the collection plate with the collection-elution plate to recover the peptides and elute peptides adding 50 μL of 70% ACN.
9. Dry down eluted peptides in a vacuum concentrator system and kept at −80 °C.

3.2 LC-MS/MS Analysis

1. Dried-down peptide samples are reconstituted with 2% ACN-0.1% FA and quantified by NanoDropTM spectrophometer before LC-MS/MS analysis.
2. 100 ng per sample is analyzed by reverse-phase LC-MS/MS using an Evosep LC coupled to an Exploris 480 mass spectrometer (Thermo). Peptides are resolved using a performance Evosep column E1112 (C18, 1.9 μm beads, 15 cm × 75 μm) using the SP20 whisper method (58 min gradient, 40 °C).

3. Sample data is acquired in a data-independent mode (DIA) with a full MS scan (scan range: 400–900 m/z; resolution: 12000; maximum injection time: 50 ms; normalized AGC target: 300%) and 50 periodical MS/MS segments, applying 10 Th isolation windows (0.5 Th overlap; resolution: 15000; maximum injection time: 22 ms; normalized AGC target: 100%). Peptides are fragmented using a normalized HCD collision energy of 30%, acquiring data in profile and centroid mode for full MS scan and MS/MS, respectively (*see* **Note 9**).

3.3 Identification and Quantification

DIA data files are analyzed using Spectronaut (Biognosys) by directDIA analysis (dDIA). Sample raw data is first processed by Pulsar Spectronaut search engine generating a pseudo-spectral library, searching MSMS spectra against Uniprot-Swissprot Homo Sapiens database (MS1/MS2 calibration and main search tolerance are set to dynamic). Maximum precursor ion charge is set to 4, and fragment selection to intensity-based. Carbamidomethyl (C) is selected as a fixed modification, and Oxidation (M), Acetyl (Protein N-term), Deamidation (N), and Gln- > pyro-Glu as variable modifications (3 maximum modifications per peptide). Enzyme is set to trypsin in a specific mode (two missed cleavages maximum). The target-decoy-based false discovery rate (FDR) filter for PSMs, peptide, and protein groups is set to 1%.Once the Pulsar search is performed, the Spectronaut DIA search engine algorithm starts the DIA search using the default settings and filtering the precursors and protein groups by a 1% Q-value.

3.4 Drug Repurposing Workflow

In this case, the drug repurposing workflow requires to prior identify differential proteins by applying statistical analysis and noting a fold change data to the results.

3.4.1 Connectivity Map

1. Open the webpage containing connectivity map software (https://clue.io/). Registration is required).
2. Go to the "Tools" section and select "Query."
3. Name your project and select the search type "Gene Expression."
4. Query parameters are set as the following: Touchstone, Individual query and 1.0.
5. Introduce the gene name corresponding to the CSF proteins with significantly altered levels between neurological conditions (*see* **Note 10**).
6. Go to the "Query history" section. Here, the uploaded dataset will appear and it may take a few minutes to finish the analysis.
7. Select one or more analyses and select query results desired (*see* **Note 11**):

(a) Heat map.
(b) Detailed List.

3.4.2 Connectivity Map Results

This tool has two output formats that can adjust to diverse applications:

1. *Heat map.*
   - Select one or more analyses in the "Query history" section.
   - On the top right, select "Heat map" in the query results section.
   - In the result page displayed on the right, a "Quick tools" box appears to customize the results shown by the tool (*see* **Note 12**).
   - The visual characteristic of the analysis can be modified in the "View" section on the top left by selecting "Options."
   - Export data and hierarchical clustering in the "File" section.
2. *Detailed List.*
   - Select one or more analyses in the "Query history" section.
   - On the top right, select "Detailed List" in the query results section.
   - In the result page displayed, in the "Connections of reference perturbations to Index," a selection of the cell lines desired to be included in the analysis can be done.
   - On the top left, the output can be filtered by the nature of the modification: compound, gene knock-down or gene over-expression.
   - Export the detailed list to a .txt format (*see* **Note 13**).

4 Notes

1. Use TCEP Bond-Breaker™ solution with neutral pH to avoid pH bias in the reducing buffer. Solutions are acidified by powder TCEP.
2. CSF consists of a total volume of approximately 135 mL in humans. CSF contains a low number of cells (0–4 cells/μL), low protein concentration, and various salts. The protein concentration in CSF from the lumbar region is about 15–45 mg/dL.
3. Traumatic taps occur in 14–20% of standard lumbar punctures, causing blood contamination in the CSF sample. This may affect CSF proteome and provide incorrect CSF protein biomarker results. It is also possible to use specific filters to remove

all membrane and cellular debris as, for example, Millex-HV filters. Another option is to incorporate a precipitation step using acetone or a methanol/chloroform liquid–liquid partitioning procedure.

4. Sample protein concentration must be checked to ensure sample quality, and calculation of trypsin quantity for tryptic digestion. Unmodified trypsin can undergo auto-proteolysis, resulting in chymotrypsin-like enzymatic activity. Use modified trypsin to avoid auto-proteolysis.
5. Adding ortho-phosphoric acid is essential to trap the proteins, denaturing proteins completely. The pH has to be ≤ 1. Add additional phosphoric acid to reach pH ≤ 1 if required.
6. Rotate the S-Trap micro units 180 degrees between the centrifugations of binding and wash steps. This is especially important when using a fixed-angle rotor because the spin column does not experience homogenous flow. Additional wash (es) may be performed and should be performed if contamination is observed by mass spectrometry.
7. Do not apply less than 10 μg of trypsin per well for effective digestion.
8. Apply 5 Hg of vacuum to filter through the well matrix plate.
9. The MS method described here, in combination with the 20SP whisper method gradient, offers a 2.15 s cycle time, allowing 6–7 quantification points per peak, a reasonable number for label-free quantitation. For other Evosep methods in whose gradient and extension peptide peak length are different, it is required to adjust the MS settings to obtain suitable points per peak.
10. Only 150 upregulated or downregulated proteins can be considered as input in each experiment. Many studies may be performed simultaneously or select the most up or downregulated proteins in your dataset. Eliminate invalid or valid genes that have not been and replace them with others considered valid by the CMap software.
11. Cmap analysis results in a list of compounds, gene knockdown, and gene overexpression that could potentially mimic or revert the pathological signature associated with the disease.
12. To filter the results, each type of modification may be selected as well as in what cell line you want to see the analysis. This could be interesting as the researcher could only select the cell lines coming from women or men if sex variable has been previously considered in the CSF differential proteomic phase.
13. Based on the CMap framework, the connectivity score (CS) is a standardized statistic parameter ranging from −100 to +100. A highly negative score indicates that the drug-proteins

signatures are opposing in which genes that are upregulated (increased levels) in neurological phenotypes are downregulated (decreased levels) by the compound and vice versa.

Acknowledgments

The Clinical Neuroproteomics Unit is member of the Spanish Olfactory Network—ROE (grant RED2022-134081-T from the Spanish Ministry of Science & Innovation) and is supported by grants PID2023-152593OB-I00 funded by MCIU/AEI/ 10.13039/501100011033 / FEDER, UE to ES and JFI and 0011-1411-2023-000028 (from Government of Navarra—the Department of Economic and Business Development-S4). PC-C is supported by a predoctoral fellowship from the Public University of Navarra (UPNA). ML-M is supported by a postdoctoral fellowship from Miguel Servet Foundation-Navarrabiomed. EA-C is supported by "Programa MRR Investigo 2023" in the framework of the European Union recovery and resilience facility.

References

1. Grabowska ME, Huang A, Wen Z, Li B, Wei WQ (2023) Drug repurposing for Alzheimer's disease from 2012–2022—a 10-year literature review. Front Pharmacol 14(September):1–13. https://doi.org/10.3389/fphar.2023.1257700
2. Rodriguez S, Hug C, Todorov P, Moret N, Boswell SA, Evans K et al (2021) Machine learning identifies candidates for drug repurposing in Alzheimer's disease. Nat Commun 12(1):1033. https://doi.org/10.1038/s41467-021-21330-0
3. Molecules S, Lamb J, Crawford ED, Peck D, Modell JW, Blat IC et al (2006) The connectivity map: using gene-expression signatures to connect small molecules, genes, and disease. Science 313(September):1929–1935. https://doi.org/10.1126/science.1132939
4. Subramanian A, Narayan R, Corsello SM, Peck DD, Natoli TE, Lu X, Gould J, Davis JF, Tubelli AA, Asiedu JK, Lahr DL, Hirschman JE, Liu Z, Donahue M, Julian B, Khan M, Wadden D, Smith IC, Lam D, Liberzon A, Toder C, Bagul M, Orzechowski M, Enache OM, Piccio GTA (2017) A next generation connectivity map: L1000 platform and the first 1,000,000 profiles. Cell 171(6):1437–1452.e17. https://doi.org/10.1016/j.cell.2017.10.049
5. Lleó A, Cavedo E, Parnetti L, Vanderstichele H, Herukka SK, Andreasen N et al (2015) Cerebrospinal fluid biomarkers in trials for Alzheimer and Parkinson diseases. Nat Rev Neurol 11(1):41–55. https://doi.org/10.1038/nrneurol.2014.232.eurol.2014.232
6. Dammer EB, Shantaraman A, Ping L, Duong DM, Gerasimov ES, Ravindran SP et al (2023) Proteomic Network Analysis of Alzheimer's Disease Cerebrospinal Fluid Reveals Alterations Associated with APOE ε4 Genotype and Atomoxetine Treatment. medRxiv [Preprint]. https://doi.org/10.1101/2023.10.29.23297651
7. Walker KA, Chen J, Shi L, Yang Y, Fornage M, Zhou L et al (2023) Proteomics analysis of plasma from middle-aged adults identifies protein markers of dementia risk in later life. Sci Transl Med 15(705):eadf5681. https://doi.org/10.1126/scitranslmed.adf5681. Epub 2023 Jul 19
8. Do AN, Ali M, Timsina J, Wang L, Western D, Liu M et al (2024) CSF proteomic profiling with amyloid/tau positivity identifies distinctive sex-different alteration of multiple proteins involved in Alzheimer's disease. medRxiv [Preprint]. https://doi.org/10.1101/2024.03.15.24304164

Correction to: An LC-MS/MS Multiplexed Method to Quantify 14 Sphingolipids in Human Cerebrospinal Fluid (CSF)

Yadira Perez-Paramo, Rathna Veeramachaneni, Emily Parkhurst, Dawn Dufield, W. Rodney Mathews, and Veronica G. Anania

Correction to:
Chapter 18 in: Enrique Santamaría and Joaquín Fernández-Irigoyen (eds.),
Cerebrospinal Fluid (CSF) Proteomics: Methods and Protocols,
Methods in Molecular Biology, vol. 2914,
https://doi.org/10.1007/978-1-0716-4462-1_18

The original version of this book was inadvertently published with error. Prof. Yadira Perez-Paramo's name has been corrected in Chapter 18. The correct name is: First name: Yadira and Last name: Perez-Paramo.

The updated version of this chapter can be found at
https://doi.org/10.1007/978-1-0716-4462-1_18

Enrique Santamaría and Joaquín Fernández-Irigoyen (eds.), *Cerebrospinal Fluid (CSF) Proteomics: Methods and Protocols*,
Methods in Molecular Biology, vol. 2914, https://doi.org/10.1007/978-1-0716-4462-1_23,

INDEX

Enrique Santamaría and Joaquín Fernández-Irigoyen (eds.), *Cerebrospinal Fluid (CSF) Proteomics: Methods and Protocols*, Methods in Molecular Biology, vol. 2914, https://doi.org/10.1007/978-1-0716-4462-1,

N

O

P

R

S

T

MIX
Papier aus verantwortungsvollen Quellen
Paper from responsible sources
FSC® C105338

If you have any concerns about our products,
you can contact us on
ProductSafety@springernature.com

In case Publisher is established outside the EU,
the EU authorized representative is:
Springer Nature Customer Service Center GmbH
Europaplatz 3, 69115 Heidelberg, Germany

Printed by Libri Plureos GmbH
in Hamburg, Germany